Electronic Drives

Electronic Drives

Robert S. Carrow

TAB Books
an imprint of McGraw-Hill
New York San Francisco Washington, D.C. Auckland Bogotá Caracas Lisbon London
Madrid Mexico City Milan Montreal New Delhi San Juan Singapore Sydney Tokyo Toronto

McGraw-Hill

A Division of The **McGraw-Hill** *Companies*

hc 1 2 3 4 5 6 7 8 9 DOC/DOC 9 0 0 9 8 7 6

Library of Congress Cataloging-in-Publication Data

Carrow, Robert S.
Electronic drives / by Robert S. Carrow
p. cm.
Includes index.
ISBN 0-07-011611-3
1. Power electronics. 2. Electric driving. I. Title
TK7881.15.C37 1996
621.46—dc20 96-16470
CIP

McGraw-Hill books are available at special quantity discounts to use as premiums and sales promotions, or for use in corporate training programs. For more information, please write to the Director of Special Sales, McGraw-Hill, 11 West 19th Street, New York, NY 10011. Or contact your local bookstore.

Acquisitions editor: Roland S. Phelps
Editorial team: Joanne Slike, Book Editor
Andrew Yoder, Managing Editor
Lori Flaherty, Executive Editor
Joann Woy, Indexer
Production team: Katherine G. Brown, Director
Toya B. Warner, Computer Artist
Jeffrey Miles Hall, Computer Artist
Rose McFarland, Desktop Operator
Nancy K. Mickley, Proofreader
Design team: Jaclyn J. Boone, Designer
Katherine Lukaszewicz, Associate Designer

0116113
EL3

To my wife, Colette, who is now probably the most
knowledgeable domestic engineer (homemaker)
on the subject of electronic drives in the world.
Without her help, this book would not
have been possible.

Contents

Introduction *xiii*

1 What is a drive? *1*

Why use an electronic drive? *2*
Drive proliferation *2*
Which drive is best for the application? *3*
Myths, misconceptions, and misapplications *5*
The drive market *7*

2 Motion control *11*

Motion control methods *15*
The evolution of variable-speed control *16*
M-G (motor-generator) sets *16*
Eddy current clutches *17*
Closing remarks *22*

3 Basic electricity, electronics, and power transmission *25*

Electrical basics *26*
Electrical formulas *27*
Frequency and amplitude *30*
Resistors and resistance *32*
Capacitors and capacitance *33*
Relays *34*
Inductors and inductance *35*
Power semiconductors *37*
Transistors *39*
Diodes *40*
Fusing and circuit protection *40*

Changing electrical power into mechanical power *42*
Work *43*
Horsepower *44*
Torque *45*
Speed, torque, and horsepower relationships *48*
Inertia *51*
Friction *54*
Gear reduction, gearboxes, and gearing *54*
V-belts and variable-speed pulleys *57*
Chains and sprockets *58*
Clutches and brakes *59*
Couplings *60*
Ball and lead screws *61*
Hydraulics *63*
Pneumatics *64*
Noise *65*
Efficiency *65*
Service factor *66*
Maintenance *67*
Closing remarks on power transmission *68*

4 Understanding electric motors ***69***
Motor cooling *71*
Enclosure types for motor cooling *72*
Motor protection *78*
NEMA ratings *80*
Motors (ac) *83*
Motors (dc) *89*
Servomotors *96*
Stepper motors *99*
Closing remarks on electric motors *100*

5 Direct current (dc) drives ***103***
Drive power bridges (dc) *107*
Form factor and ripple *111*
Drive feedback arrangements (dc) *112*
Regenerative versus nonregenerative dc drives *115*
M, or loop, contactor *117*

Speed scaling 118
Traction drives 119
Chopper drives 121
Drive features (dc) 121
Drive selection (dc) 124
Maintenance of dc drives 125
Troubleshooting and repairing dc drives 126
Common drive faults: Causes and solutions 130
Closing remarks on dc drives 132

6 Alternating current (ac) drives 135
Electronic ac drive basics 137
Electronic ac drive types 140
How electronic ac drives work 147
Energy savings with electronic ac drives 152
Variable torque and constant torque 157
Drive selection (ac) 158
Other ac drive functions and features 167
Maintenance of ac drives 171
Troubleshooting and repair of ac drives 172
Common drive faults, possible causes, and solutions 175
Flux vector drives (ac) 178
Troubleshooting and diagnostics of flux vector drives 182
Load commutated inverters (LCIs) 184
Cycloconverters 186
Closing remarks on electronic ac drives 187

7 Servo, stepper, and specialty drives 189
Servo drive systems 189
Stepper drives 201
Linear motor systems 206
Spindle drives 207
Closing remarks 207

8 Drive peripheral equipment, control, and communications 209
Power wiring and supply power considerations 210
Control and signal wiring 216
Sources of industrial electrical noise 224

Transformers and line reactors *225*
Enclosing electronic drives *230*
Switchgear *233*
The drive control *234*
Drive control schemes *238*
Drive communications *241*
Closing remarks on peripheral drive equipment, drive control schemes, and drive communications *243*

9 Drive closed-loop control and feedback devices ***245***
Open- and closed-loop control *246*
Speed and current feedback methods *248*
Optical encoders *250*
Resolvers *254*
Magnetic pickups *255*
Peripheral feedback and control devices *256*
Transducers *257*
Temperature control and pressure sensors *258*

10 Drive integration and machine control ***261***
Drive integration *262*
Machine systems and applications *266*
Tension control *267*
Coordinated drive machine control *268*
Remote control *269*
Braking and regeneration *271*
Why do drives stop running? *274*
Drive training and uptime *275*
Closing remarks on drive integration *278*

11 Electronic drive applications ***279***

12 Harmonics, power factor, and other electrical line phenomena ***313***
Drive harmonics *313*
Linear and nonlinear loads *317*
Data and equipment required to measure harmonic distortion *323*
Radio frequency interference (RFI) *325*

Power factor *326*
Electrical noise *328*
EMI and RFI *329*
Grounding *330*
Power fluctuations *331*
Dealing with electrical disturbances *332*
Shielding *334*
Optical isolation *335*
Signal conditioners and filters *336*

13 Electronic drive technology's future ***339***

Appendices

A Standards organizations and addresses ***347***

B Common symbols used in drive diagrams ***349***

Bibliography and suggested further reading ***351***

Index ***353***

Introduction

ELECTRONIC DRIVES ARE RELATIVELY NEW TO INDUSTRY. They have been around in some form for decades but really hadn't proliferated until 10 or 12 years ago. At that time, nobody knew much about them or the technology. Even today, many individuals have serious misconceptions about drives and their applications. This book is for them: the engineers, designers, technicians, machine operators, maintenance—anyone who deals with electrical equipment.

Electronic Drives has been written to provide a basic understanding of the product, the industry, and how to apply the product. Rather than a chalkboard's worth of equations and formulas, it contains just enough calculations to solve a problem quickly in the field. The book is intended as a reference to guide the reader through electronic drive selection and application. It is intended to dispel any fear and anxiety about choosing an electronic drive over mechanical methods. This book is heavy on applications and practical information accumulated from experience in the field. I have spent many years applying ac, dc, and specialty drives—and have the "scars" to prove it.

Early on, when the industry and technology was young, many drives were misapplied, and many types suffered high "infant mortality" rates. They never worked out of the box, or else when connected to the wrong type of load, they couldn't turn a motor. These situations gave drives and the drives' industry a bad name that it is still trying to shrug off. However, from those days, we have all grown together, getting smarter about the products, working the bugs out of them, and applying them where they are best suited.

This book focuses on electronic drives, as almost all drives today have some microprocessor content and most use high-frequency switching to perform their power conversion. We will discuss both ac and dc types. Specialty drives such as servos, stepper drives, and so on actually fall into one group or the other. In today's electrical scheme, a motor either has some type of alternating current or direct current power running to it.

Each chapter contains graphs, charts, and tables. In addition, there are various logs and charts available to the reader that can be used wherever a drive is to be installed or wherever one might be installed in the future. These charts or logs are in spreadsheet form so that a sound analysis can be made.

Just hearing the word *drive* causes five different people to have five different interpretations. Chapter 1, "What is a Drive?," tries to get the big question answered right away so that we can continue on to later chapters that detail specific drives. In this chapter, we discuss drive proliferation, the industry and market in general, and compare ac to dc product broadly. Discussion of some of those common myths and misconceptions about drives, which eventually lead to misapplications, is included.

Chapter 2, "Motion Control," discusses older technology as far as drive speed control is concerned. Traditional methods, many mechanical, of changing output speeds in a process are covered. Drive and startup costs compared to expected performance are also discussed. Concepts of torque, horsepower, speed, and position are introduced as they pertain to electronic drives.

Chapter 3, "Basic Electricity, Electronics, and Power Transmission," covers the fundamentals that govern the world of drives. Electrical components, electrical-based calculations with examples, and Ohm's law are all covered. Theories of resistance, inductance, capacitance, and so on are discussed. Because frequency is the basis for much of the ac drive's operation, it is covered in detail. Work, power, and torque, as well as speeds and inertias, are then expanded upon. Whenever drives are used, they are often trying to do something electronically that was traditionally done mechanically. This is where a lot of trouble starts. Understanding the basic principles of power transmission is important, and this chapter covers the basics. Various power transmission components such as gearboxes, couplings, clutches, and brakes are discussed.

To understand how a drive controls an electric motor, you must first understand how the electric motor functions. Knowing how it is built and why it can be controlled at all are principal topics of Chapter 4, "Understanding Electric Motors." Squirrel cage and ac induction motors, dc brush-type, permanent magnet, servo, stepper, and other specific-duty motors are discussed. Motor enclosure types are also covered. The electronic drive and the motor actually become one in the circuit, relying on each other for application survival.

Chapter 5, "Direct Current (dc) Drives," begins our first discussion on an actual drive type. Hardware, both analog and digital versions, dc drive theory, rectifiers, and special dc types of drives specific to certain applications, such as choppers and traction drives, are discussed. Troubleshooting and replacing some component parts, such as SCRs, is covered. Installation and drive setup is also discussed in this chapter, along with general application information.

Chapter 6, "Alternating Current (ac) Drives," takes us from the dc drive right to its cousin, the ac electronic drive. Analog and digital types are discussed. Various converter and inverter designs are covered, along with flux and voltage vector, LCI, and cycloconverter designs. Troubleshooting and diagnostic aids are provided for most of the types of ac drives discussed. Inherent features and specific application functions of the ac drive are covered one by one. Lastly, a look at energy savings and the ac electronic drive is provided.

Chapter 7, "Servo, Stepper, and Specialty Drives," discusses that class of drives used for high-performance, low-horsepower, and positioning-type applications. Servos and steppers are used in machine tool and robotic applications and thus have their own mystique. This chapter attempts to remove this mystique and give the reader a simplistic view of the drive. Other special drives such as spindles and linear motors are then discussed.

Chapter 8, "Drive Peripherals, Control, and Communications," covers all the other associated equipment and control schemes found in drive application. Transformers, line reactors, switchgear, cable and wiring, enclosures, and other associated hardware are all discussed. Proper application and installation methods, along with some troubleshooting of this equipment, are discussed. Common drive control techniques and various communication methods are also covered.

Chapter 9, "Drive Closed-Loop and Feedback Systems," covers all those little insignificant drive application devices that can shut an entire factory down whenever they're malfunctioning. Encoders, resolvers, magnetic pick-ups, transducers, tachometers, CTs and PTs, Hall effects, and shunts are all defined and discussed. Methods for attaching them to a motor, connecting to a drive, and setting them up are also covered.

Putting all the ac and dc drive components together into a machine working system is covered in Chapter 10, "Drive Integration and Machine Control." True integration involves putting other drives together into a machine environment or simply getting a single drive online. Many disciplines are discussed, including remote control schemes, tension control, and braking and regeneration. The question, "Why do drives stop running?," is also tackled.

Chapter 11, "Electronic Drive Applications," is, perhaps, the heart of the book. More than 30 common drive applications are discussed. A mix of ac, dc, servo, and other drive types are explored in their actual usage. Each application note includes the features of the application, the pitfalls, and a detailed drawing showing its implementation. As will be shown, ac and dc drives are applicable to many of the same machines and applications.

Chapter 12, "Harmonics, Power Factor, and Other Electrical Line Phenomena," are topics that are integral to electronic drive technology. With so much equipment in facilities being microprocessor-based and sensitive to line disturbances, power-converting products such as drives get a lot of attention—perhaps for the wrong reasons. Harmonic distortion and power factor are complex topics, and this chapter attempts to highlight and simplify them. EMI and RFI, grounding and shielding of wire, and filtering are discussed.

Chapter 13, "Electronic Drive Technology's Future," takes us a few years down the road. Actually, many issues discussed are happening today. Electric cars, magnetic levitation trains, drives on motors, and other drive-related topics are being either worked on or implemented. Their proliferation and extended use will be felt in a few years. The ac and dc drives are once again compared.

The subject of electronic drives is vast and can actually encompass several disciplines. This book covers all the major concerns. Electrical, electronic, computer, mechanical, and process people all have a stake in the electronic drive market. Understanding the electronic drive better will make us all more well-rounded. Who knows where and when a drive unit might pop up? Maybe someday in your toaster, microwave, car, or bathtub!

What is a drive?

THE DRIVE, OR DRIVER, IS BEST DESCRIBED AS THAT component that causes extended motion from the prime mover. Because this is an electronics book, the emphasis will be on electronic drives. For clarity, the term *electronic drive* is used throughout; *electronic* because of the microprocessor basis to most drive products in use today.

To the electrician, the word *drive* usually means an electronic-type unit. It is often the mysterious "black box" in the plant that very few workers know anything about. What's in that black box? What does it do? Why it is needed? And what to do if it quits running? Usually there is one person in the plant who learns the drive product in detail. This person becomes the "drive guru." All seek him or her, day and night, when the drive needs fixed.

The ac and dc drives have each been given many unique names over the years. The dc drive has been around for decades and has the more traditional nomenclature, for instance, the SCR or thyristor drive, or driver. It has even been called the M-G set, since a motor-generator set is necessary to provide the dc energy to control a dc motor. Others call the dc drive the rectifier or the converter, specifically describing its electrical functionality.

Although the ac drive has been around a shorter period of time, it has been given many more nicknames. Like the dc drive, it can be called the motor drive or driver. Or it can be called the variable-frequency drive (VFD), the variable-speed drive (VSD), or the VF drive. The volts-per-hertz drive is an appropriate name for the ac drive. Another, the inverter, is only half correct. The ac drive has a converter and an inverter component; therefore, the proper terminology should be converter/inverter. My personal favorite is the "freak" (for frequency) drive.

The drive is boxlike in appearance. The so-called black box can contain many electronic parts. The drive itself will be confined to the area around the power device heatsinks. In this package area there will also be other components necessary to the operation of the drive. Today's drive has a microprocessor-based main control board, and sometimes there are peripheral boards such as gate firing control boards, input and output cards, and other attachable boards specific to the drive's application. Often the actual drive is placed within another enclosure, usually metal. Sometimes there is more than one drive within this enclosure, as well as other electronic devices. This book will take an in-depth look at all types of drives, their individual components, drive systems, and the like.

Why use an electronic drive?

Originally, the need to control a dc motor prompted the development of the electronic drive, followed by the need to change speeds of dc and, later, ac motors. Today, the reasons are many. An ac electronic drive can be used as a soft-start device for an induction motor. Often when an ac motor is line-started, the inrush current to the motor is 6 to 7 times its rating. Amplify this condition by the number of starts in an hour and you have a severely stressed motor. Reduced-voltage starters exist but are not comparable in cost to an electronic drive.

The electronic drive is justified by its efficiency. It can save valuable energy dollars by reducing the actual power necessary to run a motor at less than full speed. (The payback period is very real and is elaborated on later in this book.) The electronic drive also can protect the motor from dangerous overload conditions. It will monitor the current to the motor and shut down after a predetermined period of time in order to save the motor.

Although electronic drives are mostly used for industrial applications, as time goes on, more drives will be utilized in our offices and even in our homes. The costs of such units continue to come down, making them all the more practical. The technology will also change in the future; some major breakthrough will, no doubt, further increase usage all over the planet. Right now, electronic drives are very alive and well. And they're getting more and more reliable!

Drive proliferation

In a global, competitive economy, many manufacturers and entrepreneurs of electronic drive products have emerged. Spin-off disciplines and associated industries have been instituted. Electrical, mechanical, and computer personnel have all become familiar with the electronic drive. New jobs have been created. Drive application engineers, field service individuals, and system integrators are now in abundance. The drive's industry is big, and it is growing.

Electronic drives are used almost everywhere. In fact, they are almost a commodity item. Someday it might even be possible to go to the hardware store and pick up a 1-HP drive. The ac drives are used on pumps and fans. The dc drives are used in many continuous processes. Servo and stepper drives are used on robots and other machines. The ac and dc drives are digital these days, but many analog units are still in service.

Presently, drives are being used in commercial washers and dryers, heat pumps, pool and spa pumps, and a variety of applications unheard of 10 years ago. Merge these new applications with the thousands of other common applications and we have practically a commodity. There is a good chance that electronic drive repair might one day be a course by itself at technical schools. Presently, its subject is covered in almost every electronics course in the country. It will not be long until drives are utilized in many residential products. Costs and acceptance will be key factors to its proliferation into these markets.

Which drive is best for the application?

Throughout this book, all types of drives will be analyzed—from ac to dc, servo to stepper, vector to traction, and so on. The question should not be which is best, but rather, which is best suited for the user and the application. Many factors enter into the decision: environment, load, cycling, motor preference, familiarity with a technology, cost, reliability, and so on. My suggestion is to explore all the possibilities. Ask many questions and prepare a spreadsheet similar to figure 1-1. This spreadsheet will help in selecting the best drive for your application. In addition, this book should give some guidance toward proper selection and successful installation.

Figure 1-2 provides a cost-versus-drive technology curve—not as gospel, but as a guide. Various horsepower sizes and different drive technologies such as ac, ac vector, dc, servo, and stepper are compared. Keep in mind that any one of these drive types could cost more than the other for a given application where special enclosures, additional switchgear, system integration (engineering or programming), and other items are needed. This comparison assumes like horsepower sizes, includes the motor, and compares hardware only.

Drive Comparison Spreadsheet							
Nominals: HP = Voltage = Motor = ac, dc, other________							
Drive No.	Current Rating	Output Type	Overload Capacity	Enclosure Type	Cost $	Service	Other

■ **1-1** *Electronic drive comparison spreadsheet. Use to evaluate various drives and manufacturers.*

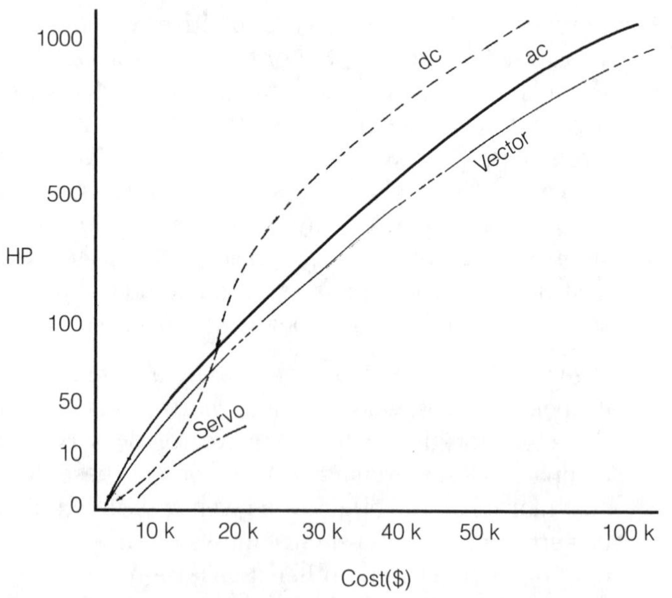

■ **1-2** *Cost versus drive type curve. Compares drive packages (motor and drive) of various technologies.*

Myths, misconceptions, and misapplications

Electronic drives have gotten a bad reputation with many individuals over the years. In part, this is due to the technology being a young science. Anytime a new product bursts onto the scene, there will be many more failures with that product early on than there will be later in the life of that product. Everyone is learning from the experience. However, these experiences leave many (users in particular) gun-shy. Nobody wants to be the guinea pig! To avoid being "burned" a second time, some might seek other options, especially the tried-and-true mechanical solution, even though the problem that surfaced the first time is now fully corrected.

There are also many misconceptions concerning ac and dc drives. Some of these have led to applying drives incorrectly, especially ac drives. For example:

- ☐ *If this electrical device (an ac drive) can be installed in an application and the speed can be reduced, then why use a gear reducer?* Because the gear reducer provides more output torque as it reduces the output speed. The torque is what drives the load. A certain horsepower drive cannot make up for a dramatic difference in torque. Thus, the drive was able to yield speed reduction, but the concept of torque was overlooked. These concepts are described in greater detail in Chapter 3. Also, the concept of directly driving a load versus using a gear reducer is shown in figure 1-3.
- ☐ *My motor is rated for 25 HP; therefore, I want a 25-HP drive*. This may or may not be practical. The first question to ask is: What is the application type, constant or variable torque? Motor manufacturers derate motors for constant torque applications, yet still nameplate them for a particular horsepower rating. This could, and does, create problems later in starting loads and simply running. Misapplying drives as variable torque when they should have been constant torque is a problem that still exists today. The solution? Simply ask. Then size the drive accordingly.
- ☐ *Can an equivalent HP ac drive replace the same HP eddy current clutch?* There are losses associated with the eddy current clutch, which must be accounted for when sizing a replacement ac or even dc drive. Best to place a service factor on the replacement drive and upsize.
- ☐ *An electronic drive can handle all my process requirements; therefore, I don't need a programmable logic controller*. Electronic drives with their input and output

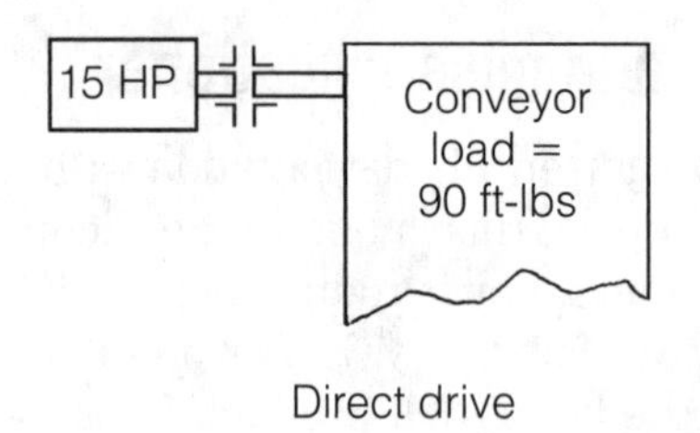

FACTS: Motor has a maximum torque output of 45 ft-lbs. Conveyor needs to be slowed but needs 90 ft-lbs to run at all.

$T = Hp \times 5250/RPM$
$= 15 \times 5250/1750$
$= 45$ ft-lbs, we need a gearbox and 2:1 ratio.

$T = 15 \times 52350/875$
$= 90$ ft-lbs.

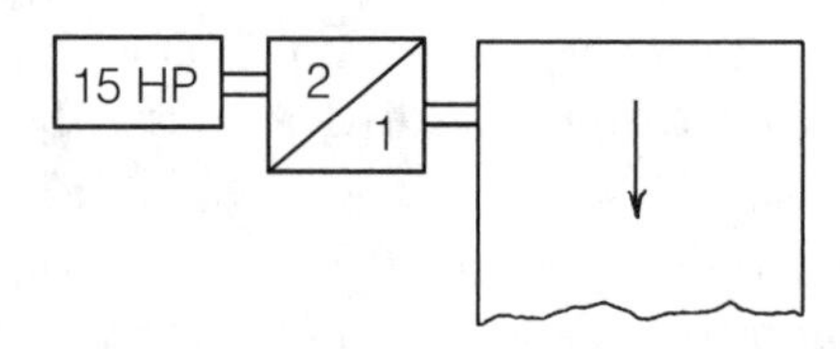

Gear reduction

■ **1-3** *The gear reducer as a torque multiplier.*

(I/O) modules can do quite a bit of process control. However, there is a limit. The drive's main purpose is to control an electric motor. The microprocessor could "time out" and cause the machine to stop running whenever the CPU can't handle all that it is requested to do.

☐ *The drive is rated for constant torque and the speed range is such that the drive can run down to 10% speed. Any worries?* Yes. Is the motor able to cool itself at those low speeds? The drive will be delivering a substantial amount of output current to the motor. If the motor's fan is attached to the shaft, then the actual fan rpm's will not be enough to cool it properly. An auxiliary blower might be necessary. The drive will not need any modifications.

☐ *The drive can stop or brake the load.* True, only if the drive is given adequate time to dissipate the energy generated by the motor or if the drive is fully regenerative. The drive protects itself when it cannot handle fast stopping by faulting.

☐ *If we make this drive and motor a 100-HP system, then we won't have to worry about running this 10-HP load.* Not necessarily so. A dc and even an ac motor like to be loaded. They run much more stably under proper load conditions. For instance, lighter load will wear the brushes on the dc motor inconsistently. It is always proper to match horsepower and current requirements whenever possible. Oversizing can be as much a problem as undersizing. Besides, a 100-HP system costs more than a 10-HP system.

- ☐ *Since dc has been around longer than ac, it is the better technology*. Not necessarily so. For energy savings (variable and centrifugal torque applications), ac drives are much more attractive. The ac might even be a better solution than dc on some constant torque applications.
- ☐ *My application only requires reduced-voltage starting not variable speed.* Today, the ac drives give both and a whole lot more. Soft starts, limited inrush currents, variable speed control, energy savings, and motor protection are just a few other electronic drive advantages over reduced-voltage starters.
- ☐ *We have to have premium efficiency motors with the variable frequency drives for maximum energy savings.* Maybe not the best recommendation. Premium efficiency motors will yield the best savings over standard motors when running full speed, across the line. However, whenever an electronic ac drive is incorporated, the drive's volts-per-hertz pattern is actually saving the energy. The extra expense of the premium efficiency motor might not be necessary.
- ☐ *If a servo drive can perform any application that an equivalent ac or dc drive can, why not just use them all the time?* Cost and complexity will curtail that idea. The servo costs more, might require programming, and might be more complicated to maintain. It is also more sensitive to line interference.

From the examples shown, it is obvious that many early drive misconceptions and misapplications stemmed from poor sizing. A general awareness might be beneficial to many, especially managers, so that the complexity and power of the drive product can be understood and adequately considered. Attention to details will save serious startup problems and possible downtime later.

The drive market

The drive's market is a multibillion dollar industry, and it continues to grow. Also, many by-products and by-product industries have emerged. The industries served are virtually any. Petrochemical, automotive, food and beverage, heating, ventilating, air-conditioning (HVAC), material handling, pulp and paper, textile, machine tool, film and packaging, steel and aluminum, and many others are using ac and dc drives as this chapter is being written. It is also estimated that more than half of the total applications are variable torque and involve fans and pumps. Each market industry has its own drive niches and requirements for electronic drives.

The drive aftermarket is very broad. The motor industry has actually benefited from the proliferation of drive products. New designs of motors have emerged, such as the inverter duty motor. Other designs of motors for low-speed vector applications have been introduced. In addition to drive specialists, system integration has grown into a multibillion dollar business; not only do drives and controls get integrated, so do computers and other electrical components. Power semiconductor manufacturers have benefited, as have microprocessor concerns. Additionally, engineering and consulting firms are now retained for their expertise in drive applications. A new breed of service engineer has come out of the drives business. Solving motion control problems and troubleshooting drives becomes paramount.

As for the technology itself, it is constantly evolving. In fact, it is evolving so fast that by the time the investment is made and equipment is installed and running, a newer product is emerging. This means that the drive had better perform right away in order to pay back the investment. An analysis of different technologies and their useful life versus their initial costs is shown in figure 1-4.

Drive electronics is becoming one of those necessary evils. In order to automate and keep up with competition, a facility must incorporate drive products wherever they can. Throughout this book all types of drives are discussed, exposing their advantages and disadvantages. Hopefully, it will help you decide appropriately!

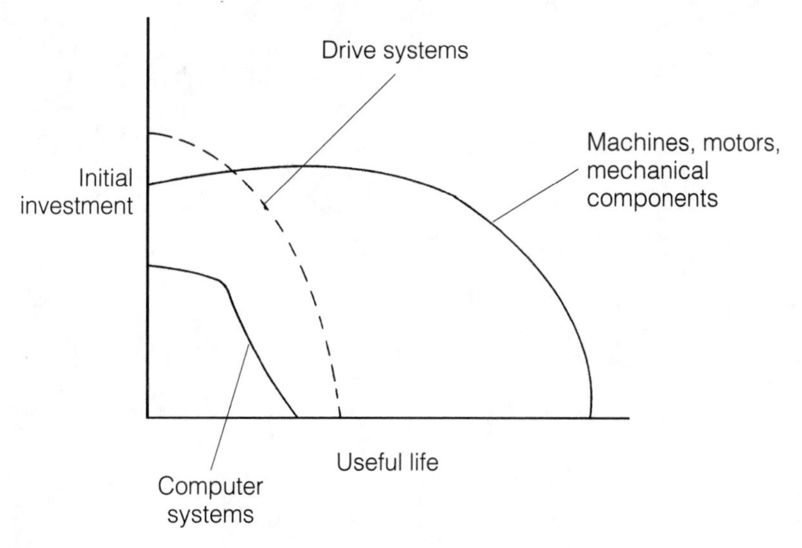

■ **1-4** *Initial drive costs versus useful life.*

2

Motion control

BEFORE ELECTRONIC AC AND DC DRIVES, THERE WAS still motion control. This motion control was in the form of starting and stopping machines by using steam and pneumatic means along with hydraulic methods. Even today, there is still quite a bit of nonelectronic drive control going on around us. The fact remains that if some machine in the factory is moving, if air is being blown, or if water is being pumped, then there is motion. And with motion comes the need for motion control.

Motion control can be as simple as opening and closing a valve or as complex as programming a six-axis robot to perform multiple circular interpolation-based moves simultaneously. The starting and stopping of electric motors is motion control, as is an air-actuated plunger knocking bad parts from an assembly-line conveyor. Many methods and disciplines are used for motion control. However, for our purposes, motion control has really only attained the level of a separate discipline within the last 10 years. With ac, dc, and specialty drive proliferation, specialists in motion control have emerged.

These motion control specialists are likened to automation engineers. They might be mechanical, electrical, or computer engineers by schooling, but they must have multiple-discipline backgrounds in order to fully understand all the different motion control applications. They also have to understand different products and how to apply them, since they are always trying to provide a solution to the motion control application. Certification programs are in place today to test and certify individuals as motion control specialists. The specialist must combine experience, education, and specialized study in a manner that makes him or her the person that we go to in a motion control application.

Electronic drives have caused motion control to rise to a higher level of technology. With powerful microprocessors and elaborate software platforms, drives can perform complicated tasks today that used to require multiple pieces of equipment. Because most of today's drives are programmable, many more tools are available to the creative motion control specialist. And because the technology of motion control is constantly changing, the motion control specialist must keep up with the technology.

However, as much as we strive to engineer a project correctly, because so many variables are involved, there are going to be mistakes in overlooked data; problems with new, high-tech equipment; and performance expectations not met. This has led to the field of *reengineering*. How often have we used the phrase, "There was not enough time to do the job right the first time, but time was made to do the job over"? This is particularly true with retrofitting or modifying old machines, reusing most of their original parts. These are the jobs that carry the hidden surprises.

The motion control engineer, or the resident drive or controls expert, is given the task of putting advanced electronic controls on an old machine, and possibly replacing some older mechanical parts, too. Mixing electronics with mechanics is always interesting. Electronically commanding a 4,000-pound hunk of metal to move from point A to point B in 2 seconds doesn't mean it will! More often, we find out why something didn't work out as planned after implementing the original plan, then trying again and again.

Motion control can be broken down into individual categories: speed, or velocity, control; torque control; current control; position control; and digital input and output control. For our purposes, we will look mainly at motion control as it pertains to controlling electric motors. Control also implies regulation. This regulation is the drive controller's function. Speed and current regulation make or break the application's performance. The amount of regulation required will also determine the initial costs for drive equipment, as illustrated in figure 2-1 for typical degrees of speed regulation.

In simple terms, speed regulation implies that an expected percentage, or actual motor rpm's, can be measured at the motor. If speed regulation is 3% for a 1,750-rpm motor, we can be off by 52½ motor rpm's. Can the application tolerate this? If not, a better, more expensive approach might be in order. What is usually ambiguous is under what conditions this speed regulation is based. The loads obviously have the main impact. Are the loads steady-state or changing rapidly, severe or light? These conditions have to be considered and compared with the predicted speed regulation in order to achieve optimum system performance in the application.

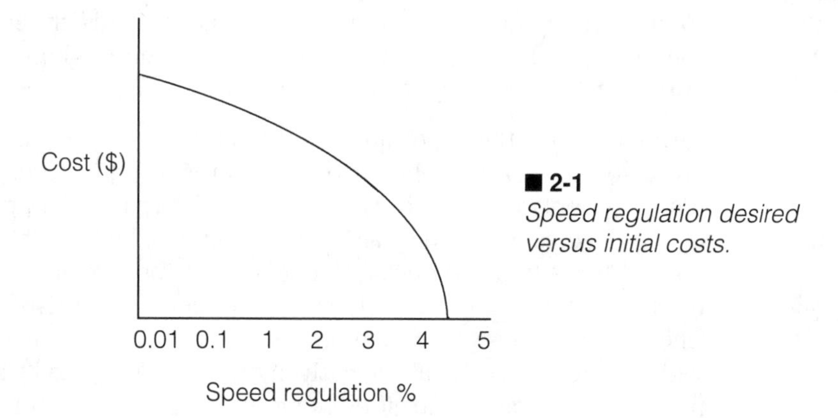

■ **2-1**
Speed regulation desired versus initial costs.

Speed control sounds simple, but it isn't. Let's take a motor, drive, and some electricity and run that motor at half speed. Accomplishable, yes, but what really happened to get the desired result? First, someone had to physically mount the motor and drive somewhere. The motor has to be connected somehow, mechanically, to the load or machine it is driving (hopefully, there are no problems in the machine). Then all the proper wires had to be pulled to the right terminals at motor and drive. Next, the means of control had to be considered, which meant more wire had to be pulled. At this point, we are at least ready to apply power. Many times the motor is uncoupled from the load and run to see if shaft rotation and direction is correct. We also check for commanded speed with actual speed whenever possible. If there is a tachometer, is it connected properly and scaled accordingly within the drive? We also must make sure that we are in sync with the tach feedback.

Then the fun begins. The load is reconnected to the motor and the motor is "bumped," or given a run command at some lesser speed to check that the load can be moved. After the bump is considered a success, we begin running up to full operating speed with load. Next, we run through multiple load changes and conditions, try going in reverse (if the drive can accommodate this), and check for any other machine nuances that might affect the speed. All this just to run a motor at half speed off of an electronic drive.

Running the motor is one thing, driving the load (which is the whole reason for applying the equipment) is another. Driving the load or machine is actually work done. Work means production, but to do work, the motor has to produce torque. Torque is needed to move the motor's shaft and to move that which is attached to the motor's shaft. An electric motor will demand more electrical current in order to produce more torque.

Again, stressing the importance of constantly driving the load at a given speed, the electronic drive and motor will give torque and current priority over speed. This means that in a particular process, a motor's actual rpm's will "sag" so that there will be enough usable current to continue producing torque to drive the given load. Many electric motors will continue to seek more current to drive a load. This is more true with dc motors than with ac motors; however, with an electronic drive ahead of the motor, we are assured that the motor will not damage itself by requesting too much current. The drive monitors this.

With motion control products being so powerful, flexible, and programmable, positioning-type applications have become very practical. Older technology directed us to positioning motor shafts by turning power on and off to the motor or by using clutches and brakes to achieve some semblance of shaft position control. If we control the position of the motor shaft and the rest of the mechanical system is stiff, then we have successfully positioned the machine.

The old way meant that accuracies were not as critical; nothing was repeatable or consistent, and components would drift. With electronic controls, positioning became more plausible. Also with programmable and digital electronics came more exacting methods for closing the position loop outside of a motor and drive system. Servo and stepper drives are true positioning-type drives. Chapter 7 discusses servo and stepper drives at greater length.

Finally, many processes require that other actions happen. These actions might be motion-related, or they might involve doing something else, such as merely turning a light on at an operator station. This is *digital input and output (I/O) control*. With today's electronic drives being furnished with both physical modules for the landing of low-voltage signal wires and the ability to (with software) scan and respond to an input, we have the makings of more than a motor controller; we have a motion controller or even process controller. Starters can be instructed to turn a motor on at full speed at a time and duration determined by the programmed electronic drive. Various input signals from an operator station can tell the drive when to run, in which direction, how fast, and so on. Likewise, the drive can carry a sophisticated program that can virtually run a machine, with all the motors and peripheral equipment, all by itself. It just needs power and a "go" command.

High-speed microprocessors have enabled the drive industry to achieve major strides in control. Coupled with the advancements in power semiconductor technology, drive designs have become dramatically smaller in package, faster in processing, and more intelligent. Compared to the CMOS, relay logic controls, and even op amps of yesterday, the technology of today is changing quickly—perhaps even too fast. For example, a drive manufacturer introduces a new design or concept, has some success with it, and then their competition improves on that design. The first manufacturer then makes the next improvement and the cycle continues.

What happens is that the end user of the drive product tries to standardize with a certain brand of drive by the same manufacturer, but to no avail. The manufacturer has changed the design so much that the product is virtually brand-new. The technology has been changed so much that the technicians and designers have to learn all over how to troubleshoot the drive product. Spare parts in inventory are obsolete and the frustration continues.

Thus, the motion control technology is not standing still. Electricians must be constantly trained on drive products as well as applications. Many traditional users continue to implement tried-and-true analog systems, which is why some old types of technology for motion control still are utilized today. Many individuals don't want to have to learn and relearn when they can achieve a similar result with older equipment. However, electronic, programmable motion control in some form or another must be achieved in the factory in order to compete on a national and world-class level. Also, the initial costs and benefits are beginning to far outweigh the negatives of learning a new technology.

Motion control methods

Motion in the factory is mainly actuated by either pneumatics (air), hydraulics (liquid), steam, or, most frequently, electrical power. The least common, steam, is sometimes used in larger plants to drive turbines to generate electricity for internal use. Pneumatic-, hydraulic-, and steam-actuated motion is controllable. However, there are disadvantages associated with each, which is why electronic motion control of electrically powered devices is most common.

Pneumatic and hydraulic systems tend to be high-maintenance. Hydraulic systems tend to leak and are usually dirty, making them unsuitable for industries needing clean plant environments, such as the food industry. Pneumatic systems often get water in the air line and are noisy. Pressure losses equate to poor performance. Today, pneumatic and hydraulic uses in motion control are either specialized or supplemented to machines mainly controlled electrically. Using a pneumatic or hydraulic solution is often appropriate only because of initial costs.

Many older plants have air lines readily available and a compressor in place. Initially, an air-driven solution is more cost-effective than a new electrical scheme but probably not in the long run. Costs for maintenance and possible loss of production must be considered. Interestingly, the compressor motor is electrically powered. Likewise, any hydraulic system needs some type of pump motor, which is usually electrically powered.

The evolution of variable-speed control

With the introduction of the induction motor, various methods emerged to reduce their speeds for an application. Early on, electrically powered machines used these constant-speed ac motors as the prime mover. Some used a dc version of the motor but needed an electrical rectifier. Obviously, as we have probably known for years, ac motors, which ran at full speed, were energy wasters when the speed was restricted mechanically. This technology also did not allow for any closed-loop velocity or position control. Since the motor ran full speed, gear reduction had to be incorporated in order to run the machine at an appropriate slower speed.

Mechanical solutions, such as brakes and clutches, initially gave plant workers some means of control. The ac motors were a necessity in the factory, even though controlling them was another issue. Even motor-generator (M-G) sets, used to control dc motors, incorporated a constant-speed ac motor as the prime mover.

M-G (motor-generator) sets

The motor-generator scheme was an initial method of controlling the speed of the dc motor. The M-G set was actually comprised of an ac motor and a dc generator. The ac motor was coupled to a generator to produce direct current electricity, which could be controlled to change the speed of a dc motor. The prime mover could also be gasoline- or diesel-powered to run the generator, thereby still eventually powering the dc motor. This was ideal for remote and isolated locations. The mainstay in industry for years, the M-G set provided direct current power to dc motors. Figure 2-2 shows an M-G set configuration.

With new ac and dc control technology, M-G sets are not very popular anymore. They are very inefficient and use more energy than is needed to perform the job. Also, finding replacement parts is difficult (if not impossible).

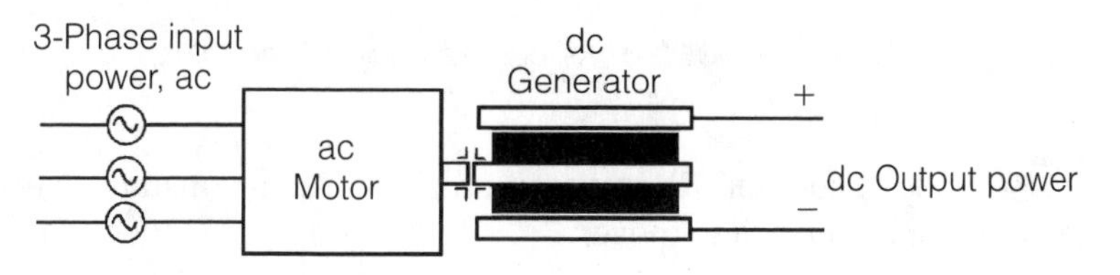

■ **2-2** *Motor-generator set.*

Out of the M-G set evolved the static controller scheme for dc motors. Static devices were developed for mainly dc motors to achieve variable-speed control. Static devices could be equated to any power conversion system that did not require a motor driving a generator to achieve the desired electrical output. In other words, nothing mechanically or dynamically was used to convert ac to dc or dc to ac. The power conversion would be handled through strictly electrical means. Power-converting devices took in ac electrical energy in one form, and dc energy emerged out in another form. Vacuum tubes (or electron or thyratron tubes) were used to change the dc voltage to the dc motor's field or armature.

These systems took up a lot of space on the factory floor and also gave off a lot of heat. Today, purchasing replacement parts for these systems is often difficult. Many are being replaced with digital dc drives or ac drives.

Eddy current clutches

Figure 2-3 shows an eddy current clutch. As the name implies, the eddy current principle is in use here. Eddy currents, induced within the conducting material by the varying electrical field, cause the desired effect of changing output speed. Soft-starting and high-torque output, especially at low speeds, made this device a workhorse. The main disadvantage is that the clutch system is very inefficient and must be cooled, either by water or air. Even with the cooling, a lot of energy is lost as heat. In its heyday, the eddy current clutch was probably the most widely used variable-

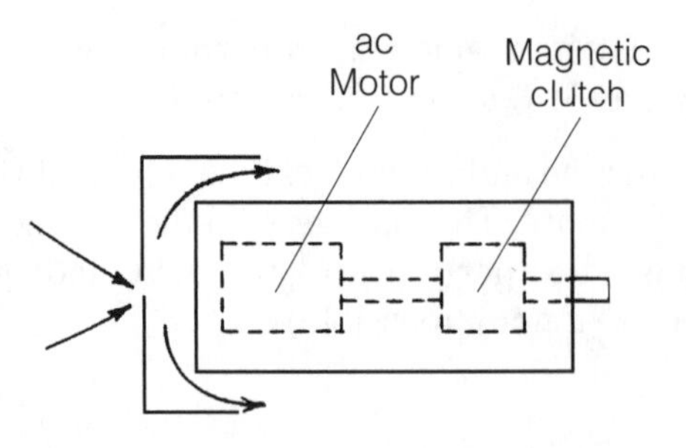

■ **2-3** *Air-cooled eddy current clutch.*

speed solution. However, it still required a constant-speed ac motor as the prime mover.

Along the way, other entrepreneurs came up with various techniques for reducing the speed of electric motors. Most focused on ac motors, and many approaches were mechanically based. Some were industry- and application-specific, while others incorporated hydraulics. In many cases, the ac motor still had to run at full, constant speed, thus making the device very inefficient. Not until electronic ac drives emerged did we start getting away from these high-maintenance, multicomponent energy wasters.

One such device is the variable-pitch pulley, sometimes called the V-belt drive (figure 2-4). Motion commences with a full-speed, ac motor, which is attached to one of the pulleys. By adjusting the distance between pulleys and by changing the actual depth of the pulley, which can be opened or closed, the output can be changed. The V-belt moves up and down within this adjustable groove.

This speed controller is employed throughout industry and is still used today, but in dramatically reduced numbers. The reasons for its demise include limited speed range turn-down, belt slippage during acceleration, general V-belt wear and breakage, and limited horsepower sizes. Also, soft starting of the load is not practical. The V-belt was good in its time because it was a simple, inexpensive design.

Another common device used to slow the speed of a process but more useful for soft starting is the fluid coupling (figure 2-5). This device has internal impellers that turn from the motion of a prime mover. Fluid is centrifugally forced throughout the coupling, eventually creating motion and, finally, full speed. The fluid coupling is used mainly for soft-start control, but some speed control is attainable by controlling the flow of liquid, usually some type of oil-based product. It can reverse a load without damage to mechanical parts internally.

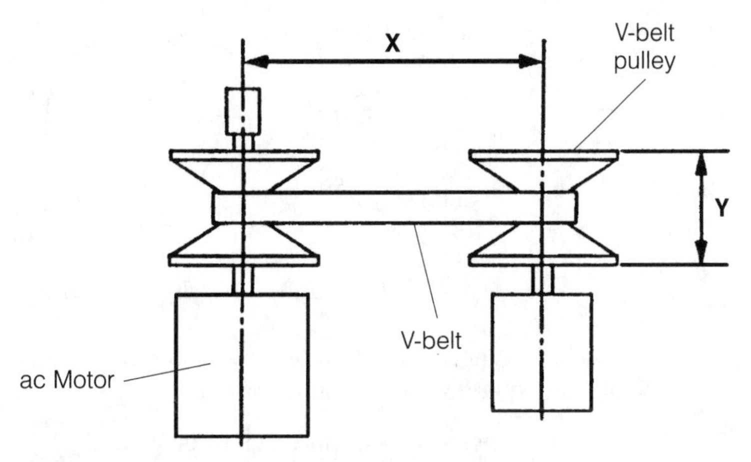

■ **2-4** *V-belt drive or variable-pitch pulley.*

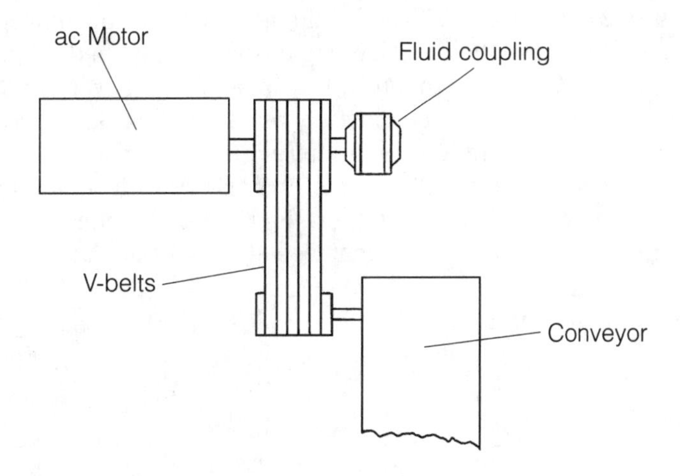

■ **2-5** *Fluid coupling.*

Fluid couplings are inefficient devices because of their constant-speed motors and various heat losses. In addition, fluid couplings are prone to leaks and require routine maintenance (e.g., changing the oil). They are available in very high horsepowers and are generally lower in cost, initially.

A variation of the fluid coupling is the fluid-based speed variator. This device, illustrated in figure 2-6, could be called a fluid coupling in-a-box. By configuring internal jets and impellers a certain way, the fluid can flow against those parts producing motion. The fluid-based speed variator offers soft starts, reversing, and a wide speed range. It can run very low speeds and provide constant output torque. However, it gives off a lot of heat, needs a periodic oil

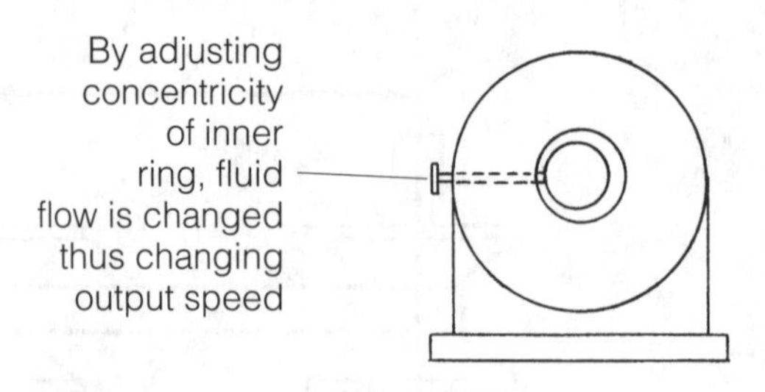

■ **2-6** *Hydrostatic fluid variator.*

change, costs more initially, has horsepower limitations, and needs a constant-speed ac motor on the input.

Elaborate, expensive transmissions offer a similar speed control without using a fluid as the basis for motion. These transmission boxes, sometimes called PIVs, have many high-precision gears housed within. The machining of these gears and the sheer quantity make these devices expensive. A simplified PIV is shown in figure 2-7. Unfortunately, the speed range is very limited. Also, if a gear breaks, other mating gears may break, and repairing them costs a fortune. PIVs require oiling and maintenance and are not very efficient. An ac motor must provide motion as the prime mover. Additionally, there is no real soft start capability and reversing can only be achieved after the transmission is stopped.

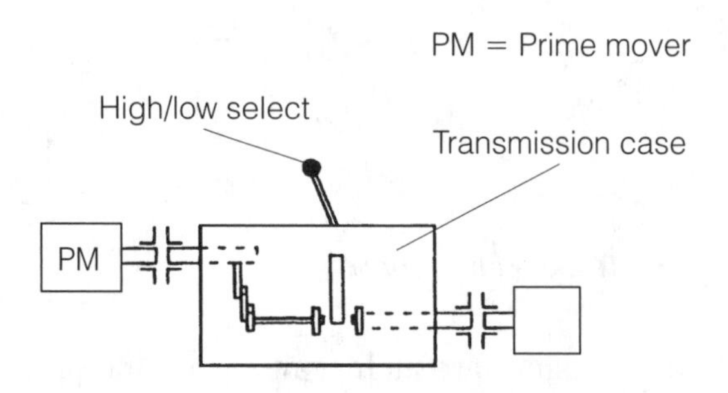

■ **2-7** *Variable speed transmission, PIV.*

For fan and pump applications, methods for slowing air and fluid flow range from simple to complex. However, all the methods still are dramatic energy wasters because of the affinity laws of speed and power. This is the biggest reason that more electronic drives are incorporated into these kinds of systems every day. The common methods used to slow airflow are inlet vanes and dampers in the ductwork system (figure 2-8). In pumping systems, the flows are usually restricted by opening and closing valves in the piping

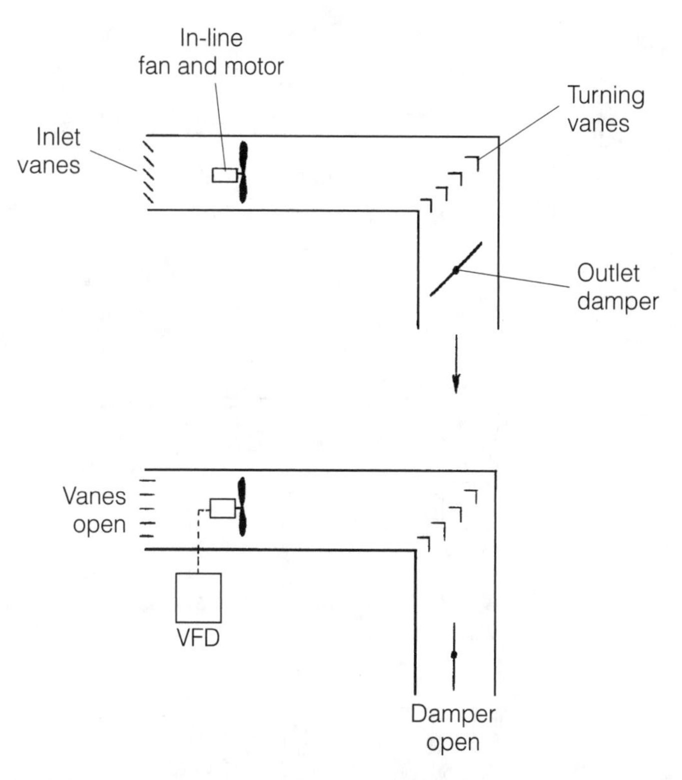

■ **2-8** *Fan system with inlet vanes and outlet dampers with VFD, vanes and dampers are left wide open.*

system. The full-speed ac motor and valve arrangement is shown in figure 2-9. Further discussion on energy savings and these types of applications appear in later chapters.

One other method applied to large fan systems is the use of variable-pitch blades. By mechanically changing the pitch of the blade, a different surface area is introduced into the system (figure 2-10). This has the same effect as a parachutist falling faster or slower depending on body (surface area) position and wind resistance. The fan will turn slower or faster based on the position of the louvers on each blade. This approach does not allow for great deviations in speed but is useful for high-horsepower fan systems. Coupled with a two-speed motor, the speed range is greatly enhanced. However, energy savings are still not evident, and special, complex fan blades are expensive.

Throughout this analysis there have been allusions to electronic drives. Speed, torque, and motion control have to be accomplished in some way. Previously, the choices were limited to mechanical or

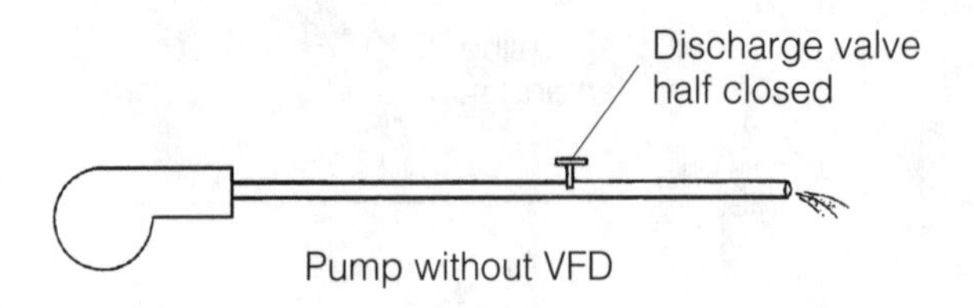

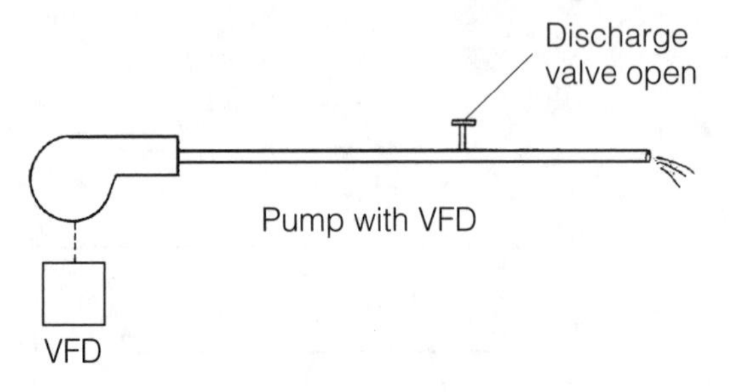

■ **2-9** *Pump system with discharge valve with VFD, valve is left open.*

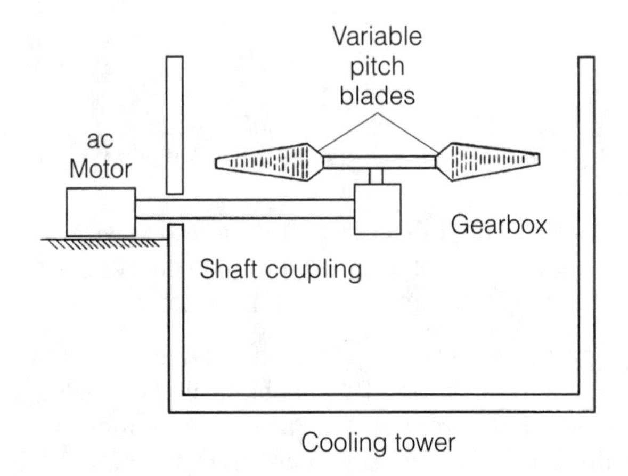

■ **2-10** *Variable-pitch fan blades.*

electric devices. Now there is an electronic choice: the drive controller; *electronic* because of the computerized nature of converting the electricity to control a motor. For motion control, electronic drives are the wave of the future.

Closing remarks

Motion control is the essence of production on this planet. Used in the slowing or speeding up of a system, it also entails the starting and stopping of a process, along with controlling when and how other events in the process occur. Motion control is whatever the designer or engineer wants it to be. At the heart of motion control will be found some type of electronic drive.

With advancements in power semiconductor technology, newer feedback technology, and the more frequent use of microprocessors, major strides have been made in motion control, primarily in the past decade. Sometimes it seems as though we are advancing too fast, and many technicians and plant personnel will agree. But with technology comes progress, and over time, after bugs and kinks are worked out, new technologies are finally accepted.

Beyond speed and torque control have emerged a plethora of unique control concepts regarding position control, I/O control, and many more specialty drive applications. System integration, competition, faster semiconductors, better software platforms, and a general acceptance of the present technology will further advances in motion control.

Electronic ac and dc drives can be found in schools, factories and industry, hospitals, and practically anywhere there is an electric motor. Whether ac or dc, the drive is becoming more and more accepted as the method of slowing the speed in the process. Reliability has improved dramatically. Misapplication of electronic drives is dwindling (with the help of books such as this). The benefits are starting to far outweigh the negatives. And with costs and competition bringing the electronic drive to many, proliferation will continue.

3

Basic electricity, electronics, and power transmission

IN ORDER TO UNDERSTAND AC AND DC ELECTRONIC DRIVES and how to apply, repair, and troubleshoot them, you must be knowledgeable in mechanical, electrical, and computer engineering. Electronic drives mix these disciplines in almost every application. After all, the drive is actually nothing more than a robust electrical circuit, complete with resistors, inductors, capacitors, and power semiconductors. The drive controls an electric motor, which is mechanically coupled to a load and has to move, do work, and produce torque (all mechanical concepts). Our electronic drive is also a computerized device. Today's drives are fully digital, microprocessor-based controllers, and therefore have computer software, firmware, and hardware inherences.

The engineer or technician must quickly become a drive specialist. By learning a machine and its electronics, he or she becomes the company's resident expert on that specific type of equipment. A certain controller, made by a specific manufacturer, is programmed and maintained by the person in the plant who is most familiar with it.

Companies are constantly attempting to cross-train their technical workforce because of the rapidly changing technology. More than one individual has to be knowledgeable about the same electronic drive. These individuals have to have a tremendous understanding of the basics in many disciplines in order to effectively apply, troubleshoot, and repair drive products. For example, an electrical engineer might be asked to size a motor for a part of a machine because it is, after all, an electrically powered device. At this point, the electrical engineer must quickly find out what torque, inertia, and their relationships to speed really mean. He or she might also have to select a coupling and a gearbox.

The on-the-job experience doesn't end with an off-discipline assignment here and there. Today, most all pieces of electrical equipment have some type of microprocessor content to them. Whether the microprocessor is in the drive itself or in a sensor feeding the drive, the engineer or designer has to understand computers. This can involve anything from the engineer actually programming the product to simply knowing which microprocessor is used in the drive.

Electrical basics

In drive and motor systems, electricity and magnetism are the basic elements behind the scenes. Once thought to be two separate forces, Albert Einstein's theory of relativity showed that the two have much in common. Electricity is produced from basic electric charges. Magnetism is created by charges in motion and reacts with moving charges. Electromagnetic forces and magnetic flux are all a result of the presence of electricity. Electricity is a form of energy that consists of mutually attracted protons and electrons (positively and negatively charged particles).

Therefore, with every electronic drive, electrons are flowing at some level. These electrical functions may or may not be apparent. The raw, incoming power to a drive for rectification is one level. Communications power is a form of low-level electricity traveling over conductors to pass data. Motion, on the other hand, is caused by an electric *prime mover*, or motor generated by higher levels of electrical energy and converted into mechanical energy. It all originates with the power-producing utilities and comes to us in two forms: ac (alternating current) and dc (direct current).

From substations outside the facility, the electricity is distributed throughout, usually via transformers. These devices actually provide the usable voltages, 460 V, 230 V, and 115 V being the most common. In addition, lower voltages such as 24 V and lower are still required for most control schemes. Most of the drive equipment commercially available in the United States requires a supply voltage previously mentioned. Some voltages, such as 575 V, are used in various parts of the United States and Canada, but are not as common. Interestingly, odd voltages do exist within certain facilities and industries (e.g., the mining industry uses 1,000-V power quite extensively). Typical frequencies are normally 60 cycle/hertz in North America and 50 cycle/hertz outside North America.

Electrical formulas

The three basic elements in an electrical circuit are voltage, amperage, and resistance. Voltage, usually shown as V or E, is the force that causes electrons to flow. Amperage, or current, measured in amps, is expressed as I or A and is the actual flow of electrons. Resistance, R, measured in ohms, is the opposition to current flow. These three elements make up Ohm's law, from which many basic electrical circuit calculations can be made (see figure 3-1). Ohm's law is adequate for dc circuit analysis and for some ac circuit analysis. Three-phase power, however, tends to be a bit more complicated. Figure 3-1 shows a pie chart that is useful when trying to remember the following equations:

$$V = AR, \text{ or } volts = amps \times ohms$$

$$A = \frac{V}{R}, \text{ or } amperage = \frac{volts}{ohms}$$

$$R = \frac{V}{A}, \text{ or } ohms = \frac{volts}{amps}$$

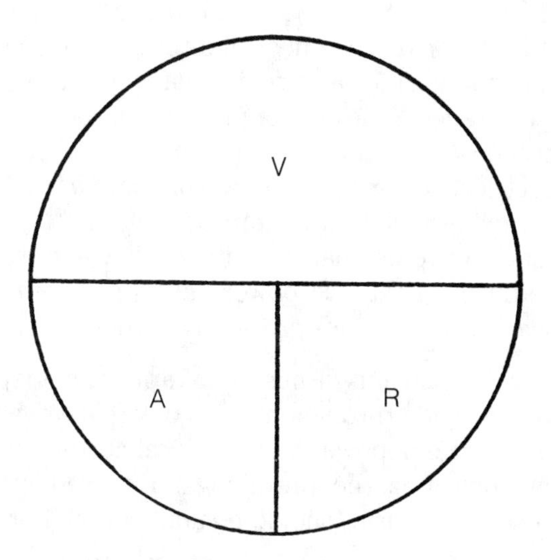

■ **3-1** *Ohm's law pie chart.*

When using the pie chart, the V value is shown above the A and R values. This represents V being divisible by either A or R when solving for either A or R. Likewise, to find V it is necessary to multiply A times R. The pie chart is an effective and easy way of remembering the aforementioned equations and their relationship to each other. Ohm's law allows the technician to quickly determine whether or not there is an immediate discrepancy.

There are many derivations to Ohm's law, as well as other laws of electricity, and the technician must understand all of the principles involved. Complete college curriculums take the electrical engineer through all facets of circuit design and analysis.

Several analyses and transformations are used to prove and disprove certain issues relating to drive equipment. However, for our purposes, we will use some basic rules of thumb to help keep drives and motors online. Also, we will look at certain analyses, postulates, and transformations only to indicate how and why they are important to the application of electronic drives.

One such transformation is called the La Place transformation. It is a mathematical integration used in harmonic analyses and other issues related to electricity, magnetics, and gravitational concerns. Another is the Fourier analysis, which is also useful in harmonic and waveform investigation. These theorems are used frequently in relation to drive equipment. Harmonic distortion due to power rectification by drives lends itself well to these calculations and theorems.

With electronic drives, power is defined by the type of drive in use, either ac or dc. We must differentiate between equations for dc and equations for ac circuits mainly because of steady-state conditions, such as dc and for practical purposes, single-phase ac, and the nonsteady state conditions, such as three-phase ac power. Electrical power P, or the rate of doing work, is measured in watts and is expressed for dc circuits as P (watts) = $V \times A$. The ac circuits have to be analyzed differently when three phases of power are involved. The ac power has to be averaged and RMS (root-mean-square) values provided in order to get proper results.

In drives, power is discussed in terms of horsepower. However, the point is made throughout this book that drives should be sized by current. Horsepower and electrical current are related, and since horsepower is the prevailing term used in motor discussions, horsepower basics should be understood. For a given dc circuit:

$$horsepower = \frac{volts \times amps \times efficiency}{746}$$

Since production is a direct relationship to work in the industrial plant, electrical power usage is also a gauge. Facilities employing electronic drive and motor products today are trying to get the absolute best output for the electricity that they purchase. Older mills and factories had enormous electrical bills—and some still do. Now the mode is to have equipment as near to 100% efficient as possible. This is factored into automating any plant, new or old.

The equation

$$efficiency = \frac{746 \times output\ horsepower}{input\ watts}$$

can be used for ac circuits and is a good indicator of where a particular process or piece of equipment is relative to its cost and productive output. Where energy savings are concerned, an electronic ac drive is justified not only by its payback from running a motor with less energy, but also because it uses every bit of available electricity throughout its circuit when running that motor. Heat loss is wasted energy. It is always proper to evaluate the entire transformer-drive-motor package for a total efficiency of the system.

The incoming power to the plant is usually three-phase alternating current (ac). Figure 3-2 lists various common formulas for power in ac circuits. There will also be some single-phase power available, especially at lower voltages, mainly for lighting and office circuits. The bulk of the work done in the factory is powered by three-phase systems. Motors are designed and built around three-phase supplied power, although there are many single-phase motors in use. Consider which is the driving force from a design

single-phase horsepower = *volts* × *amperes* × *efficiency* × *power factor*/746

single-phase power factor = *input watts*/*volts* × *amperes*

single-phase efficiency = 746 × *horsepower*/*volts* × *amperes* × *power factor*

single-phase amperes = 746 × *horsepower*/*volts* × *efficiency* × *power factor*

single-phase kilowatts = *volts* × *amperes* × *power factor*/1000

three-phase horsepower = *volts* × *amperes* × 1.732 × *efficiency* × *power factor*/746

three-phase power factor = *input watts*/*volts* × *amperes* × 1.732

three-phase efficiency = 746 × *horsepower*/*volts* × *amperes* × *power factor* × 1.732

three-phase amperes = 746 × *horsepower*/1.732 × *volts* × *efficiency* × *power factor*

three-phase volt-amperes = *volts* × *amperes* × 1.732

three-phase kilowatts = *volts* × *amperes* × *power factor* × 1.732/1000

efficiency = 746 × *output horsepower*/*input watts*

■ **3-2** *Common ac electrical formulas.*

standpoint, in other words, why three-phase over single-phase? One important factor is the conductor size. If a single-phase motor with hundreds of horsepower had only one wire running to it, the diameter of that wire would be immense. Large wires are very difficult to work with and route in the factory.

Thus, three-phase systems, which incorporate three individual wires plus a ground conductor, are the most common. The current is spread out, or averaged, over all the wires. The three-phase incoming power can be rectified into dc power in order to run dc motors. The dc systems are discussed in more detail later in this book. Also, dc is often the only power choice in remote locations when oil or diesel generators are producing the power. Obviously, there is much to know about electricity and electrical calculations.

Frequency and amplitude

Frequency is the amount of electrical pulses that are transmitted over a given period of time. Frequency is expressed in hertz (Hz), or cycles per second. For example, most power in the United States is in the 60-Hz range. This means that every second there are 60 pulses of electricity through a given point in a wire. Figure 3-3 shows a typical waveform depicting the frequency portion of the wave and the other necessary portion, called *amplitude*. This wave is actually in sine wave form, commonly referred to as the *fundamental*.

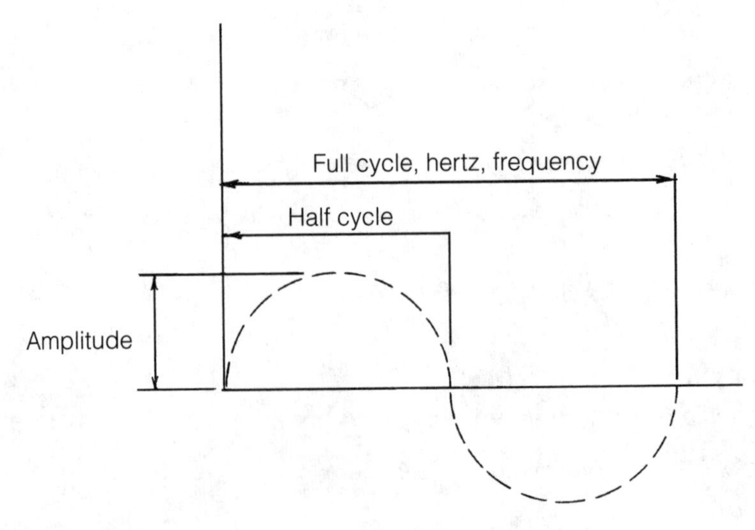

■ **3-3** *Frequency and amplitude, sine wave.*

For every electrical wave, there must be a corresponding amplitude in order to provide any usable power. This amplitude is often in the form of voltage. This wave has two time-based components called *periods* or *cycles*. A half period is one of the halves of the wave and is sometimes called a *half-cycle*. A half period is illustrated in figure 3-3 and is typical for ac. Figure 3-4 illustrates a square wave, and figure 3-5 shows the typical triangular wave shape, which are seen periodically with various electronic devices. Outages sometimes are expressed in cycles, or portions of a second. If the lights flicker, most likely either the amplitude for a given cycle or so dropped, or a complete cycle or wave was not even present.

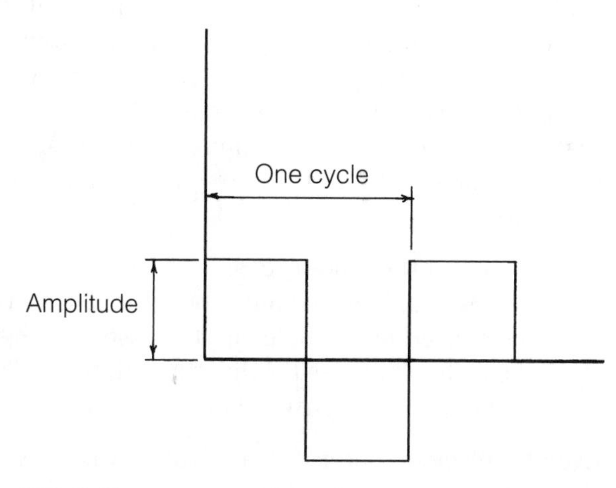

■ **3-4** *Frequency and amplitude, square wave.*

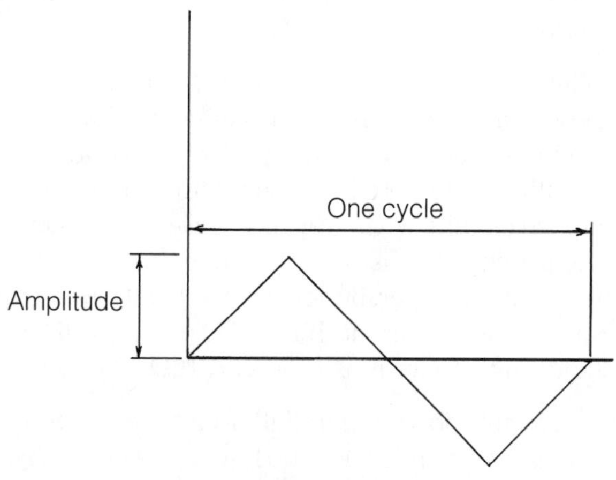

■ **3-5** *Frequency and amplitude, triangular wave.*

Resistors and resistance

The most often used electrical component in electronics is the resistor. Electronic drives are concerned with resistance from the motor control standpoint (a motor is a resistor), wire and cabling to and from the drive (copper or aluminum have resistance values), and all the resistive values within the drive circuitry (the drive's power semiconductors, resistors, and boards). Its ratings are in ohms, or R, and can take on many shapes and sizes, as evidenced by all the resistive components within an electronic drive controller.

In addition to drive equipment, electricity is flowing through many other different resistive devices in a facility. There is resistance in many areas of the plant we don't often consider. For instance, lights, computers, and most other electrically driven components have some resistance value. On a smaller, physical scale, when looking at a common printed circuit board (PCB), you can potentially find hundreds of small resistors populating the board.

Perhaps one of the more prevalent uses for resistors and drive applications is found in dynamic braking. These braking resistors might be housed in an enclosure because they are so much larger and have higher voltages flowing through them. The application of dynamic braking is discussed later in the book.

Even an electric motor has a resistive value in the overall plant electrical circuit scheme. Over time, some of these resistive values change with repeated heating and cooling of the components. Because impeding the flow of electricity is sometimes desirable, resistors play a big role in making electrical and electronic devices perform as desired.

While we're on the subject of resistance, let's look at parallel and series circuitry. Many electrical devices, such as resistors and capacitors, have to be strategically located in an electrical circuit. It should also be noted that there are subtle differences between ac and dc circuits when utilizing series or parallel schemes. Basically, the placing of resistor values in a circuit can provide solutions for the electrical personnel in the plant. For example, figure 3-6 shows a series circuit. If $R_1 = 7\Omega$ and $R_2 = 9\Omega$, then the total resistance in that circuit is $R_1 + R_2 = 16\Omega$.

Conversely, for the parallel circuit shown in figure 3-7, the same values for R_1 and R_2 exist. However, electrically, current flows dif-

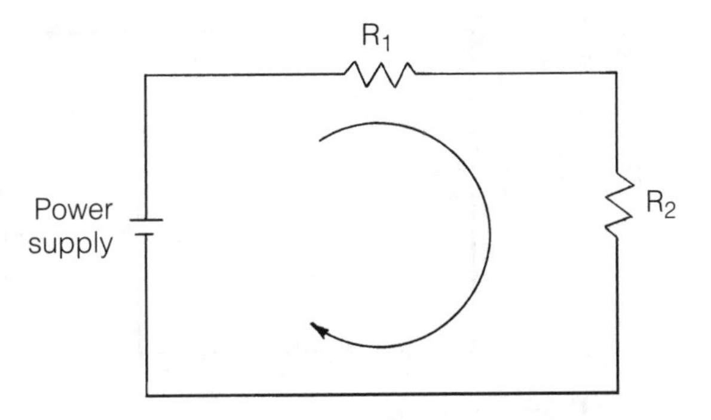

■ **3-6** *Resistance in a series circuit.*

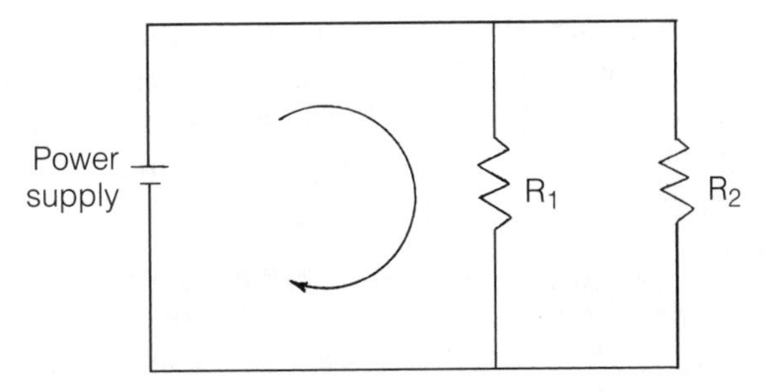

■ **3-7** *Resistance in a parallel circuit.*

ferently (the paths of least resistance), and we calculate the total resistance accordingly:

$$R = \frac{7 \times 9}{7 + 9} = \frac{63}{16} = 3.938\Omega$$

Thus, it can be seen that in one scheme we can get 16Ω of resistance, and in another, almost 4Ω.

Other components can be incorporated into the parallel or series circuits. By utilizing a capacitor in an RC (resistor–capacitor) network, the corresponding action is as a filter in an electrical circuit. This is shown in figure 3-8 and is used frequently in drives and drive applications.

Capacitors and capacitance

The capacitor is capable of storing electrical energy. It is sometimes referred to as a condenser because it consists of two conducting materials separated by a dielectric, or insulating material. The conducting materials get charged, one positively and one negatively, thus creating a potential between them. The size and type of mate-

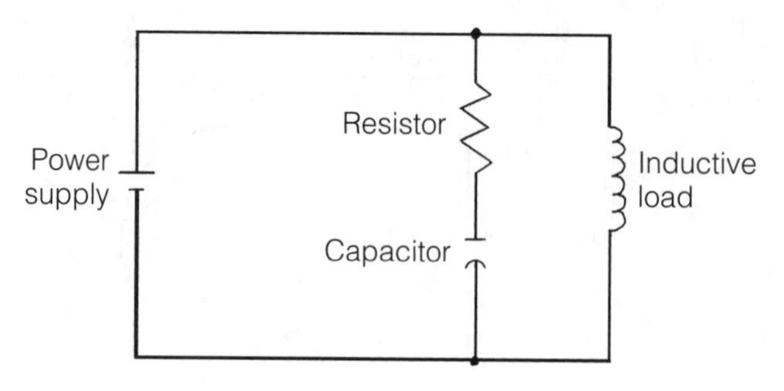

■ **3-8** *RC filter.*

rial of the conductors, the distance between, and the applied voltage determine the capacitance, in farads. (A farad equals 1 coulomb divided by 1 volt.) One farad is a large amount of capacitance. Usually, capacitances are in the microfarad range. Uses of capacitors include placing them into circuits for tuning purposes and even as batteries for stored electrical power.

Electronic drives typically have a capacitor network built-in. Depending on the type of drive, they are sometimes called *commutation capacitors*. These capacitors supply stored energy to turn inverter devices on and off.

Relays

Relays play an integral role in drive applications and drive control. Although digital control techniques are more common these days, relays have been used for a very long time in controlling machines and processes and are still used. Relays work on the principle of electromagnetism. This principle allows the relay to function as a device that can either automate or control from a remote location any electrical flow. There are basically two circuits: the relay circuit and the energizing circuit. These work together to open and close a switch. Enough energy is provided via the relay circuit to magnetize the appropriate element and, thus, complete the other circuit. This other circuit can be and usually is a higher voltage and carries more current than the relay circuit. The operation of the common contactor can be likened to that of the conventional relay.

The relay is still very much an integral part of drive control, although input and output modules within drive controllers are becoming commonplace. These I/O modules can accept direct low-voltage signals, which can directly be used in the drive's operational program. However, there are still many instances where the relay has to be implemented. Its switching is an important function in most every process, and years ago, attempts at automation were by relay logic. Relay logic utilized many interlocking relays and contactors to control machines and processes. If a machine was to start, then permissive relays had to be in place and energized to allow that start. Today, more equipment is controlled by computers and microprocessors that, in effect, do the switching in the software.

Inductors and inductance

Another component often found in conjunction with drive products is the inductor. Inductors are devices used to control current, and the associated magnetic fields relative to the currents, over a given period of time. Inductance is measured in units called henries, which are equal to 1 volt/1 amp per second. Typically, an inductor, called a transformer or reactor, consists of a coil of conducting material with a specific size and shape. This material is coiled around a core, most often of soft iron (sometimes the inductor is called a choke for this reason). Transformers are often used to slow the rate of rising current and for noise suppression, and are an integral part of the industry. Without these specialized inductors, there would not be available any usable electricity in lower voltage levels.

A transformer's simplified construction and symbol is shown in figure 3-9. Basically, transformers work on the principle of mutual inductance, hence, they are classified as inductors. In order for a transformer to work, there must be two coils, positioned so that the flux change that occurs in one of the coils induces a voltage across each coil. Typically, the coil that is connected to the electri-

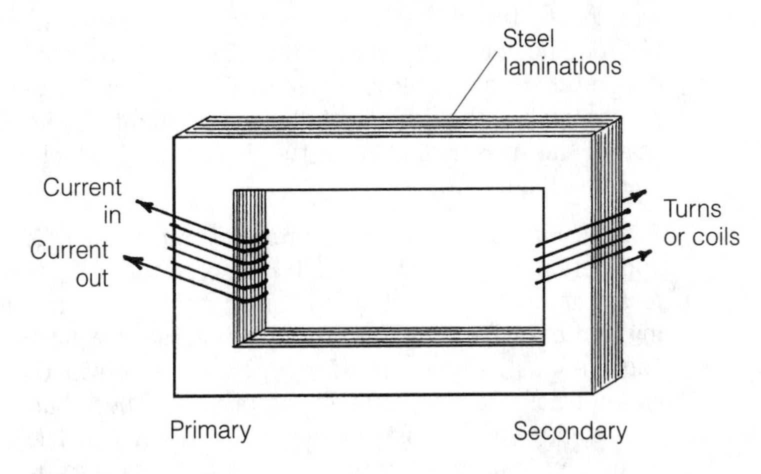

■ **3-9** *The transformer.*

cal source is the primary, and the coil that is applied to the load is the secondary. Transformers are rated in volt-amps (VA) and kilo-volt-amps (kVA). There are dry and oil-filled types, with the size sometime dictating which type to apply. There are also iron-core types, air core types, and units called autotransformers.

Autotransformers are a type of transformers that have one winding common to both the input and the output. Instead of having isolation between the coils and employing the traditional two-circuit principle, one winding is used and higher kVAs can be transformed. Often the cost of the autotransformer is the attractive feature, especially when voltages, which are to be matched, are not that far apart to begin with.

Transformers can be used as isolation devices and step-up or step-down devices. (Transformer applications with drives is discussed further in Chapter 8.) Often it is desirable to include a transformer with similar primary and secondary voltages, mainly to isolate one section of the electrical system from another. This way, there is no direct physical connection and conditions such as ground faults can be prevented from traveling throughout an electrical system and destroying components along the way. Also, the step-up or step-down transformers used to match voltages in a given system will offer isolation. Much of today's electronic equipment requires 460-V, 230-V, or 120-V power in order to function properly. Since it is impractical to run 460-V power many miles in large-diameter cables, higher voltages are carried to substations where transformers reside. From there, matching voltages to other equipment is provided.

Rate of rise with respect to temperature is a constant that is attributed to transformers and other peripheral equipment subjected to higher electrical currents. The rate of temperature rise is defined as the allowable rise in temperature over the ambient temperature when the device is fully loaded. Ambient temperature is that of the surrounding air at the device itself, which acts as the heatsink for cooling.

When comparing two or more different transformers for the same application, a rule of thumb is to compare the weights of the different units. If one has less iron and copper (or aluminum) winding content, then it will show in the actual weight. This might come into play when a transformer has to be provided in an application where harmonics might be present. These harmonics will create eddy current losses within the transformer. The severity is hard to predict, but a transformer could overheat. To deal with this phenomenon, the K-factor value is now assigned to transformers and the amount of harmonics that it can handle.

The K-factor is sometimes inappropriately called the form factor. The term *form factor* refers to the ratio of RMS current to average current. More often, the form factor is the effect of rectifiers on motors. This phenomenon will be looked at more closely in the upcoming chapter on drive peripheral equipment. However, it can be stated that anytime there are devices performing power conversions (ac–dc, dc–ac, etc.), there will be some sort of distortion to the waveform somewhere. That is why transformers must be analyzed with regard to their K-factor.

Power semiconductors

Interestingly, as power conversion and drives are discussed, terms are often mixed, causing confusion. The piece of hardware physically installed within a drive that switches, turns on and off, conducts and doesn't conduct is the *power device*. Many types of power devices are incorporated into electronic drive rectifier and inverter circuits. They can be built as modules for ease of installation, or they can be left as individual units. New power devices are being developed and new ways of using them explored.

Without a doubt, these devices become the main component within any ac or dc drive. The inability to change electrical energy for specific motor control purposes makes the electronic drive useless. Power semiconductors and similar devices have been used in electronic drive equipment for years. Often called thyristors, these devices can be SCRs, GTOs, and their predecessor, the vacuum tube. Various transistors and diodes are discussed later.

Dating back to the late 1800s, the vacuum tube, or electron tube, was the standard method for rectifying alternating current. It can also be referred to as a diode because it acts as a valve to control the flow of electrons. The vacuum tube consists of a glass or metal enclosure from which the air has been evacuated. There are pins at the base of the unit where connections must be made to the cathode. The cathode is made from tungsten and supplies electrons via a filament that heats the cathode. From here the anode, which is a plate, collects the electrons and then the grids that control the overall emission. In industry, the ignitron was used to rectify currents in the higher ampacities.

Another version of the electron tube is called the thyratron, which is a tube filled with gas that utilizes three internal elements. The thyratron was commonly used to convert ac to dc.

With semiconductor technology and advanced solid-state electronics, electron tubes have been all but forgotten. There are many old installations still dependent on their rectification, but the reality is that there are no spare parts readily available and no one wants to support this old technology. Tube technology is a thing of the past.

From vacuum tubes evolved thyristors for power rectification. Thyristors are types of transistors in which there are several semiconducting layers with corresponding p-n junctions. The thyristor is a solid-state version of the aforementioned thyratron. The most common thyristor is the silicon-controlled rectifier, or SCR. The SCR is still a fairly common device used in power rectification. Similar in operation to a diode, the SCR can block the flow of current in the reverse direction, but it can also block the flow of current in the forward direction.

The SCR has a blocking state and a conducting state. In its blocking state, no current is allowed to flow through it. Likewise, in its conducting state it acts like a switch that is closed. The SCR receives a small current signal called the gate signal to be triggered into conduction. In the conducting state, the SCR will keep conducting until the gate signal is removed, and the current flow reduces to zero. This turn-on and turn-off function of the SCR allows for extremely good control and very small losses in terms of current leakage. The SCR, while conducting, has a very good forward voltage drop value, which means that large amounts of current can flow through it with very little energy loss.

There are several different packages for SCR devices, including stud types, complete modules, and hockey-puck versions. The hockey-puck design of an SCR is the most common. It looks like a hockey puck, is usually white, and has two leads for receiving its gate instructions. Very high values of current can be run through the SCR, making it the rectifier of choice in higher horsepower applications.

Additional ratings are assigned to SCR design. The peak inverse voltage (PIV) is typically in the 1,400-V range. The PIV rating is sometimes called the peak reverse voltage (PRV), which means that the device will only block a certain amount of voltage in the opposite direction. If this voltage rating is exceeded, the device might be destroyed. Another factor when selecting devices is the proper heat dissipation of an SCR, or any other rectifier system.

The surface area for heat dissipation, typically for the heatsink, is important.

GTOs are gate-turn-off thyristors, which are power semiconductors that have self-commutating capability. This means that once commanded, the GTO will turn on and turn off repeatedly without having to be told so. These devices are available in high current ratings and have high overcurrent capabilities, making them very suitable for larger horsepower applications. Their turn-on and turn-off times are good, and the speed at which they switch is adequate for the typical higher horsepower applications in which they are used. However, a major drawback to GTOs is that they cost 5 to 6 times more than a conventional SCR device.

Transistors

For electronic drives, the transistor has become the power semiconductor device of choice. These solid-state components have been around in some form since the early 1950s, but really have gotten popular in the past decade. Basically they are made up of different semiconductor materials, sometimes with arsenic or boron in conjunction with silicon. The way the electricity moves through the silicon is directly related to the amount of arsenic or boron contained in the transistor. What transistors provide is fast-switching capability for a relatively low cost. The three general types of transistors are the bipolar transistor, the field effect transistor (FET), and the insulated gate bipolar transistor (IGBT).

Over the years, the bipolar transistor has been used in many applications. It can be found in oscillators, high-speed integrated circuits, and many other switching circuits, such as variable-speed electronic drives. It is available in rated currents much lower than thyristor-type devices—6 to 7 times lower. These transistors can be paralleled in operation to achieve greater current-carrying capacity; however, this design costs more. This self-commutating device also cannot withstand too much in the way of overcurrent conditions. Bipolar transistor switching speeds, for its time, were very adequate—in the 2- to 4-kHz range—making the bipolar transistor a very good switching device for its relative cost, even though it had somewhat marginal turn-on and turn-off capabilities. However, with the introduction of faster switching devices, bipolar transistors are starting to be replaced.

One such device that is extremely fast switching is the MOSFET, which stands for metal-oxide semiconductor field-effect transistor. The MOSFET is a self-commutating device that is not too costly to incorporate into a system. The drawback, presently, is that MOSFETs are not available in current ratings much above 20 A. This limits its applications, even though it has switching capabilities in the 100-kHz range, good overcurrent capability, and very good turn-on and turn-off conditions.

Many electrical switching applications today are well suited for the insulated gate bipolar transistor. This self-commutating device is available in 300-A current ratings, has good turn-on and turn-off ability, and has switching speeds of 18 kHz. It is somewhat cost-effective to manufacture and can be implemented into an electrical circuit at relatively low costs. With costs and performance driving the semiconductor industry, don't be surprised if a newer, better, and faster transistor emerges within the next couple of years.

Diodes

A *diode* is a solid-state rectifier that has an anode, which is the positive electrode, and a cathode, which is the negative electrode. These nodes allow electricity to flow in one direction only. Diodes are commonly used to convert alternating voltage to dc. The diode, in effect, acts as a valve for electricity. A diode is shown in figure 3-10, with its accepted symbol and basic components. This diode might be used for voltage regulation.

A reverse-operating version of the standard diode is the zener diode, whose symbol and characteristics are shown in figure 3-11. These are sometimes called breakdown diodes because they allow reverse currents under breakdown conditions. This quality is sometimes desirable in regulating voltage.

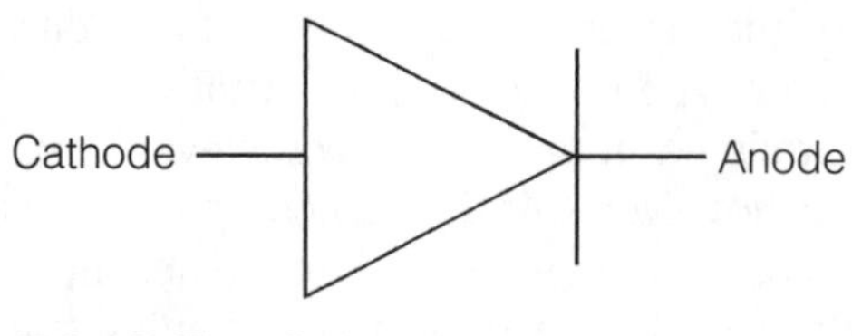

■ **3-10** *The diode.*

Fusing and circuit protection

All electrical components end up being integrated into some circuit. As costs increase for individual components, it is very important—and thrifty—to protect these higher valued devices in the circuit. A fuse costs much less than most of the other components in an electrical circuit. Thus, it is the "protector" of choice. The fuse is a

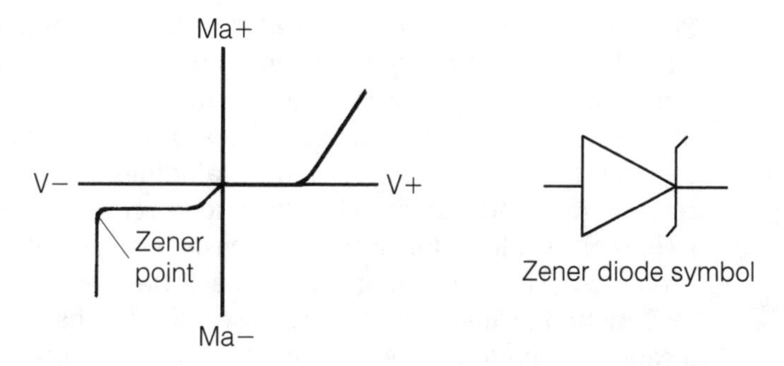

■ **3-11** *The zener diode.*

safety device that protects components from overcurrent conditions. The common construction of the fuse is a current-carrying strip of metal or wire that can melt in an excessive current situation.

In industry, the cartridge fuse is the most common type used. The cartridge is placed ahead of the component to which it is designated to protect. Most often, fuses are placed at the point where power enters a given installation, component, or piece of equipment. For further clarification on where fuses should be located, consult Article 240, Overcurrent Protection, Sections E and F, of the *National Electric Code Handbook*.

Two common reasons for an electrical component not working are the dangerous short and the irksome loose connection. Figure 3-12 shows a typical circuit with a resistive value that is part of the desired circuit. The dotted lines show a path of lesser resistance where the current will want to flow freely. In these short-circuit conditions, the current flow can increase to dangerous levels until a protective device or, in costly instances, another electrical component is destroyed. Wire and cabling can even melt, since the heat can become very intense. Often, there will be the distinct smell of something electrically burnt.

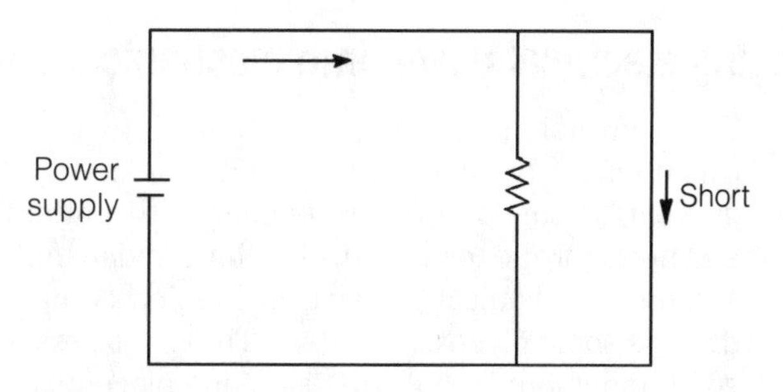

■ **3-12** *The short circuit.*

Shorts are not as frequent as predicted and reported as the problem. Out-of-the-ordinary occurrences have to happen in order for a short circuit to appear. Foreign conducting debris, the presence of water (yes, water and the molecules carried in it make it a conductor), and breakdowns in motor windings are some common reasons for short circuits. The best solution is to place inexpensive, replaceable fusing at those pieces of equipment that are best protected. The loose connection is a little more forgiving in its malfunction properties. In these cases, it's just harder to find the location of the loose connection. They don't typically create the havoc that a short circuit does.

When an electronic drive isn't working, the short or loose connections are usually cited as the problems. Finding out if these are the problems and where they are actually located, and then figuring out how to correct so it doesn't happen again, becomes the challenge. This takes a methodical troubleshooting approach, along with the proper equipment. One of the most important tools an electrician must have is a multimeter, or volt-ohmmeter, to measure values of voltage, current, and resistance in components. Also included might be the digital, stored-image oscilloscope. This device will display and analyze waveforms that can be then stored in memory for later use.

Another way of protecting devices is the circuit breaker. A circuit breaker works via the principle of a magnetic field. In an overcurrent condition, an electromagnet draws a metallic portion of the circuit out of the circuit. This opens the circuit, thus protecting downstream components. Unlike the fuse, the circuit breaker can be reset after the electromagnetic condition has passed and reused again. For more on circuit breakers, consult Article 240, Overcurrent Protection, in Section G of the *National Electrical Code Handbook*. Interrupting capacities, where to locate, and uses in parallel or as switches are covered.

Changing electrical power into mechanical power

The conversion of electrical energy into mechanical power is called *power transmission*, or PT. An electronic drive is an integral part of this process. Volts change to watts, which eventually end up as torque to move the machine and manufacture a product. Much of mechanical power transmission equipment available today has some electrical control. The plant personnel must have a good understanding of what is taking place mechanically in order to assimilate the electrical needs. This is particularly true when applying and troubleshooting electronic drives. We have to understand the mechanical aspects of what it is we're trying to move in order to keep the electronics happy.

In addition to drives, other electronic controls in the factory have become more and more a part of power transmission products. Clutches, brakes, speed variators, variable speed fluid couplings, and many other devices have either a closed-loop controller or other means of adjusting their speed. These controls are electrically fed and usually control the voltage or the current to the power transmission device. Many times they will have to connect to the drive controller in some way. To understand how these controls work, we'll need to review the basics of power transmission.

Work

Production in the factory and work are well-related. In its basic definition, *work* is a force acting through a distance. Work is equal to force times distance ($F \times D$). A force is a push or pull that causes motion of an object. The object moves in a straight line in the direction of the force applied to it. *Power* is the amount of work done in a period of time, and it is usually expressed in horsepower (HP) or in watts (W) for electrical power. Figure 3-13 is an example of work and how it is expressed. It shows that if 25 pounds are moved 10 feet, then 250 foot-pounds (ft lbs) of work has been expended.

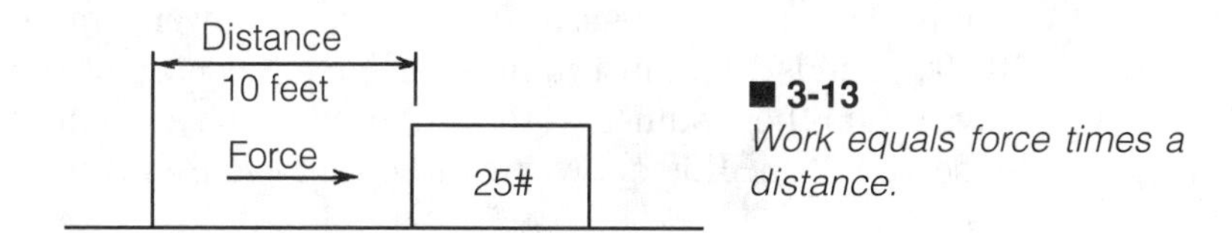

■ 3-13
Work equals force times a distance.

Often the terms work, force, power, and torque are used interchangeably in dialogue to get a point across. If everyone using the terminology had a good, basic understanding of each term, there might be less confusion and perhaps fewer problems in the factory. For instance, as we will see in more detail, torque and work are very similar. Work is a force times a distance, whereas torque is a force times some radius. Similar, yet different. It all depends on where the term is applied and what is trying to be accomplished.

The device used to convert electrical energy into torque is the electric motor. The point in the system where this distinction is made is usually at the coupling off the motor shaft. Torque is transmitted from the motor shaft to the rest of the drivetrain in order to achieve work. Figure 3-14 illustrates a typical mechanical drivetrain and its associated components. Each of the individual pieces will be further analyzed, but a basic understanding of torque and horsepower is necessary first.

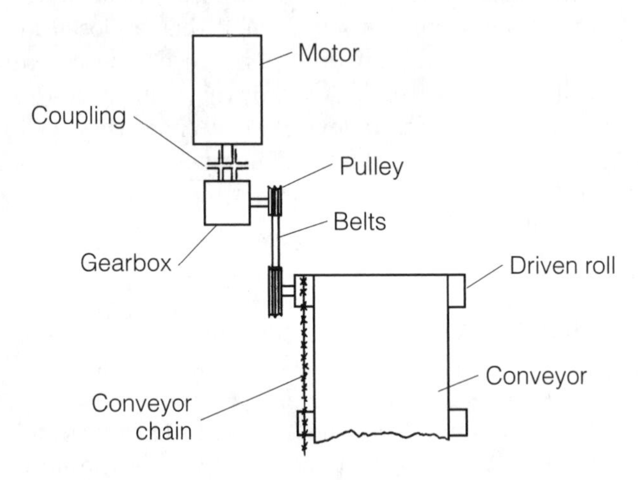

■ **3-14** *A mechanical drivetrain's various components.*

Horsepower

In the late eighteenth century, James Watt of Scotland determined that 1 unit of horsepower was equal to 33,000 ft lbs of work in 1 minute. This is equivalent to the amount of power required to lift 33,000 pounds 1 foot in 1 minute. A horse was used as the "prime mover," and these values were attributed to the amount of work a horse could do. Horsepower can be derived from the following equation:

$$HP = \frac{F \times D}{33{,}000 \times t}$$

where F = force, D = distance, and t = time.

The term *horsepower* has probably caused more costly undersizing when it comes to drives and motors than we can fathom. The fact is that when sizing the foot-pound requirements of a given application, we are really concerned about *torque*. After that, it is important to then look at the electrical current requirements of the application and how they relate to torque. The electrical equivalent of 1 HP is 746 W. To find a value for horsepower when other values are known, the following formula can be used:

$$horsepower\ (HP) = \frac{torque\ (ft\ lbs) \times speed\ (rpm)}{5{,}250}$$

Drives and motors are rated both in horsepower and in current, actually full-load current, or full-load amps (FLA). To be more exact, when sizing the actual power requirements and selecting a motor, brake horsepower (BHP) is the term used. Drives are rated on their continuous current output. In some instances, peak or starting torques and currents are important. The actual power requirements begin at the motor shaft, with the load. It is always good practice to check the torque requirements of the driven load and make sure the motor can produce the necessary output. Motor speed versus torque curves are available to help with this check.

It is safe to say that if a motor is sized properly to move a given load, and the electronic drive applied to that motor is also sized to the motor's current rating, the drive will not be a limiting factor in the application. To double-check, many engineers, where physically possible, turn the motor shaft by hand. If turning the load by their own power is possible, then most likely the electric motor will provide the torque necessary to turn the load also. This also indicates if there is any binding or unwanted friction, which might be a problem later. Of course, it can be noted that when process speeds allow, a certain amount of gear reduction can go a long way in gaining the mechanical advantage in turning a given load. Sometimes the gearbox is a lifesaver in getting an application to work. Gear reduction and gearboxes are looked at later in this chapter.

Torque

Torque is defined as a rotating force. Any twisting, turning action requiring force is torque. As is shown in figure 3-15, torque is the

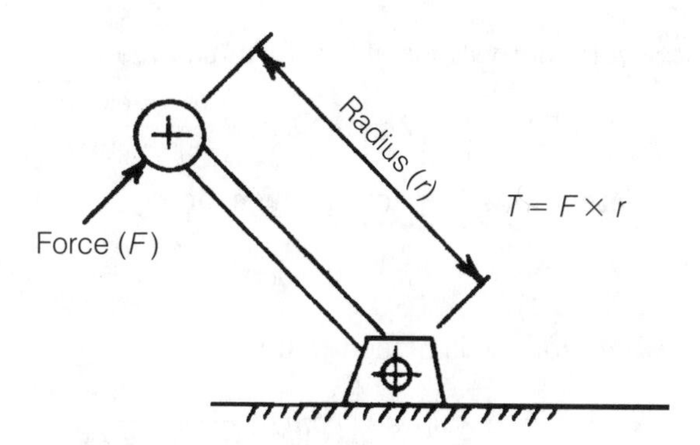

■ **3-15** *A torque lever arm.*

product of some force times a distance, or $T = F \times r$. If a force F is applied to the lever arm at a distance equal to the radius r shown, then a resulting torque is produced. Derivations of this formula can give us the amount of work as a torque acting through an angular displacement.

There are two basic types of torque: static and dynamic. Most often factory personnel are concerned with dynamic torque. Rotating apparatus will exhibit dynamic torque. It is in a state of constant movement, correction, and change. Static torque, on the other hand, is a more consistent value. It might be an issue, for instance, with a holding brake. For most of this book's purposes, dynamic torque will be most often analyzed in examples.

Torque can take on different values depending on what is required of it. To start a machine, we might be interested in *breakaway torque* than all other values of torque in the operating cycle for that machine. In another instance, we might just be concerned with *running* or *process torque*, since the application might be sensitive to torque variations once we are at rated speed. Running torque can be expressed as simply the torque required to keep the machine running at a given speed. Process torque, on the other hand, could factor into the requirements the need to compress, cut, or act on a material periodically, possibly increasing torque needs for an instant. And we cannot ignore *accelerating torque* because all too often someone wants a machine or process to cycle faster; therefore, this torque value plays a major role in sizing.

The universal formula for determining torque of a rotating piece of equipment is

$$torque = \frac{horsepower \times 5{,}250}{revolutions\ per\ minute}$$

Several derivations of this formula follow:

$$horsepower = \frac{torque \times speed}{5{,}250}$$

where *torque* is in foot-pounds, or

$$horsepower = \frac{torque \times speed}{63{,}000}$$

where *torque* is in inch-pounds, or

$$speed\ (rpm) = \frac{5{,}250 \times horsepower}{torque}$$

where *torque* is in foot-pounds, or

$$speed\ (rpm) = \frac{63{,}000 \times horsepower}{torque}$$

where *torque* is in inch-pounds.

Speed, which is sometimes referred to as revolutions per minute (rpm) can also be shown in formula form as N. A rule of thumb to remember is that at 1,750 rpm, 3 ft lbs of torque equal 1 HP. Another constant to keep in mind is 5,250, used with foot-pounds of torque. As shown above, the 5,250 constant is changed to 63,000 when torque is in inch-pounds (in-lbs). These inch-pound calculations are more typical for servo and stepper motor applications. Sometimes the sizing even gets into the ounce-inch realm.

The aforementioned formulas are for rotary motion. For linear motion the formula for horsepower is

$$HP = \frac{F \times V}{33{,}000}$$

where F is force in pounds and V is velocity in feet per minute.

This formula may be useful when trying to determine the horsepower required for a line that is moving a material a given speed (V), linearly, at a given tension (F).

Beyond the linear and rotary torque and horsepower basic formulas, there are several other useful equations to aid in sizing motors and prime movers. One is used to find the accelerating torque of a rotary device. Accelerating torque is the torque required above and beyond the torque required to drive the given load. You must add it to the (usually larger) value of load torque in order to adequately size an application with substantial acceleration rates. Accelerating torque is found by using the following formula:

$$T = \frac{(WK^2) \times change\ in\ speed}{308 \times t}$$

where:

T is the accelerating torque expressed in foot-pounds.

WK^2 is the total system inertia, including motor inertia (from the rotor) and the load's inertia, and is expressed in foot-pounds squared.

Change in speed, sometimes shown as N, is in rpm.

t is the desired time in seconds to accelerate the load.

If a gear reducer or other important inertia-containing component is part of the system, then its inertia value must also be included. Any power transmission component that must be part of this acceleration has to be factored into the equation.

Torque is definitely a more appropriate value than horsepower to determine what size prime mover is needed to move the load accordingly. Once the torque value is known, a determination of which motor to use is within reach. The problem is that strictly sizing motors by horsepower is potentially dangerous. The better scenario is to provide the motor supplier with speed and torque requirements at those speeds so as to size the motor for duty cycle and complete heat dissipation. Let the motor supplier select the motor and frame. They know their motor constructions better than we do. Some particular frame sizes of motors might be capable of more in the way of overloading than others.

A torque wrench is sometimes used to tighten nuts onto a seated surface. This device measures the amount of force in in-lbs (inch-pounds). A similar device can be used to measure the torque requirements of a particular shaft. This evaluation is good for any rotating component, but won't be a factor into the equation's acceleration or peak torque requirements. A motor is a stupid device. If a motor can't turn because its load is too great, then the current to the motor will increase to try and move that load until the supply is shut off. That is why electronic overload protection devices should always be implemented into a motor system, whether ac or dc. If a motor doesn't want to turn, the motor might be undersized, and the torque output from that motor is inadequate to perform the application.

Speed, torque, and horsepower relationships

Electronic drives are speed and torque controllers for motors. Both speed and torque have a relationship to each other and to horsepower. Figure 3-16 shows a typical curve with the designations for speed on the x-axis and torque on the y-axis. Most speed/torque or speed versus torque curves are laid out in this manner. This particular curve is for a variable torque application. The torque increases proportionately as the speed increases until full speed has been reached. This is the point where full torque capability is met.

As discussed earlier, horsepower gives an indication of how high in speed a motor can be run and still obtain usable torque. The peak

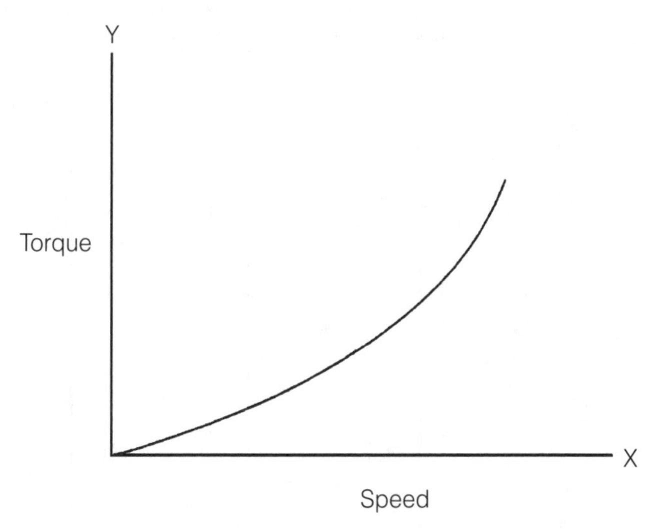

■ **3-16** *A speed-versus-torque curve.*

on a horsepower curve indicates the maximum power available and at what speed that power is attained. Selecting a speed beyond the peak yields no more usable power. A motor is said to be in a state of constant horsepower at this point. The power achieved at higher speeds is also available at lower speeds. Thus, we can save undue wear on the motor and run at lower speeds. It is also important to note that in any given mechanical or power transmission system (gears, pulleys, reducers, etc.), the value for horsepower is constant throughout that system at any given time.

Figure 3-17 shows a speed/torque/horsepower nomogram. This gives an excellent graphical comparison of the three values and how they relate. A known value, for instance, speed, is indicated on the graph. If either torque or horsepower are also known, the corresponding line through the two points (the known components) can be drawn, and the unknown value found. A nomogram can save you the time of calculating the values using the equations previously mentioned.

Charts showing constant horsepower and constant torque curves are shown in figures 3-18 and 3-19. Figure 3-18 indicates the expectations of torque when a motor approaches a state of constant horsepower. Torque will begin to diminish in a constant horsepower application after we have reached 100% of rated speed. The converse is true for constant torque, as shown in figure 3-19. If we are able to continue to increase in speed and maintain a constant torque, horsepower will continue to increase. Examples of con-

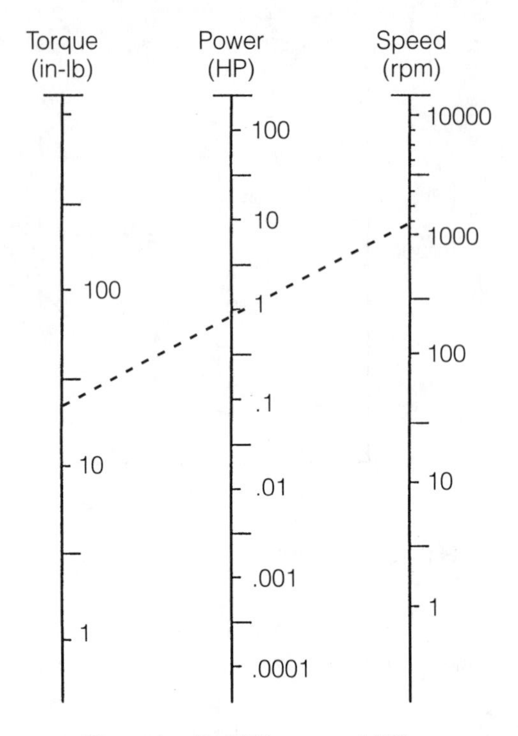

Example: At 1750 rpm, a 1 HP
system yields 3 ft-lbs or
36 in-lbs of torque

■ **3-17** *A speed–horsepower–torque nomogram.*

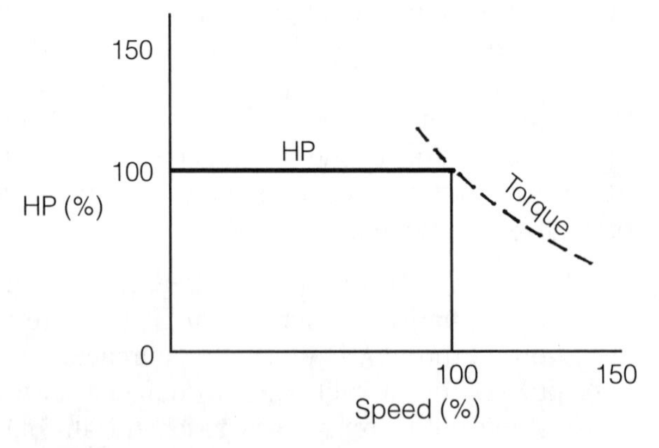

■ **3-18** *Constant horsepower.*

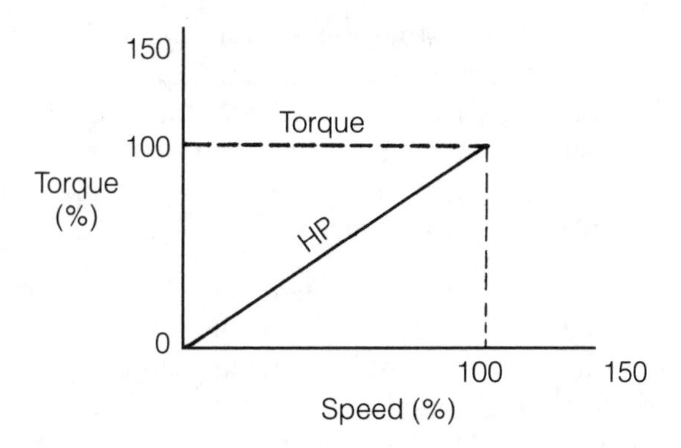

■ **3-19** *Constant torque.*

stant torque applications are plentiful in industry, including conveyors, rotary tables, and countless other constant torque applications that rely on an operating zone that does not take them beyond 100% speed and, thus, diminish the torque requirements. Constant horsepower application examples include center-driven winders, drills, milling machines, and so on. However, it is always worth repeating that torque is the component that is doing the work. And somewhere on the curve, we must be concerned with having enough torque to do so.

Inertia

Many times we hear the term *inertia* used in motor sizing and system evaluation. Inertia is often designated by the letters J or I. We even sometimes hear the term WK^2 used in conjunction or even in place of inertia. Actually, WK^2 is the weight of an object times its radius of gyration value, K, which is squared in the equation:

$$Inertia = WK^2$$

The inertia formula is sometimes called WR^2. *Inertia* is that property by which an object in motion will stay in motion until acted upon by another force. Once in motion, a machine will stay in motion until an external force acts upon it. For instance, a motor running full speed is stopped. The rotating elements will come to rest because friction and other forces will act upon those moving parts. Clutches, brake systems, and regenerative means are all effective at stopping objects with substantial inertia values.

There are two distinct types of inertia. One is applicable to rotating elements, and the other is applicable to linear motion. Because 90% of the applications involve rotating elements, plant workers are usually referring to that type of inertia. Inertia in a linear-motion application will be inherent in an elevator or conveyor type of power transmission system.

Most of the time, inertia is only a factor when changing speeds of a machine. In practice, WK^2 is the total value of all rotating components of the system in which the main components are the inertias of the rotor (motor) plus the inertia of the load in ft lbs. Sometimes the total values of all the inertias except the motor's are referred to as the "reflected inertias." All this would be true for inertias in rotating elements.

Reflected inertia is an important property. Its value can have a dramatic effect on a system's performance. Here's why: When incorporating gear reduction into a motor and power transmission scheme, inertia values are affected by the gear ratio. With a change in speed, the reflected inertia will change inversely by the square of that gear ratio. For example, if the speed increases by four times (because of the ratio), then the reflected inertia decreases by 16 times. This can work in one's favor in a power transmission system where there is a high reflected inertia and acceleration rates are important. However, this ratio can work against performance in those instances where speed increase, rather than reduction, is needed at the output of the motor. This can be overcome with a higher torque output motor. Be aware. This is one of those "gotcha" type surprises that could cost extra time or money if not considered beforehand. Figure 3-20 shows a typical gear reduction scheme in comparison with a speed increasing scheme.

The *flywheel effect* is the phenomenon whereby half of an object's mass is on the outside and half is on the inside. A large flywheel is both hard to start and hard to stop. It can work for an application to help provide the ride-through capabilities in intermittent dramatic load changes once at running speed. It can work against the application when acceleration and deceleration rates need to be increased. To understand the flywheel effect, study the relationships in figure 3-21. Two objects, one solid (A) and one basically hollow (B), have the same masses. However, object B is much harder to set in motion. Thus, the shape of power transmission components can play an integral role in an application's performance.

When considering drives and inertia in applications, ask the following questions: Does the motor have to change directions or

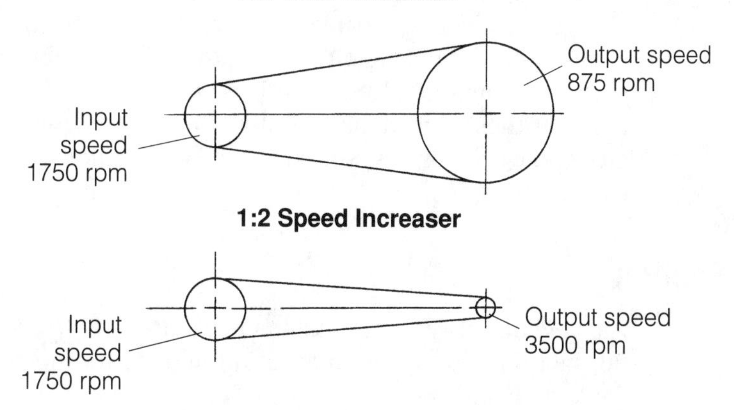

■ **3-20** *Gear reduction (top); speed increaser (bottom).*

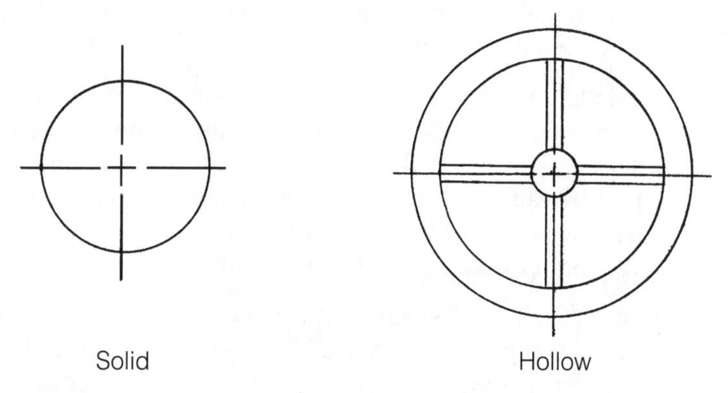

■ **3-21** *Both objects have the same mass. The solid flywheel has a smaller diameter, but the hollow flywheel has more inertia.*

speeds quickly? Does it have to start and stop frequently? What components in the drivetrain contain high inertia values? What are the inertias of the rolls that accumulate the material? This buildup of inertia has to be accounted for in the system sizing. The rule is to size a power transmission system for its worst cases and factor in some margin of safety to be assured that when the operator hits the RUN pushbutton, every component starts and stops together in a timely manner.

All objects have some mass, and, therefore, some inertia. Inertia is an important factor when seeking a desired performance for a mechanical system, especially when incorporating electronic drives. Electronic drive controls do not magically provide the "escape route" if a power transmission system will not respond as desired. They can deliver electrical current that can be converted into mechanical power, but we often run into the immovable object. It has to be determined why that object won't move. The answers to many of these questions will ultimately determine which type of electronic drive is best suited for the application: ac, dc, servo, or other.

Friction

All mechanical systems exhibit some amount of friction. When sizing motors for a given application and its torque requirements, friction should be factored into the equation. A small amount of friction is actually desirable because it can aid in a given load coming to rest sooner. Some machines even have enough friction so that during an emergency stop condition, the machine can stop instantly. Of course, mechanical and electronic braking should always be considered from a safety point of view.

Friction is actually welcome in many other power-transmitting products. V-belt drives need friction to transmit torque from one sheave to another; therefore, a variable-speed pulley relies on this friction, also. Clutches and brakes also operate on a medium of friction. Friction is not desirable in gears, since it creates heat, which, in turn, creates losses and inefficiencies.

Often the term "windage and friction" is used when discussing rotating components. When a high inertia device is at full speed, the only real torque requirements at that time to maintain speed (with the loading steady) are to overcome friction, compensate for the deflection losses due to wind, and keep the device from running away. Another term relative to friction is "stiction." This is sometimes used to describe a mechanical component's tendency to "stick" in a particular location. This might mean that the coefficient of friction is very high at this time. Two other types of friction are rolling, or rotating, friction and static, or at-rest, friction. An object at rest has to have more friction overcome than an object in motion.

Gear reduction, gearboxes, and gearing

Gearing is a very important facet of power transmission. It can provide the application, and the motor, with a tremendous mechanical advantage. As long as the speed reduction can be tolerated, torque advantages can be achieved simply by implementing gear reduction. Instead of paying more for a larger motor, it might be more economical to install a gearbox.

A good example of gear reduction as a mechanical advantage is a revolving restaurant high above the ground. The diners are seated, and the restaurant turns very slowly, maybe one revolution per hour. The motor used as the prime mover for this application could be very small in torque output, say, a 15-HP motor running at 1,750 rpm. With so much gear reduction, there is a tremendous amount of torque produced—enough to move a several thousand-pound object, such as the restaurant. Since the net revolutions per minute (per hour in this case) are so minuscule, it is possible to utilize all of this gear reduction.

Sometimes called a gear reducer, a gearbox is an assembled device that houses all the gear reduction needed for a given application. It has become the convenient device to house gears, input and output shafts, lubricant, and so on. Before fully housed gearboxes, gears and pulleys, along with belts and chains, were installed to provide the necessary reduction for speed. The gearbox evolved as a compact unit in which all the same reduction was achieved in a smaller, and safer, package. The gearbox can remain in service for several years as long as the lubricant is maintained.

Gearboxes have rated outputs and, therefore, appropriate efficiencies. If 20 HP is supplied at the input shaft, the gearbox might only have an output rating of 17.5 HP. This means that this gearbox is 87.5% efficient. The lost horsepower is in the form of heat and noise losses. However, for drivetrain system analysis purposes, horsepower is assumed constant throughout a system. Losses and efficiencies are not considered initially.

Gears made of hardened steel are often found in lubricant within the housing of a gearbox. These components are extremely critical in the transmission of power in the factory. They are often the source of a lot of the audible noise in the factory. As is seen in figure 3-22, the configuration of gears can change rotation and can provide a very high ratio of torque and speed, and vice versa. As a rule of thumb, the lower the ratio, the more efficient the gear train system.

Each introduction of another gear is called a *stage* in the gear train, as shown in figure 3-22. Each additional stage will add another 2% loss to the overall system efficiency and will add more noise. It might be safe to say that those 2% losses are made up of mainly more friction, heat, and noise. An odd number of stages in a gear train will reverse the rotation, and an even number will keep

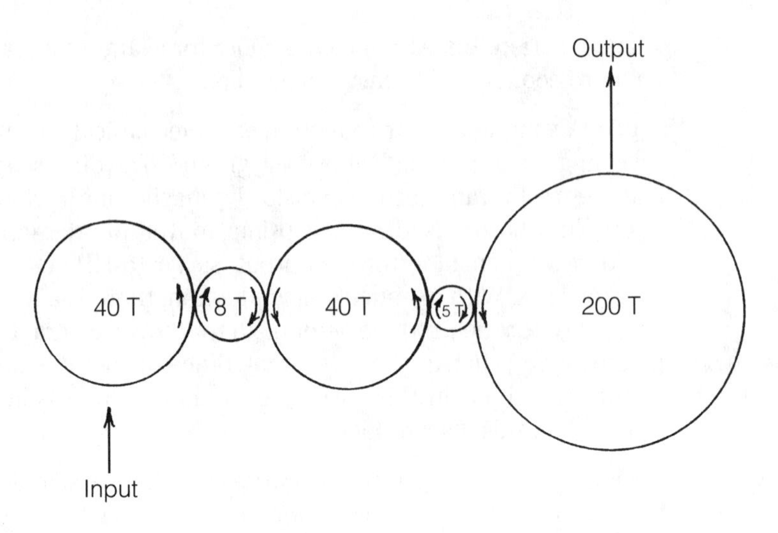

Total Gear Reduction = 8/40 : 40/8 : 5/40 : 200/5
= 5 to 1

■ **3-22** *Gear rotation and reduction.*

the rotation the same. Again, this can be seen in figure 3-22 by the rotational arrows on each gear.

There are different classifications of gears used in gear reducers. One type is called the spur gear; its teeth are parallel to the shaft and it is typically noisy. Another is called the miter gear, which is used to change direction, usually at a 90-degree angle at a 1:1 ratio. Very similar to the miter gear, another common gear is the bevel gear. One thing common to most of these gear arrangements is *backlash*. Backlash actually occurs in most mechanical systems where two pieces of metal are in physical contact with each other. Typical backlash is shown in the partial gearset in figure 3-23. The term *zero backlash* implies that there is no play between the components. While backlash can never be zero because of wear from the initial engagement, it can be minimized. However, this requires extra-precise machining of the gears, which can be expensive.

There are different types of gears and configurations of gears within a gearbox. One is called the *worm gear reducer*. Its main components are the worm and worm gear. One attractive feature of the worm reducer is that it is difficult to drive backwards. This can be a built-in safety feature with this device. One disadvantage is that the worm gear is not as efficient as the standard helical gear. The *helical gear* engages more teeth at any given time and is therefore capable of transmitting more torque, having less back-

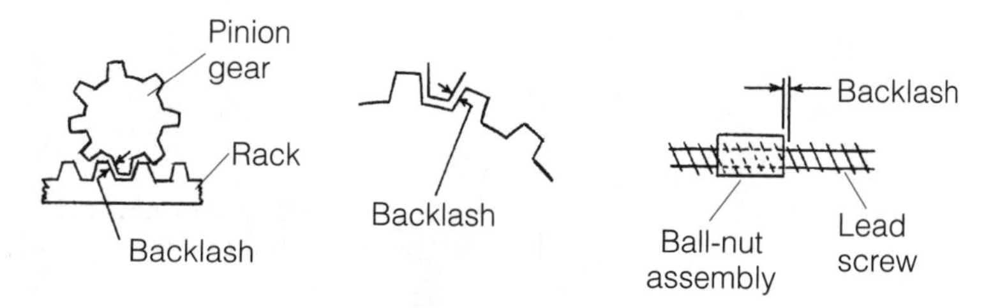

■ **3-23** *Various examples of backlash.*

lash, and being less noisy. The helical gear teeth have angles such that more teeth area is able to engage at a time.

Another gear arrangement is the *planetary gear*. As the name implies, there is a sun gear around which the other gears rotate. This package yields the highest reduction ratio in the least amount of space. The sun gear is the driven gear and the ring gear provides the output. Normally, the planetary gear design is low backlash and fairly efficient.

V-belts and variable-speed pulleys

As the name implies, V-belts are power transmission belts with a cross section in the shape of a *V*. This is so that the bottom portion of the *V* seats well into the pulley or sprocket into which it is placed (figure 3-24). The power transmitted is a direct function of how much actual surface area is common to the belt and the pulley. When there is slippage, in effect, the surface-to-surface contact is lessened. An industrial V-belt is made of high-grade rubber product and has tensile and other cords interlaced within it. Many times multiple belts are used for transmitting larger amounts of power.

Advantages of using V-belts include lower installation and replacement costs, practically no noise, and no lubrication (in fact, oil on a V-belt causes real problems with slippage). In addition, V-belts are fairly forgiving and can absorb shock loads. They are easy to install, require very little maintenance, and can be used for high-speed applications.

A disadvantage is that they aren't recommended for oily, hot, and harsh environments because they can slip. The possibility of slippage makes them impractical when trying to synchronize components. Low-speed operation is typically not recommended, either. A common practice in industry is to use a minimum of two belts even if one will do. If one belt breaks, the other can keep the machine running.

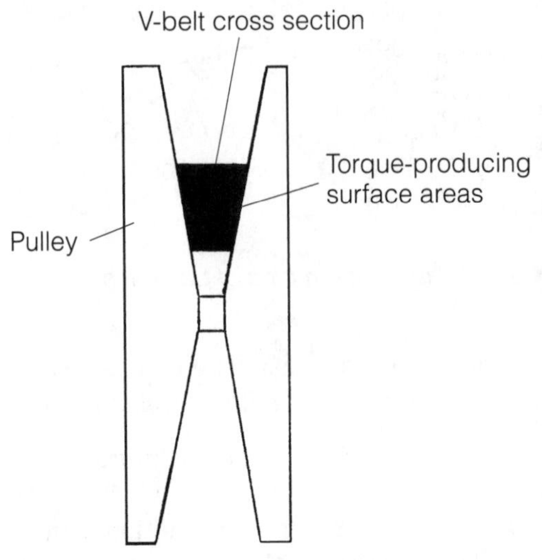

■ **3-24** *V-belt drive and V-belt cross section.*

The V-belt is the main power-transmitting component in V-belt drives or V-belt pulleys. Often called variable-pitch or variable-speed pulleys, these speed-control systems are being replaced routinely by electronic drives and motors. Although used extensively in industry for many years, with so many other more efficient and modern ways to gain variable speed, the pulley system has lost popularity. Its disadvantages, besides being inefficient, include frequent adjustments and high frequency of belt wear. It requires a constant-speed prime mover, which is usually an ac motor. Typically, these variable-speed pulley systems are not seen above 50 HP in size.

Often it is necessary to use a belt to connect a critical device to another, and no slippage is allowed. This might be the case with a pulsing feedback device that cannot be directly mounted to a motor shaft and therefore has to be mounted in parallel to the shaft. This is often accomplished with a timing belt. Timing belts are flat belts with teeth. They are highly efficient, lightweight, and eliminate slippage. However, they have substantially less torque-transmitting capability than their V-belt counterparts and should not be used for extreme power transmission conditions.

Chains and sprockets

Often referred to as roller chain, this system is used to transmit power in nasty environments. Roller chain can provide no slip (maybe a little backlash), run in high ambient temperatures and oily atmospheres, and is very efficient. It requires accompanying sprockets with teeth in order to be a complete package. These sprockets and the chain itself are more expensive than belt-drive systems. The chain system needs to be lubricated, is obviously noisy, and is not as forgiving as V-belts. Shock loads can be amplified or can even cause the chain to snap. Also, over time, the chains will actually become elongated. At some point they will have to be replaced.

Clutches and brakes

A clutch is defined as a device that engages and disengages motion. It must be used in conjunction with a prime mover, usually an electric motor. Clutches can also be actuated by air or fluids. The electromagnetic clutch utilizes a permanent magnet and an electromagnet. By providing some dc voltage via a voltage controller, the amount of magnetism is controlled, providing as much clutch engagement as necessary. Engagement material, which transfers the torque and engagement times, is critical to a clutch's performance. Nowadays, engagement material should contain no asbestos; however, there is still some in use. Friction media include ceramics, leather, and other non-asbestos material. Still, these newer materials do wear out and must be replaced.

Clutch actuation is either manual, electrical, or mechanical. Air, electric, hydraulic, centrifugal, or magnetic means are all common. Engagement time is the amount of time required from when a signal is sent to a clutch until enough contact has been established with the friction material to cause the desired motion. Engagement time is critical, especially in high-cycling applications where the load has to be picked up and let go often. Heat then becomes a real issue with the clutch.

In an air-actuated clutch, air is the medium by which pad pressure is achieved. Hydraulic clutches use oils and synthetic fluids to provide engagement. The control of the flow of these fluids is proportionate to the clutch's engagement.

Other types of clutches include the magnetic particle type, which utilizes particles that are magnetized to provide certain degrees of clutch strength. These clutches are used often to set a tension, or drag, on a rotating component. They usually are not seen in high horsepower applications. Two other types of clutches are the over-running clutch and the centrifugal clutch. The over-running clutch is used as a backstop, such as on a conveyor system, to ensure that backdriving cannot occur. These clutches can also perform simple indexing applications and will free-wheel when not engaged. The centrifugal clutch works on the principle of centrifugal force. At zero and low speeds, the clutch is not engaged to any other component. However, when speeds are such that the outer housing can centrifugally "fly" out to engage with another rotating device, the clutch is fully engaged and working. These clutches are inexpensive and are used quite a bit with two-cycle machines.

A brake is very similar to a clutch in operation and actuation, but its sole purpose is to stop all motion, most often rotary motion. The types of friction materials and actuation methods are common to both brakes and clutches. And, like clutches, there are many types of brakes. Some brakes are air actuated, while most are electrical. The strength of the electrical signal is directly related to the amount of braking, or engagement, to be applied. Similar to the clutch, heat dissipation is a major issue with the brake. When engaging and disengaging in high cyclical applications, it will be necessary to size a brake unit for not only the torque requirements of the application but also for the heat buildup. If the heat is not dissipated, the brake will begin to lose its braking capability.

There are different uses for brakes in industry. Some brakes perform a holdback function, providing levels less than full braking to apply tension, pressure, or drag to a rotating component. These applications will generate heat. Other brakes are called fail-safe brakes. Their operation is exactly opposite to normal brake usage. Normally, we command a brake to be on by using a signal to indicate that we want to stop the machine or process. The brake is normally in an OFF or nonengaged state when electrical or supply power is present. Sometimes, however, for safety reasons, it is desirable to have the brake normally disengaged when electrical power is present; in the event of a power failure, the brake goes to its other state, which is fully engaged. This ensures no motion when there is a loss of electrical power.

Couplings

There are many types and uses of couplings. A coupling connecting a motor shaft and its loadside shaft should be selected based on its use, abuse, and environment. Types of couplings include jaw type, shaft couplings, elastomeric, flexible, rigid, and metal and flexible disc couplings. Typical couplings consist of the mating halves with an appropriate bolt configuration and a bore requirement (it has to fit over two shafts, sometimes different diameters from one another). Duty cycles and service factors must be factored into the selection of any coupling. Obviously, the application mainly dictates what type of coupling to use. Common couplings used in direct drive applications include flexible and compliant types, which provide some degree of nonrigid connection so as to be forgiving in the application.

The perfect coupling should transmit full power efficiently. There are enough places in the drivetrain where electrical and mechanical losses exist; the coupling area should not be one of them. Inevitably, there will be misalignment between shafts that the coupling must connect. Ideally, the coupling will be able to compensate for some misalignment. Some couplings can handle more misalignment than others. In addition, the coupling should be able to provide a certain amount of cushion in shock-loading applications.

There is also the torque-limiting coupling, sometimes called a torque limiter. This coupling is designed to break at a certain level of torque, or stress. This prevents damage to more expensive power transmission devices such as motors and high-precision gearboxes. It is much more desirable to lose an inexpensive coupling and be out of service for a short period while the problem is identified, than to lose an expensive machine and be out of service indefinitely.

Coupling selection should also be based on easy installation, low maintenance, and simple repair. The unit selected should be relatively inexpensive and spares should be readily available in case of a breakdown. Although the coupling is a smaller, somewhat less significant component in the power transmission system, it is critical for a number of reasons. When selecting the right coupling many parameters must be considered. The obvious include torque required, maximum speed, shaft sizes, availability, and cost. The not so obvious include whether a service factor should be considered, how much misalignment is anticipated, whether more endplay is necessary because of shock load, whether the environment is corrosive, and what the distance is between shaft ends. Although this seems like a lot of considerations for a simple coupling, some upfront analysis can save downtime and headaches later.

Ball and lead screws

The ball, or lead, screw changes rotary motion into linear motion. It is a very common means of power transmission, particularly in the machine tool industry. The ball screw gets its name from the steel balls in a sleeve-like housing, called the nut. This nut moves on the grooves, or leads, of a screw. When an electric motor turns the screw, the balls actually roll in the grooves, moving the whole nut assembly (figure 3-25). Because ball-screw assemblies are most commonly found in positioning applications in machine tool systems, the motors used are often stepper or servomotors with high-positioning, high-accuracy controls.

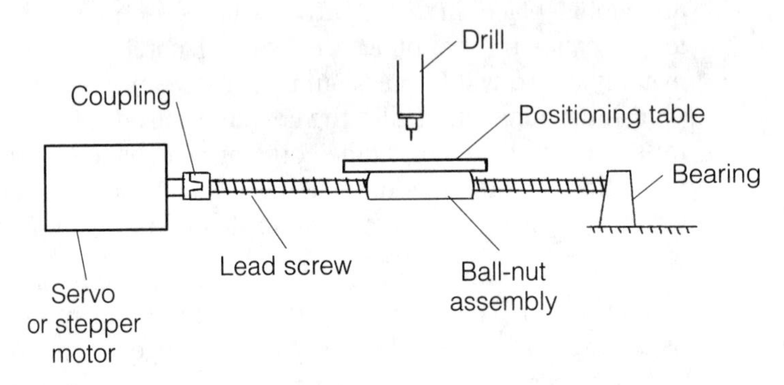

3-25 *Ball-screw assembly.*

When sizing the torque requirements of a ball-screw application, three main criteria are needed:

1. What is the load, W?
2. What is the lead-screw pitch, P, in revs per inch?
3. What is the efficiency of the lead screw? Efficiencies range from a high of 95% for a ball nut to a low of 50% for an Acme screw with a plastic nut. Consult the lead-screw manufacturer.

The results are then factored into the following equation:

$$T\ (total) = T\ (friction) + T\ (acceleration)$$

To find the torque requirement of friction, use the following formula:

$$T\ (friction) = \frac{F}{2 \times \pi \times pitch \times efficiency}$$

where T = torque and F = frictional force. Of course, inertia will play some role and must be considered.

The pitch of the lead screw is expressed in revs per inch and is usually shown in symbol form as P. Pitch is not to be confused with the lead of the screw, which is actually 1/*pitch*. The efficiencies of lead screws are fairly close, in the 80 to 90% range. The variations depend on whether or not the ball nut is metal or plastic. One type of lead screw, the Acme, is very common and inefficient, sometimes as low as 30%. It frequently has to be adjusted for wear in order to maintain performance.

Sometimes it is necessary to hold to very tight tolerances when positioning with a lead-screw system. This might dictate that there be no, or zero, backlash. To accomplish this, it is common practice to preload one ball nut against another, which minimizes looseness and end-play.

Hydraulics

Hydraulics is an often misunderstood term. In its basic sense, hydraulics pertains to fluids in motion. These fluids, usually water-or petroleum-based, can exert pressure in the piping in which they are contained; therefore, hydraulics is also concerned with the pressures and resistances to flows. However, in the power transmission world, we are concerned with mechanisms and the means of transmitting power via any fluid. Additionally, accurately controlling flows and pressures help to control a process or machine better.

Cylinders, brakes, and transmissions can be hydraulic. Like air-driven devices, pressurized fluid can be compressed or decompressed to attain motion, as in cylinders, and to gain a mechanical power advantage. However, to perform the pressurizing functions, an electric motor is normally found running the compressor. Larger systems will entail more electrical and electronic content. Pressure and flow controls are becoming more powerful, requiring the mechanic to be able to troubleshoot the electronics.

In any hydraulic system there are these five main components: (1) An electric motor, or prime mover such as an engine, to drive (2) a pump for increasing the pressure of a fluid, controlled by (3) valves in the medium for transferring fluid, some kind of (4) piping system, and of course, (5) the load itself. There is also a need to move the fluid through the piping, but the one electric motor can often be the drive source for both pump systems. As you can see, there is room for electronic and electrical control in many of these components.

The motor is often ac, and many installations require an electronic variable-speed controller, or drive. The control valves can be electrically actuated based on a transducer's feedback to a valve controller. These valve controllers can be monitoring pressure, flow, temperature, or all. In fact, flowmeter control has spawned some fairly sophisticated devices. The overall process may be controlled by a more complicated controller, such as a programmable logic controller. Thus, understanding the electronic hardware is just as important as understanding the mechanics of the process components and the process itself.

The process control involved mainly with hydraulics is a big part of industrial automation. Good control of the system is dependent on the sensors, limits, and transducers. Specific devices have been developed for the control of fluids. One such device is the magnetic flowmeter. Sometimes called magmeters, these magnetic flowmeters measure the flow of liquids in an enclosed pipe. The fluid must be able to conduct electricity in order for the magmeter to function properly. These devices are specialized process control products for industry. They are covered in greater detail in Chapter 5.

Pneumatics

Air-driven systems in the factory have a definite niche. As long as compressed air is available and distances from compressor to device is not too great, a pneumatic solution is viable. Pneumatics is that branch of physics concerned with the mechanical properties of gases. Most often the discipline is concerned with the properties of air, especially compressed air, since air must be pressurized in order to use it in factory processes. Air is pressurized via a machine commonly known as an air compressor.

There are two basic types of air compressors: dynamic and static. Dynamic compressors include centrifugal and fluid jet types. Static compressors are the most common and act as a volume-changing or displacement device. Known as positive displacement compressors, their action is much like a manual tire pump, where a moving piston within a cylinder decreases the actual volume of space in which the air resides. This action compresses the air into a smaller space, thus increasing the pressure of that air. This air is typically rated at 90 to 105 pounds per square inch gauge (psig). This would be usable air pressure in a plant where air-powered tools, air motors, and actuators must be driven by air.

Air-powered systems have several advantages. Air devices do not create shock hazards, and therefore can be applied in wet applications. Because they do not create sparks, unlike electric systems, they are fairly safe in explosive environments. A small compressor can also provide a great deal of available air when used in conjunction with a storage tank. Air systems provide a great deal of flexibility and economy.

The control of air systems is simple. Tubing, piping, valves, pistons, and cylinders are the main components, and it is important to note that there are few moving parts. Reliability is high, and from an electrical standpoint, the motor used at the compressor is one of the wear, or maintenance, items. The flexibility of an air system can be seen in its ease of upgrading. By adding another valve and some piping, more capability is introduced. Relief valves can be incorporated to guarantee the protection of a particular system. As for performance, an air-actuated piston within a cylinder can change its motion quickly in discrete steps and with very little shock.

Air systems have mainly been replaced by electrical systems wherever practical mainly because of convenience. The cost of electrical systems has come down and the performance of an electrical system is now very good with the advent of digital control. Disadvantages of air systems include extra noise in the facility, mostly from air relief, and maintenance. The air lines in a compressed air system sometimes get water in them, ultimately contributing to loss of pressure. All in all, there will always be a place for air-powered devices in the factory, but as electronic control continues to be developed, fewer air systems will be specified.

Noise

Walk into any production facility and you will probably be given earplugs or wish you had been given a set. Electromechanical operations are loud. Motors spinning at high speeds produce one band of audible frequencies. Metal banging against metal produces another. Even some electronic drive equipment adds audible noise with cooling fans and the firing of power devices. Some factories are simply too noisy. Maybe this condition can be blamed for misinterpretation of torque and horsepower terms on the plant floor—nobody can hear too well when shouting above noisy machinery. Noise is an indicator that the process could be more efficient. The presence of noise means that energy is being lost due to sound energy.

Efficiency

Efficiency is defined as the amount of work output divided by the work input. This value may be attributed to a single piece of equipment, or it can be calculated for several components comprising a full system. To get a clearer picture, sometimes it is more important to know what the entire machine's efficiency is. By virtue of its design, one component might be somewhat inefficient, while all other components in the system are high in efficiency. The one inefficient component does not bring the entire machine's efficiency that far down. However, it is always worth investigating alternate methods and designs to further increase efficiency. Power costs are constantly rising and if machines run continuously, substantial savings can be attained.

Efficiencies can be broken into mechanical and electrical segments. Some components in a machine might even exhibit properties of both disciplines. Energy losses are usually in the form of heat; however, noise and surges of power upon starting and stopping should also be factored into the overall equation. As we strive for perfection, we must accept that nothing is 100% efficient, even though many manufacturers claim to be approaching this value. Will superconductive components cause this to happen? Can we have a situation where we are over 100% efficient? If it is to happen, then competitive industry will drive it to fruition!

Service factor

When selecting power transmission components, it is customary to size the component with a factor of safety, or service factor. This is to ensure that for a given application a catastrophic failure does not happen. This, of course, is theoretical and does not take into account the fact that there are many components in a power transmission line, and one might have been overlooked. Also not taken into account are flaws, defects of material, and external forces, which might cause failure. If money and size are not an issue, then perhaps a component can be designed and selected with a service factor 10 times its required value. That would be safe!

Realistically, experience and manufacturers of power transmission equipment have given us guidelines for choosing service factors. A service factor of 1 is the base and is also the absolute minimum. Service factors are used in selecting couplings, brakes, gears, gearboxes, drives, motors, and almost every single component in a power transmission scheme. With 1 as the base, it is practical to establish values for service factor. For example, centrifugal fans, liquid mixers, and variable torque pumps will be given a value of 1. Conveyors and feeder equipment applications will be 25% higher, or 1.25. Machine tool, heavy material mixers, mills, cranes, and elevators will start to require service factors of 2. Safety of personnel and users comes into the picture, as well as the severity of the application and its duty cycle.

Service factors will differ with the type of equipment and component being applied. A variable speed drive will have different service factor needs than a V-belt. Also, the type of prime mover is a consideration. Smooth prime movers such as electric motors, steam, and gas turbines are predictable. Diesel and gasoline engines are sometimes not smooth and thus require additional service factor attention. All in all, the service factor is the designer's "ace in the hole." Often, certain manufacturers will, upon fielding a request for additional service factor, select a frame or model based on their experience with the specific application and use. This sometimes makes it difficult for the designer to discern between similar equipment offerings. Double-checking with the manufacturer, comparing weights in some instances, and doing physical comparisons will usually provide the answers.

Maintenance

Most mechanical systems need constant attention. Parts are physically wearing on or against other parts. Lubrication is necessary, but it must be monitored. Lubricants get contaminated with metallic particles and dirt; they also act as coolants for machinery. Today's synthetic lubricants can hold up longer than before, but there is still the need for the periodic oil change. Scheduled maintenance for equipment is often hard to adhere to, but in the long run, it will keep machines, which are normally huge capital investments, running longer.

Many times a machine runs 24 hours a day, 7 days a week without stopping, except to accommodate tool and product changes. Thus, there is no convenient time, in management's eyes, to shut down. Yet when a coupling breaks or a motor overheats, there is usually time to panic! Mechanical and electrical systems all need periodic maintenance. Normally, suppliers of machinery and equipment can provide a schedule for maintenance and specify what components need more attention than others.

A current trend is to minimize components and to get maintenance-free equipment in the plant. This is becoming increasingly practical with more electronic equipment now available to perform work that previously took several mechanical components to accomplish. But don't be fooled; electrical equipment needs scheduled maintenance, too. Heat, dirt, and moisture buildup will shut the plant down just as fast as a broken coupling!

Also, a new breed of service person is emerging. The electrician must be cross-trained on mechanics, and the mechanic must be cross-trained on electrical systems. Even operators of machines must know a little about every part of their machine. They are usually not looked upon favorably when they are constantly calling for a service person every time there is a problem. The person who is well-versed in multiple disciplines will be in position to get compensated accordingly, and will be better prepared for change and new technology.

Many electronic drive controls are equipped with on-board diagnostics, and some can even pinpoint a problem within themselves. Diagnostic features between pieces of equipment are beginning to be very similar, especially among drive products.

Closing remarks on power transmission

Power transmission is the "brawn" of industrial production and the drive controls are the "brains." So many times an application is undersized or mechanical problems create electrical failures. As these technologies have been merged, the need for cross-understanding of all disciplines is increasingly intense. An electronic drive cannot make up for insufficiencies in a mechanical system, although it does provide some flexibility.

This chapter's intent was not to be the standard for power transmission, but rather a quick reference for those necessary formulas and definitions that the plant and facility people will come into contact with each and every day. Hopefully, it will settle an argument or two and maybe pinpoint a problem.

Various facts have been presented, but it is always a sound idea to get another's opinion on a subject, especially when safety is an issue. Also, as technology becomes increasingly complex, you should rely on the vendor's expertise and competence. After all, the vendors and manufacturers of high-tech products are constantly upgrading and trying to outdo their competition. So put the onus on them to provide good, sound facts and judgment on critical issues and take some of the responsibility, especially with regard to design and sizing of a motor and drive system.

4

Understanding electric motors

IN ORDER TO UNDERSTAND HOW ELECTRONIC DRIVES control a motor and how best to apply them, you must have a thorough understanding of electric motor basics. After all, there is no need for any type of electronic drive controller unless there is a motor. A complete understanding of how the motor functions, how it is built, and what its limitations and shortcomings are factor greatly into how well the selected drive performs. When troubleshooting the drive, if we have a basic idea about what is going on within the motor, our chances of resolving the problem quickly are enhanced. This chapter's intent is to analyze all common electric motors and relate them to electronic drives.

The electric motor is everywhere. It moves air and fluids via fans and pumps in every city. In today's factory, the prime mover is probably an electric motor, either ac or dc. This electric motor converts electrical energy into mechanical energy, which nets into work. Imagine life without it: there would have to be quite a few steam lines, diesel generators, or turbines to act as the prime movers. This just isn't practical. Mr. Faraday should be thanked many times.

Most electric motors incorporate a rotating, round element into their construction. This is called the rotor because it rotates. The rotor can be referred to as the inside of the motor. It can be found in dc as well as ac machines. In the electric motor, the complementary component to the rotor is the stator. The stator is the stationary part of the motor and can be referred to as the outside of the motor. Figure 4-1 shows the rotor and the stator. Regardless of whether powered by ac or dc voltage, the electric motor must have a rotating element and a stationary element in order to produce motion and torque. Additional components must be connected mechanically to continue the motion. Usually, more rotating items such as pulleys, sheaves, lineshafts, rolls, and others are connected via belts, chains, and couplings.

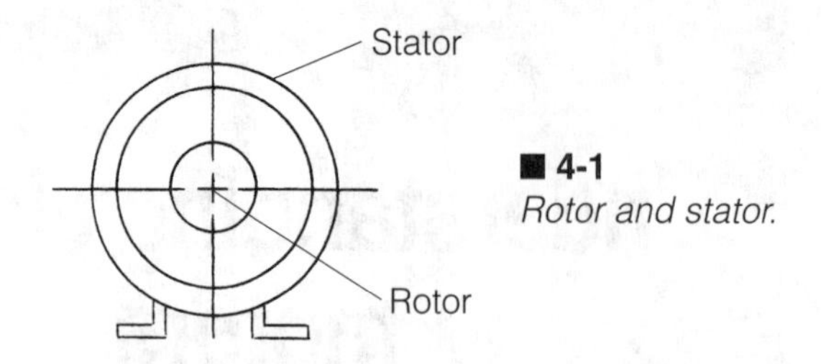

■ **4-1** *Rotor and stator.*

Figure 4-2 shows an ac induction motor in simplified form. Figure 4-3 shows a dc brush-type motor. The similarities are many: the rotating and stationary components, front and back bearings, the motor's base, terminal or conduit box, end plates (sometimes called bell housings), and the motor frame. Depending on the type of motor and its desired operation, construction will vary and so too will material types. Motor stators can be made from cast iron or even aluminum. Other motor stators can be several steel laminations welded together.

In any case, when selecting a motor, stator construction material should be carefully considered, especially with regard to heat. After all, an electric motor is nothing more than a heatsink. It will get

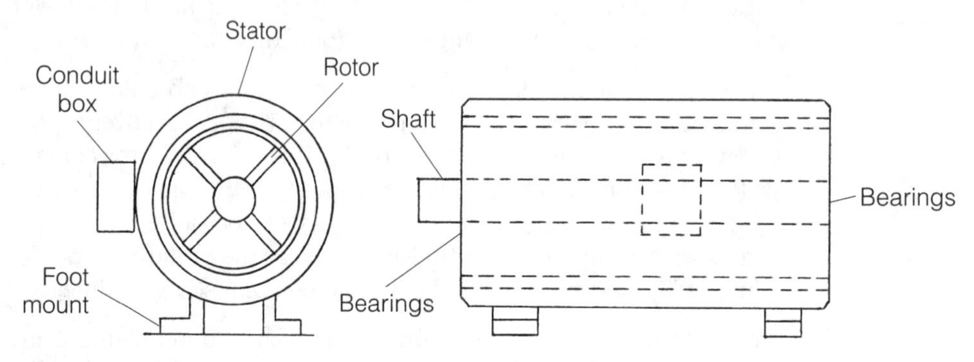

■ **4-2** *Simplified ac induction motor.*

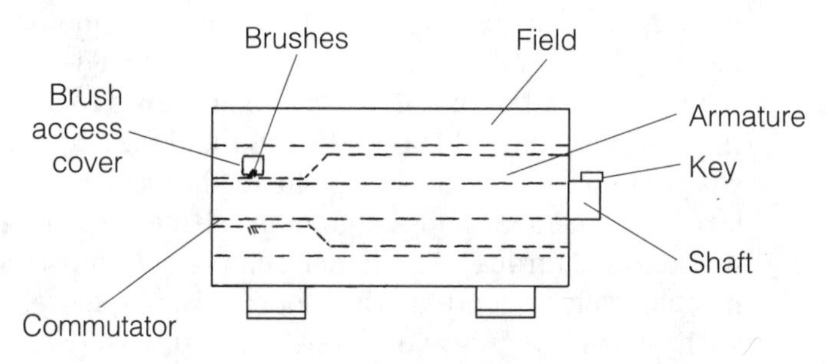

■ **4-3** *Simplified dc brush-type motor.*

hot as current runs through it. And as it heats and cools, the life of the motor lessens. When comparing two motors of identical rated horsepower, the actual weights of each can help determine which motor can withstand heavier loads over a longer period of time. This weight comparison can also aid in determining which motor is providing the best "amount of metal for the price." When applying the motor to electronic drive output, motor manufacturers will provide a frame that can, conservatively, handle the maximum currents and particular drive's output waveform.

Motor cooling

Applying electronic drives to motors means that the speed of that motor is going to be changed. And if changed, depending on the motor loading, special attention had better be given to how the motor is to be cooled. Addressing cooling needs of the motor starts with the application type: constant or variable torque, how low in speed the motor is to run, and what type of motor exists or will be selected. Whenever electrical current is passed through the electric motor there is a buildup of heat. The amount of heat produced is a function of the work, or loading, done by the motor; the type of waveform of the actual electrical signal to the motor; and the eventual changes due to bearing wear and friction.

Premium or high-efficiency motors provide even better use of the electrical energy so as to get more work than losses out of the circuit. Special attention to rotor bar designs and cross sections (see figure 4-4), better conducting materials, and attention to air gaps are some of the areas where better efficiency (i.e., handling the heat buildup) can be gained out of the electric motor.

A motor that runs fully loaded or sometimes overloaded will naturally require a greater amount of current. This could affect its heat content, especially with respect to the duty cycle and the speed range. Another factor that affects motor heating is the incoming power signal itself. In the calculation of motor losses, a pure sine wave will provide a known value. However, when that signal is subject to spikes, noise, or other line disturbances, the motor suffers as losses increase, and less of that incoming power gets used for flux and torque production. Over time, bearings wear, as do other driven components in the drivetrain. This might cause the motor to perform extra work, resulting in extra heating. Getting the heat away from the motor is very important.

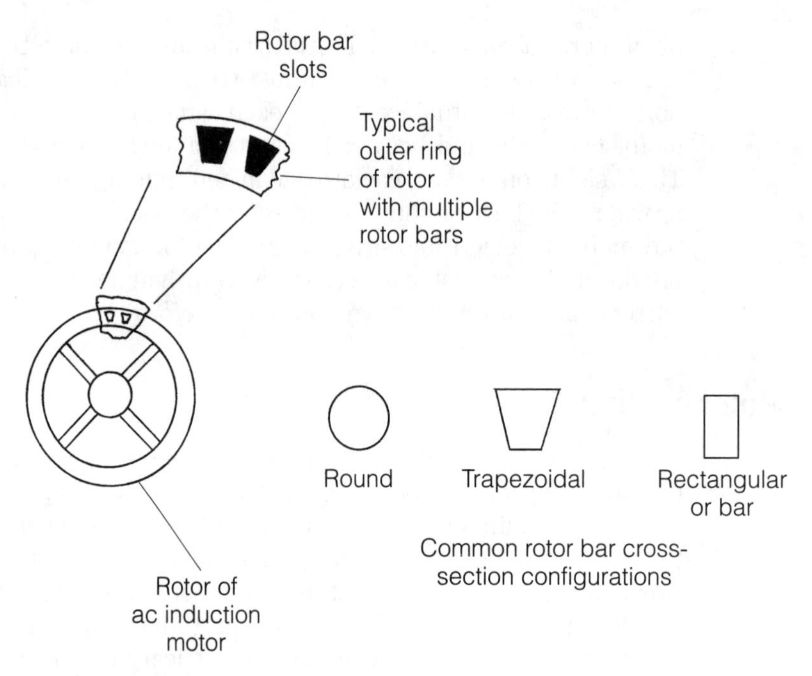

■ **4-4** *Rotor bar designs and cross sections.*

Enclosure types for motor cooling

Many motors are sized for a particular application, or horsepower rating, so that the heat produced from the current can be accepted by the metal content of the motor. Normal convection and radiation dissipate the heat with the aid of an internal mixing fan. These motors are classified as "open drip-proof" (ODP, shown in figure 4-5) or "totally enclosed nonventilated" (TENV, shown in figure 4-6).

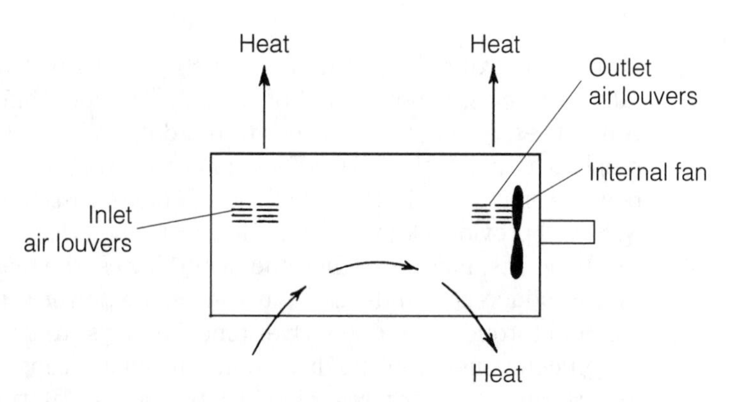

■ **4-5** *Open drip-proof (ODP) cooling.*

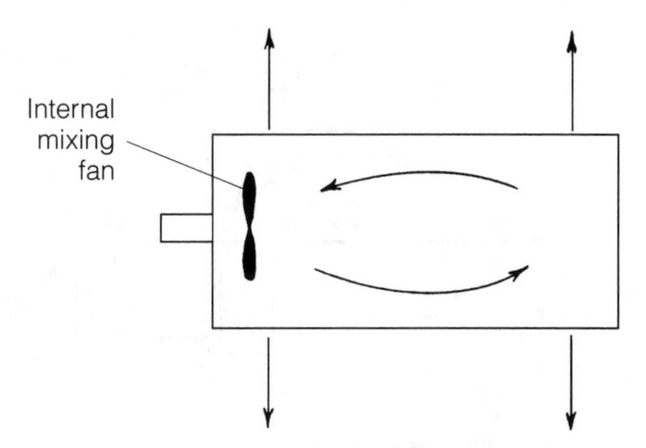

■ **4-6** *Totally enclosed nonventilated (TENV) cooling.*

Other electric motors incorporate a fan blade that rotates at the same revolutions per minute as the motor shaft. This fan blows air across the outside of the motor, cooling it as it runs. However, if a variable-speed drive is used, the lower in speed the motor is made to run, the slower the cooling fan will run also. This can result in a dangerous amount of heat buildup in the motor. Fortunately, the drive will protect the motor by tripping on an overtemperature fault. These motors are called "totally enclosed fan-cooled" (TEFC, shown in figure 4-7).

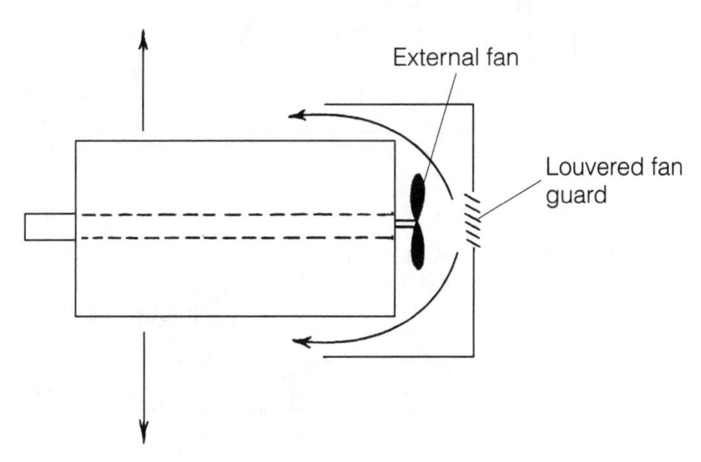

■ **4-7** *Totally enclosed fan-cooled (TEFC) cooling.*

Other types of motor cooling and enclosure types are shown in figures 4-8, 4-9, and 4-10. These are the "totally enclosed water-to-air cooled" (TEWAC), "totally enclosed air over" (TEAO), and the "totally enclosed unit-cooled" (TEUC) motor, which utilizes an air-

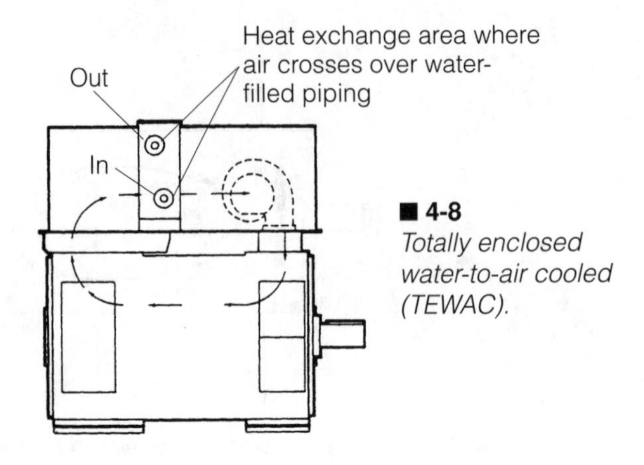

■ **4-8** *Totally enclosed water-to-air cooled (TEWAC).*

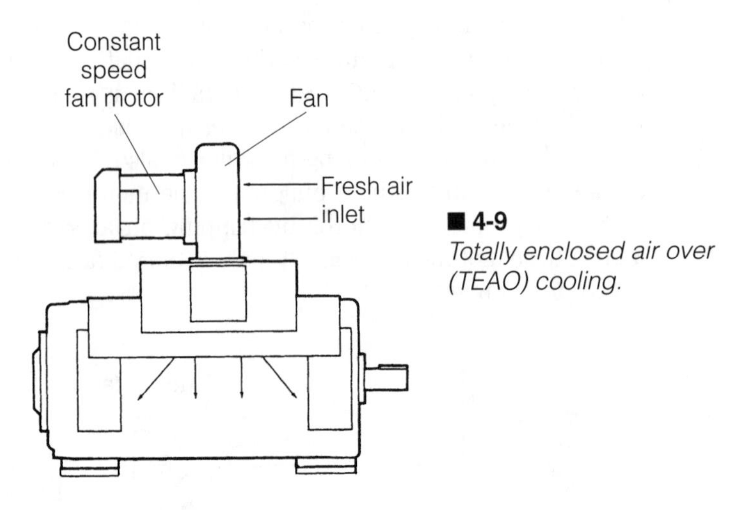

■ **4-9** *Totally enclosed air over (TEAO) cooling.*

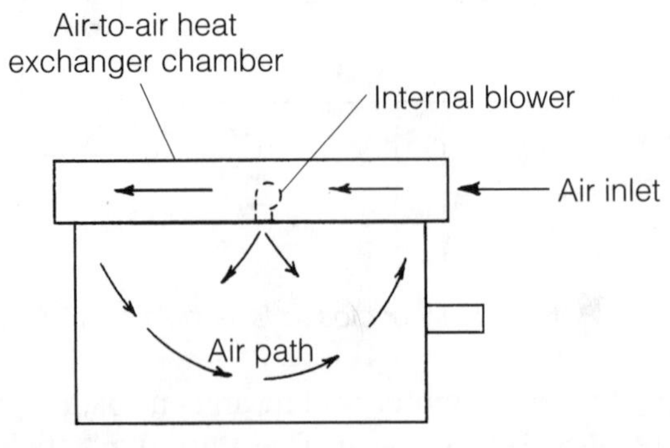

■ **4-10** *Totally enclosed unit-cooled (TEUC).*

to-air heat exchanger unit for cooling. Obviously, the more elaborate the cooling system for the motor, the more expensive the actual unit will be.

In any of the aforementioned cases, if the motor heating is too great in comparison to the rate of heat dissipation, as with ODP or TENV designs, or if the motor rpm's are not adequate to move the heat away, as in a TEFC design, auxiliary cooling measures must be taken. If not, the motor will either damage itself or stop running if motor thermal protection is built in. Either condition is undesirable and upfront care can be taken to avoid these circumstances. Therefore, knowing ahead of time how low in speed the motor will actually run and under what type of load is very important.

One approach is to size the motor with a service factor. Another is to simply go up in horsepower, which is how motors are sized. This might put a motor into a larger frame designation, thereby making it weigh more and allowing it to handle a greater amount of heat. A service factor of 1.15 means that the motor has 15% more capacity when operating conditions are normal for voltage, frequency, and ambient temperature. This 15% extra capacity means that the motor is built and sized to handle operation from an electronic drive, the duty cycle is severe, or the loading and speed range is moderate.

Two scenarios often emerge. One that occurs often is that a motor is selected for a particular load and duty cycle. Someone decides to play it safe and ask for a service factor of 15%. Then another player bumps the horsepower rating up by a factor of two. Pretty soon the application has a motor well oversized for the loading, causing energy to be wasted—not to mention the high initial cost of the motor. The other scenario is the exact opposite: a motor is selected for a variable torque load with a 3-4 to 1 speed range. The next thing that happens is that the motor is running full current down to 10% speed. We now have a heat problem!

Fortunately, in those conditions where the motor was applied for the loads and speeds not originally specified, the other course in motor cooling, the auxiliary blower, can be added later when a heating problem emerges. Basically, an auxiliary blower is a separate fan and motor mounted on the main motor (figure 4-11). This auxiliary fan motor is much smaller and runs full speed all the time, providing moving air across the larger motor to take the heat away and cool it. All that is required here is a starter, installed usually at the drive, and a smaller, full-speed motor with the fan.

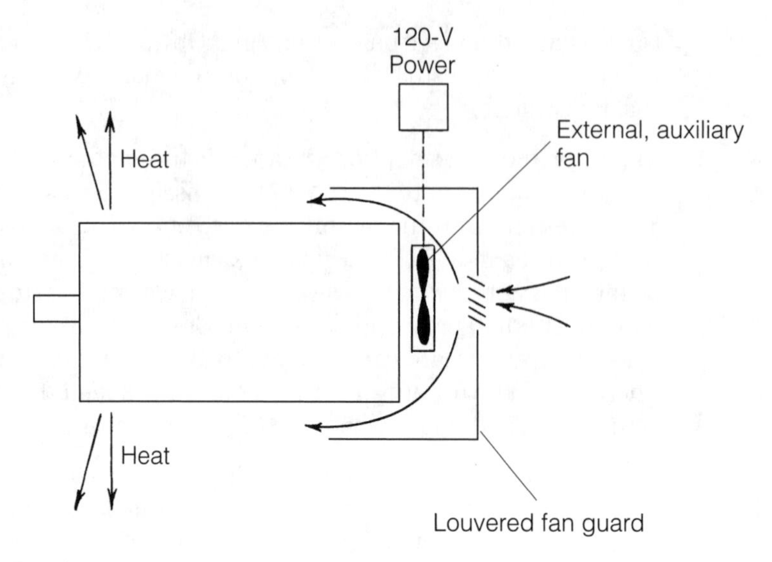

■ **4-11** *Auxiliary blower scheme.*

Another method of cooling is to duct cooled air to a motor. Figure 4-12 shows a drip-proof fully guarded, separately ventilated (DPFG-SV) scheme. This scheme is attractive when the environment around the motor to be cooled is harsh, and a separate auxiliary fan motor is not attractive. In this manner, clean, cooled air can be provided. Other means of motor cooling, such as water or liquid cooling, are possible, but are usually more expensive. Often the environment where the motor is to reside dictates the type of cooling method chosen.

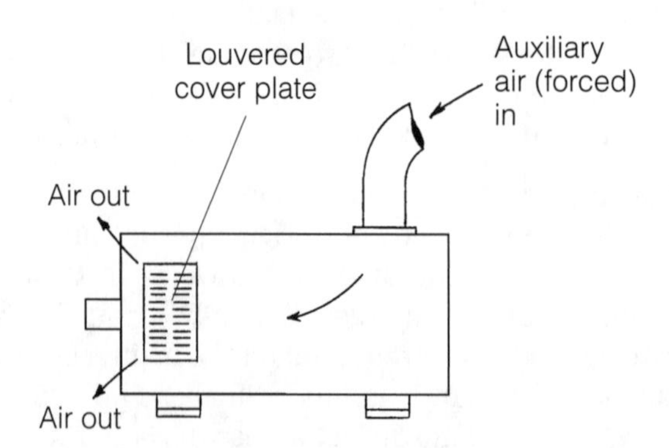

■ **4-12** *Drip-proof fully guarded, separately ventilated (DPFG-SV).*

Explosion-proof motors get us into a whole new realm of motor enclosures. An explosion-proof motor is designed with a housing construction to contain any explosion within the motor caused by any hazardous, ambient atmosphere entering the enclosure and being ignited. An explosion-proof motor must prevent the equipment from triggering a fire or explosion. The ambient hazardous areas can be classified in accordance with the National Electric Code and are shown in Table 4-1. Motor enclosures for explosion-proof purposes can be very expensive, especially for dc motors.

■ Table 4-1 Hazardous area definitions.

Class	**(or type of fuel)**
Class I:	Gases and vapor
Class II:	Combustible vapor
Class III:	Fibers
Division	**(possibilty of fuel being present)**
Division 1:	Present or likely to be present in normal operation
Division 2:	Not present in normal operation
Group	**(specific type of fuel and gases/vapors of equivalent hazards)**
Group A:	Acetylene
Group B:	Hydrogen
Group C:	Acetaldehyde, ethylene, methyl ether
Group D:	Acetone, gasoline, methanol, propane
Group E:	Metal dust
Group F:	Carbon dust
Group G:	Grain dust

Because of their brushless construction, ac motors are a better choice for some hazardous locations. This nonsparking facet of the ac motor gives it an advantage over the brush-type dc motor. However, depending on the severity of the hazardous environment, the ac motor's enclosure might have to be modified for the ambient classification. Purging motors with fresh air and keeping the inside of the motor under a positive pressure (not allowing hazardous gases to enter) are both common practices used in industry.

Whenever the potential exists for a fire or explosion, the designer should consult with the motor and drive manufacturer for the safest approach. By minimizing sparking conditions and overload potentials, and keeping electricity from direct contact with hazardous gases, catastrophes can be greatly reduced.

Motor protection

In addition to cooling the motor, attention must be given to protecting the motor. After all, the motor is an investment that should last several years when properly maintained. Motors should be equipped with thermostats, which cause the supply power to cease once a predetermined motor temperature is reached. Many times a motor is running off of an electronic drive, which can monitor the current being sent to the motor. The drive can shut off power to the motor if it detects an overload condition exceeding a certain length of time. Additionally, thermal overload relays can be installed in the circuit to add further protection for the motor.

Figure 4-13 shows one of three different circuits with corresponding motor protection. In figure 4-13 the electronic motor/speed controller (most often a drive, but sometimes a starter/motor protector) has an electronic overload circuit built in. It can monitor the current going out to the motor and shut off power to the motor in the event of excess current sensed over a fixed period of time.

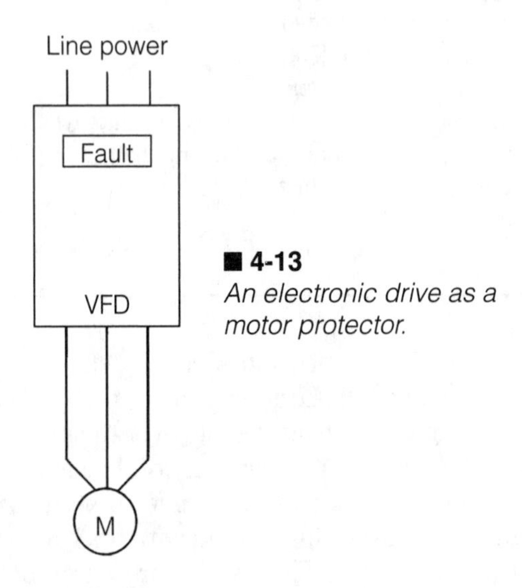

■ **4-13**
An electronic drive as a motor protector.

Another method of motor protection is by thermal overload relays, shown in figure 4-14. The adjustable, overcurrent device is set at a predetermined current value to cause disconnection of the motor from the power supply when that condition exists. This device does not protect itself, but rather protects the load, which is the motor.

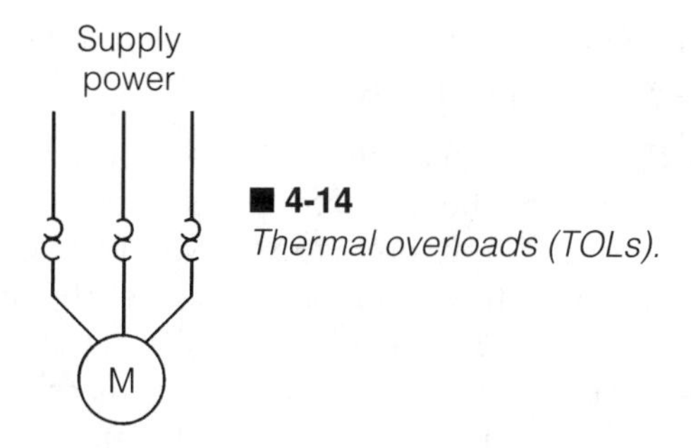

■ **4-14**
Thermal overloads (TOLs).

The last motor thermostat approach is shown in figure 4-15, which is the simplest and least expensive form of motor protection. When opened, this contact shuts power off to the motor. Sometimes these contacts are used in addition to other motor protection means (i.e., going into an electronic drive's fault circuitry). Extra protection can never hurt.

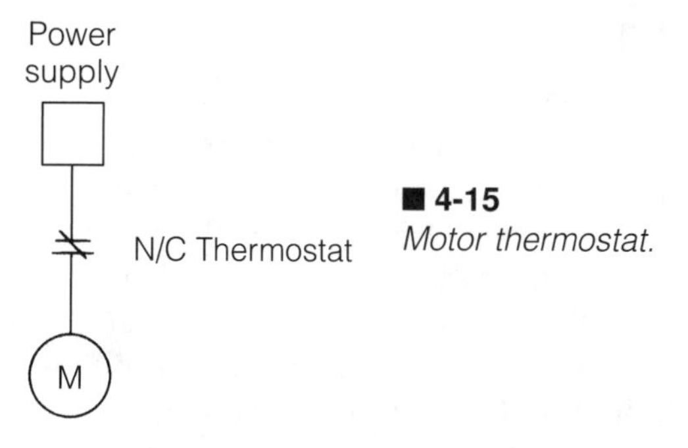

■ **4-15**
Motor thermostat.

The motor exemplifies a product whose design has evolved based on the need for extra protection and extended motor life. Care is taken when the windings, which are basically turns of copper or aluminum wire, are incorporated into the motor package. Each turn, or coil, of wire must be placed precisely, or else a premature failure in the motor will occur. Often, a motor is required to be form-wound, which means that even extra special placement of the turns in the windings must occur.

The term "random wound" is used to describe the more common method of winding. As the windings are being wound, insulating tape is applied to further isolate the windings from each other. Then the entire network of coils is coated with a varnish. This seals the insulation on the windings and adds to the life of the motor. Remember, any deviations or imperfect turns of wire can cause extra heating, and over time, will degrade the motor's torque-producing capabilities and eventually cause overheating.

Another important piece of the motor package is the bearing arrangement. The front and the back of the rotor must rest on sets of bearings. Often, premature wear at the bearings in a motor can cause a machine to shut down. Most of the time the early failure of the bearings can be attributed to the actual mounting of the motor itself. If there is side loading or if the motor is coupled in a manner such that the shaft is somewhat off-center, unwanted forces begin to act on the bearings, wearing them out faster. Heavy-duty bearings should be requested when the motor mounting is overhung or side loaded. Additionally, good maintenance and lubrication of the bearings will lengthen their life. Some motors can be supplied with oil mist systems, which perform lubrication continuously. These systems are prevalent when motors must rotate at extremely high speeds.

NEMA ratings

The National Equipment Manufacturer's Association (NEMA) has provided for some standardization in the motor industry. This standardization includes physical, mechanical sizing and electrical characteristics. For instance, three-phase, ac induction motors have specific ratings for speed and torque. Each of these designs has a different characteristic for starting current, locked rotor current, breakaway torque, and slip. The most common designs are NEMA A, B, C, and D. Each has a distinct speed-versus-torque relationship and different values of slip and starting current.

The most common is the NEMA Design B motor. Its speed/torque curve is shown in figure 4-16. The NEMA B motor's percentage of slip ranges from 2 to 4%. It has medium values for starting and locked rotor current, and a high value of breakdown torque. This type of motor is very common in fan, pump, lighter duty compressors, various conveyors, and some lighter duty machines. The NEMA B is an excellent choice for variable torque applications.

The NEMA Design A motor, whose curve is shown in figure 4-17, is similar in many ways to the NEMA B motor. It typically has a higher value of locked rotor current and its slip can be higher.

The other two classifications of NEMA design motors are the NEMA C (figure 4-18) and the NEMA D (figure 4-19). NEMA C motors are well suited to starting high-inertia loads. This is because they have high locked rotor torque capability. Their slip is

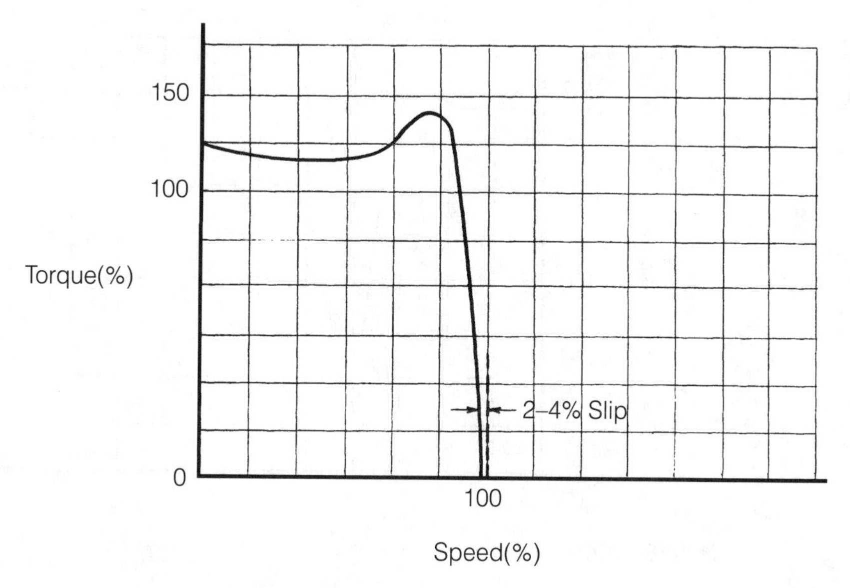

■ **4-16** *NEMA B motor curve.*

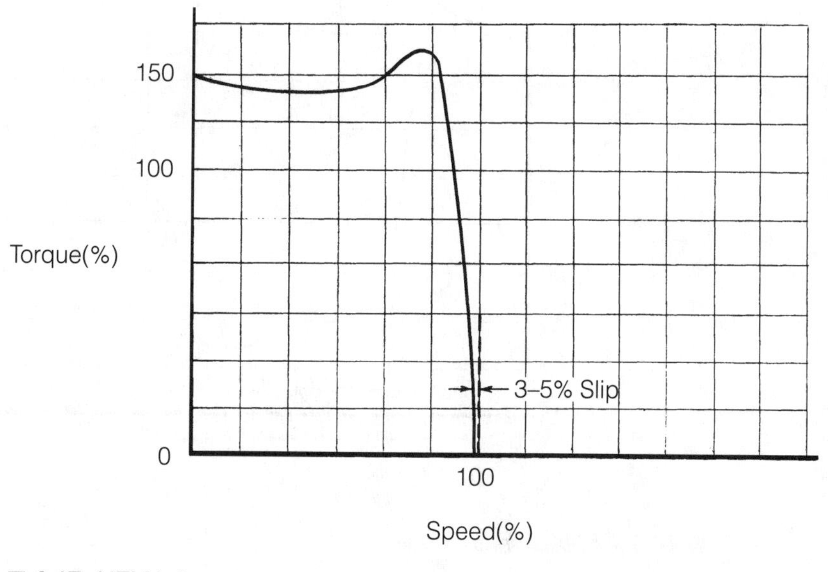

■ **4-17** *NEMA A motor curve.*

around 5%, and their starting current requirement is average. The NEMA D motor is found in heavy-duty, high-inertia applications. It has high values of slip, very high locked rotor torque capability, and its base speed is usually below 1,800 rpm. Figure 4-19 shows the speed-versus-torque curve for the NEMA D motor. Typical ap-

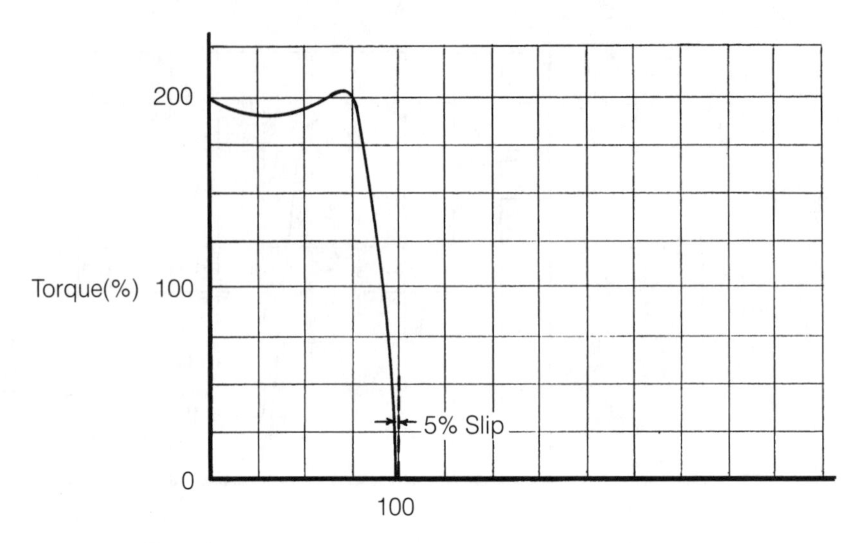

■ **4-18** *NEMA C motor curve.*

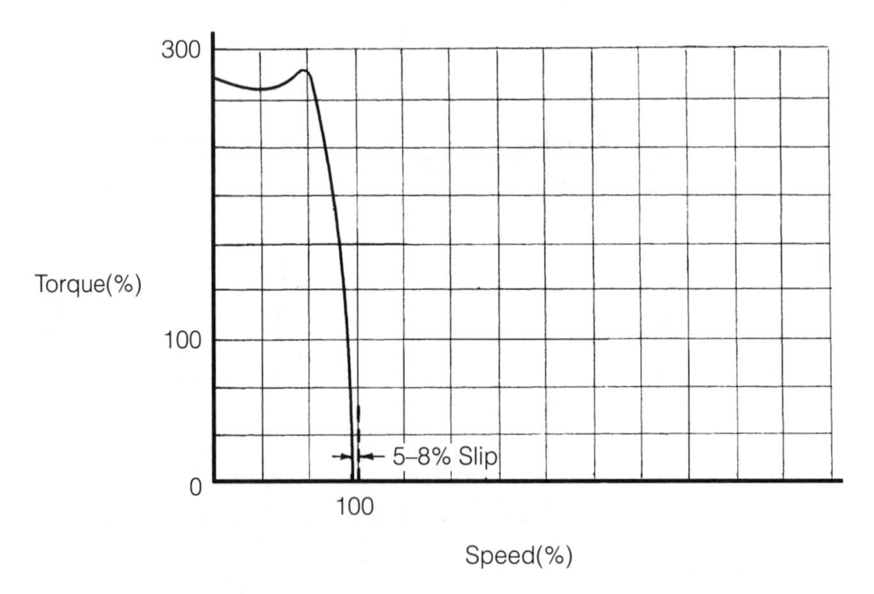

■ **4-19** *NEMA D motor curve.*

plications include punch presses, shearing machinery, cranes, and hoists. Whenever applying an ac drive to NEMA C or NEMA D motors, attention must be given to the slip and full load current ratings of these motors in order to get a proper size match (drive current rating and motor current rating).

Many times a motor is supplied based on its frame size. Motor manufacturers have their own designations when the horsepowers exceed a certain level. For three-phase ac motors, this is usually beyond the NEMA sizes. NEMA has provided for some standardization in motor sizes up to approximately the 449 frame. Above this, manufacturers of motors can classify their motor frames with their own designations (referred to as "above NEMA ratings").

For a given horsepower rating of a motor and for a given base speed, there are standard frames that most manufacturers adhere to. For instance, there is a certain foot to centerline of shaft dimension and a certain frame diameter (figure 4-20). Dividing the first two digits by 4 usually indicates the foot to centerline of shaft dimension. For example, 284 frames have a 7-inch dimension, because $28 \div 4 = 7$. This helps the machine designer in locating the motor physically to the machine and in comparing a motor sized by one manufacturer with one by another.

Other issues to be considered are based on the duty cycle, service factor, ambient conditions, and torque requirements. Many times, different frame sizes will be offered by different manufacturers for the same application and horsepower. Fully analyze what all the impacting factors mean and which are most important. This upfront analysis can save money, downtime, and headaches later.

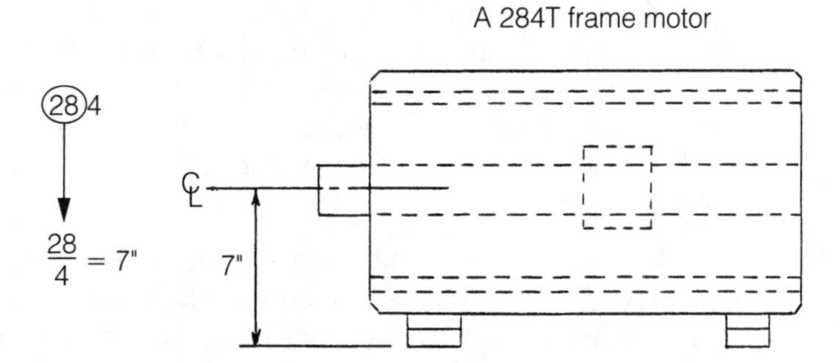

■ **4-20** *284 frame—shaft to centerline.*

Motors (ac)

All electric motors utilize either some form of electromagnetism, induction, or repulsion. These principles of physics are the basis for motion in the factory today. In order for these phenomena to occur inside an electric motor, electrical power must be supplied to make the necessary electrical changes so that motion can begin and be sustained. This power is either in the form of ac or dc.

From this point, there are multiple ways of achieving the desired motion. This section will deal primarily with ac-supplied electric motors. But first we must go back to the 1830s to evaluate Michael Faraday's discovery of electromagnetic induction.

Faraday's law of induction has been demonstrated in our factory production for over 160 years. He discovered that by varying the magnetic field around an electromagnet, a current would be produced and the magnet would actually move. By controlling the opening and closing of the circuit, he was able to gain control of the magnetic flux. This is the fundamental principle used in an induction motor today. Faraday's law also concludes that an induced electromotive force, or EMF, component and the magnetic flux's rate of change are proportional. Induction motor discussion often revolves around EMF, which is expressed as voltage.

Induction motors and repulsion motors are very similar in operation, and there are even overlaps in the function and names of various types. For example, there are repulsion motors, induction motors, and repulsion induction motors. Basically, the induction motor works on the principle of changing the state of electromagnetism around a magnetic field to achieve motion (induction). Likewise, the repulsion motor is close in operation. It contains two magnetic fields of like polarity, thereby opposing each other and causing motion. A repulsion induction motor is the hybrid; it contains brushes, a commutator, and a wound rotor. The repulsion induction motor uses the brushes and commutator to get hard-to-start loads going. Once it reaches 60 to 70% speed, the brushes are lifted off of the commutator by centrifugal force, and the motor continues running as a squirrel cage type.

At this stage, a more detailed exploration of the construction and characteristics of both the induction and repulsion motors is necessary. Because the induction motor is the most common, we'll begin with it. Most induction motors utilize ac incoming power to operate. Most are three-phase supplied and are sometimes called polyphase motors. The electrical action of this type of motor, particularly the squirrel cage type, is likened to a transformer with a shorted secondary (see figure 4-21). The motor is much like a step-down transformer because the primary is equivalent to the windings of the motor. This is where the three-phase ac power is fed. The secondary is likened to the rotor, or armature. The fewer the secondary turns and the higher the gauge of the wire equates to higher value of current induced by the primary, with the maximum current created when the secondary is shorted.

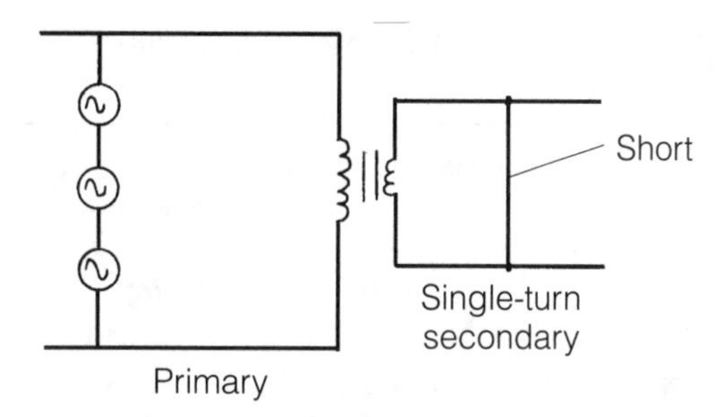

■ **4-21** *Induction motor as shorted secondary.*

The ac induction motors are also classified by their slip characteristics. Further evaluation of this characteristic can be found in speed-versus-torque performance. When a load is applied, an ac motor has a tendency to slip. This phenomenon allows the motor to request more current in order to continue driving the given load at the desired speed. Some motors have more slip than others. Typical values of slip are 2 to 3% of synchronous speed, which is common with NEMA A, B, and C motors. Some values can get as high as 5 to 6% for special high-starting torque applications. These higher slip values would be found in NEMA D and other design motors.

Another crucial issue with ac induction motors, particularly when applying them to electronic drives, is insulation. Motor insulation consists of a nonconducting material that basically separates current-carrying components within the motor from each other. Often this is in the form of insulating tape, which is applied prior to any dipping or coating of the windings. Motor insulation has been given ratings based on temperature maximums versus normal life expectancy of the motor. These values are shown in Table 4-2.

Insulation ratings are a good gauge when applying motors to variable-frequency drives. The fact is that variable-frequency drives do promote extra heating in the motor because of a nonsinusoidal waveform. Depending on its severity, this extra heating can lessen the life of the motor, although this phenomenon is somewhat overexaggerated. Many factors enter into the life expectancy of a motor, including starting currents, quantity of starts in its life, shock loading, and so on. The way motor manufacturers deal with motors on variable-frequency drives is to promote class F or greater insulation systems into the motor for these applications. It is always preferable to get the better grade of insulation whenever possible and practical. This can only help to lengthen the motor's life, whether or not an electronic drive is controlling it.

■ **Table 4-2 Temperatures for various classes of insulation.**

Insulation class	Temperature rating	
	Degrees F	Degrees C
Class A	221	105
Class B	266	130
Class C	Over 464	Over 240
Class E	248	120
Class F	311	155
Class H	356	180
Class N	392	200
Class R	428	220
Class S	464	240

The most common type of induction motor is the squirrel cage motor (figure 4-22). A workhorse of industry, this motor is found in most ac applications in a wide range of horsepower ratings. The rotor is built up from steel laminations. Each lamination has a set of slots around its perimeter. Long rotor bars are inserted into these laminated slots. The rotor bars are usually made from copper and have various cross sections to increase impedance by creating eddy currents. Traditionally, squirrel cage motors have had a disadvantage in actual started torque because of their fixed rotor design. Different rotor bar designs can provide the extra impedance to better control the rotor flux at low speeds. At high speeds this is not an issue.

When the stator windings have been supplied with ac power, current is induced in the rotor. The difference in polarity between the stator and rotor causes motion, which can occur in a three-phase motor because the phases are displaced by 120 degrees.

A single-phase motor, on the other hand, cannot start by itself. In order to get rotating there has to be a starting winding with a

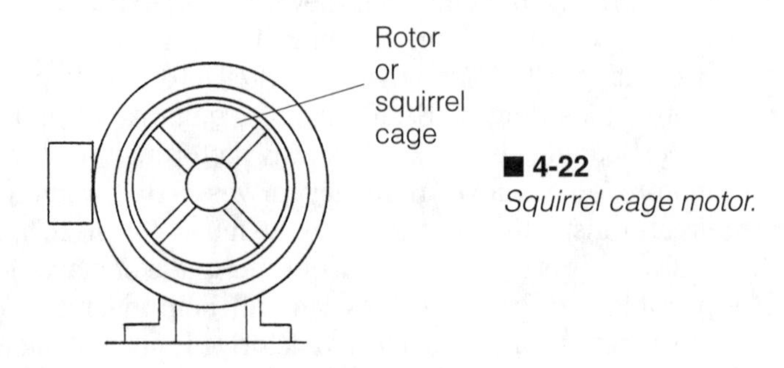

■ **4-22**
Squirrel cage motor.

switch. This configuration is sometimes called a split-phase induction motor. Shown in figure 4-23, the split-phase induction motor also requires a run winding. It is powered by single-phase, usually 120-V power. This type of motor typically is in the fractional horsepower range and has more commercial applications than industrial. Its starting and running operation is similar in principle to that of a repulsion-start-induction motor. The split-phase induction motor starts off of the starting winding, and once it reaches 70% speed, changes over to the run winding. The switch is centrifugal, which differs from the centrifugal brush system used in the repulsion motor.

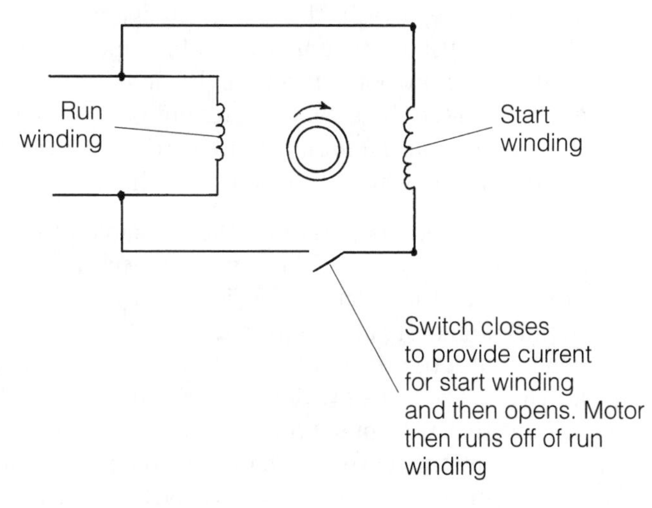

■ **4-23** *Split-phase induction motor.*

Other types of induction motors include the capacitor motor, the shaded pole motor, the wound rotor motor, the synchronous motor, and the polyphase motor. These motors are not used often, if ever, with electronic drives. However, they do exist, and the request is often made to fit them with an ac drive. Therefore, knowing how they function is important to the applicability of a drive. The capacitor and shaded pole types are single-phase units, limited in size. The capacitor motor works on the principle of storing energy for starting and can be obtained in horsepower sizes up to 20 HP. The shaded pole motor utilizes salient poles (sometimes called shaded poles), which are copper loops acting as the starting winding. Shaded pole motors are very inexpensive and don't get above ⅛ horsepower.

Wound rotor and synchronous motors are used in more industrial applications. These types of induction motors can be built in horsepower sizes from fractional to several thousand horsepower. The wound rotor motor contains many groups of copper coils. The stator windings equal the number of poles so that a rotating magnetic field is produced. The wound rotor motor is the older brother of the squirrel cage induction machine and can provide high starting torque.

Synchronous motors run at a constant speed regardless of the load changes. They run from a fixed frequency ac source at the same speed as the source frequency, as dictated by the number of poles in the rotor. The poles "synchronize" with the rotating magnetic field. Many synchronous machines have low synchronous speeds and therefore have 10, 12, and even higher counts of poles. These motors are used in compressor and pump applications quite often. They can be provided in very high horsepower configurations and are well suited for steel and aluminum mill environments. Often the synchronous motor must be started with a dc generator so that the poles can eventually lock in with the field.

Repulsion motors operate on the principle of opposing magnetic fields to attain motion. They need an armature, a field, and means of switching the polarity. This is done by means of a commutator. Repulsion motors also require a set of brushes, which ride on the commutator to provide the means of transferring electricity to the armature. From the previous description of a repulsion motor, it is evident that the common dc motor is, in fact, a repulsion motor. Since dc motors have been a major force in the factory for so many years, more on these machines will be covered later in this chapter.

Often when a motor of larger size is installed, evaluating that installation relative to its associated, mechanical drivetrain components and its electrical components is necessary. This is called a torsional analysis and is usually done before the actual installation. Design criteria will help to predict whether or not there will be potential problems with vibration, resonance, and shock conditions. All of these conditions can be detrimental to the life of an installation. This analysis can also help to suggest whether a certain type of coupling should be used, if vibration isolation is required, or if the electrical pulses will be a problem.

Basically, the analysis requires data about the application: loading, speeds, motor mounting, type of electronic controls used, and so on. Then extensive calculations are performed to determine as much about the operation of this system as possible and to predict where harmful operating points might be. Steps can then be taken to either eliminate or minimize the effects of these hazardous situations. For example, consider an electronic drive controller that varies the frequency to the motor to change speeds. If allowed to remain at that setting, at a certain frequency there could be an amplification of oscillations, or resonance. This could cause couplings or motor shafts to break.

The electrician's best friend is the motor nameplate (figure 4-24). When starting or troubleshooting an ac drive and motor installation, knowing what you're working with is imperative. Most nameplates contain much of the data required to troubleshoot, set up a drive controller, and even wire the motor. (New motors are usually supplied with complete test data, but this is often lost over time and not available.) The frame, horsepower, full-load speed and current, and even wiring diagrams are sometimes provided on the nameplate. If the motor can be wired for either 230 or 460 V, the connection information is usually imprinted on the nameplate. The nameplate is riveted onto the motor frame and rides along with the motor for its entire life. Even if painted over, the data is engraved onto the plate, thereby making it still readable. This motor data is extremely important to the electronic drive application engineers if they are to implement the drive on that particular motor.

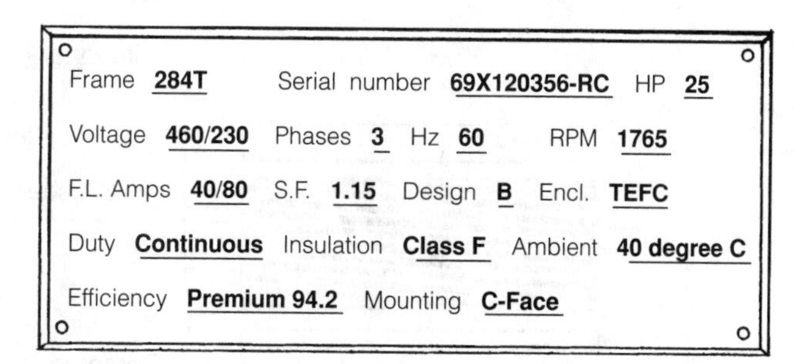

■ **4-24** *ac motor nameplate.*

Motors (dc)

The jury is still out on dc motors and their future. The dc motor has lost a lot of its prominence in the past few years. With ac induction motor technology getting stronger and with its manufacturing and maintenance costs a factor, the dc motor just isn't used in the factory like it was years ago. However, it still has a place, and there are still applications that require it. One important use is that of running off of a portable power supply such as a battery. Another common practical application of dc motor technology is in the winding and unwinding of material where holdback torque is necessary. Other applications that we can call dc motor strongholds are cranes and hoists, traction motors requiring high starting torque, elevators, and large mill systems.

A major threat to the dc motor is ac vector technology, especially as these systems find cost-effective methods of handling regeneration. The dc motors have traditionally given high starting and ride-through torque capabilities, along with excellent speed regulation. However, the ac flux vector drive is able to deliver similar performance with an ac motor.

As previously noted, the dc motor is a repulsion type. Figure 4-25 shows a dc motor in simplified block form. The main components are the armature, field, commutator, and brushes. These components provide for an electrical to mechanical scheme whereby dc power is supplied to the brushes, which transfer it to the commutator. The commutator then reverses the polarity in the armature, which opposes the magnetic property of the field, thereby causing motion. This is the basic premise for dc motor operation. It was and still is a straightforward operating principle. The equivalent circuit for the dc motor is shown in figure 4-26.

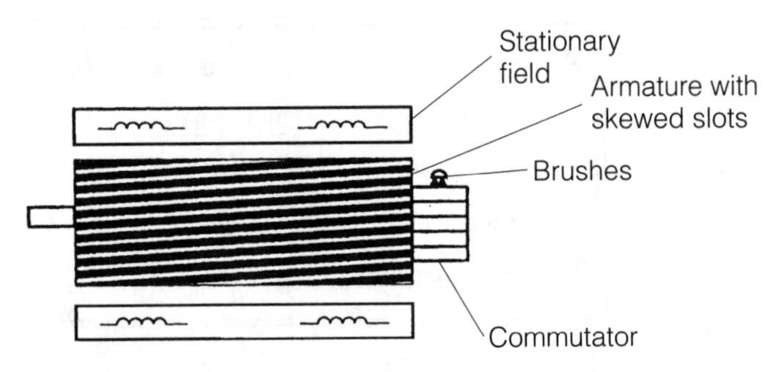

■ **4-25** *dc motor—simplified block form.*

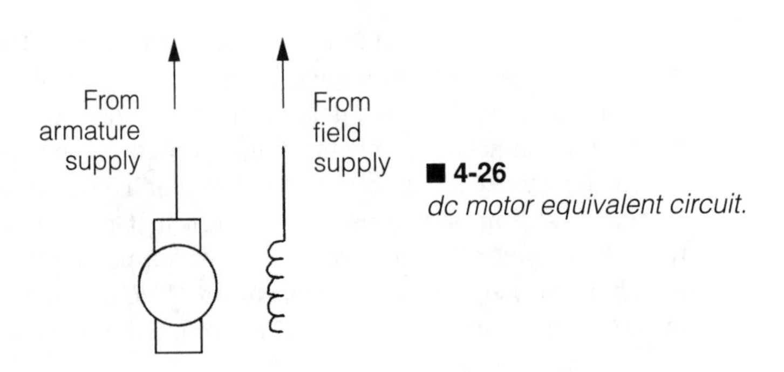

■ **4-26**
dc motor equivalent circuit.

The current to the field windings maintains the field strength. This is half of the equation to the dc motor's commutation. The other is the armature winding's current and its smooth transitions from coil to coil. This current must reverse at each coil as a new pole is encountered. This process is actually the commutation, and the timing of this reversal must be synchronized by the brush position at the commutator itself. If the timing is off, flashing and arcing can occur. Therefore, it is always imperative to have clean and properly aligned brushes with respect to the commutator gaps in order to provide smooth, effective torque output from any dc motor.

The typical frame construction of a dc shunt-wound motor is more square than round (permanent magnet dc motors are usually round). This makes for a shorter distance from the feet to the centerline of the motor. Also, this design minimizes magnetic losses, thus getting greater use out of available current. The frame is typically made of iron laminations epoxied together with an oxide. The shunt field coils are mounted on the inside of the laminated frame, thereby becoming the field.

The armature is mounted to the shaft within the housing of the frame and is supported at both ends by bearings. The armature is again made of iron laminations, this time in a round configuration to fit compatibly within the rounded field windings. The armature coils are fitted inside slots, sometimes skewed to reduce cogging, and insulated. The ends of these coils are at the commutator, which is connected to the armature.

It should be noted that a skewed armature, although better for handling cogging conditions, is harder to manufacture. The amount of skew, or the amount of slot pitch offset, makes winding the armature a little more difficult and therefore more costly. The commutator is a round, copper-based device with many individual wedge-like pieces. The surface is machined smooth to allow for a good, clean contact between the commutator and the brushes. The brushes, which are made from carbon, ride on the commutator to pass current to the armature windings. The brush contacts are spring-loaded and the number of brushes required in a given motor is determined by the current rating of the particular motor.

There are different types of dc motors: permanent magnet, series wound, and parallel wound, also called shunt wound. The permanent magnet dc motor is typically found in the smaller, fractional horsepower range. It usually has a lower armature voltage rating than shunt-wound motors, usually 90 and 180 V. The magnets are actually part of the stator and equate to the field. Thus, no power is required at the field. These magnets are usually ferrite type. Ferrite magnets are fragile and are susceptible to breakage if the motor is mishandled. Also, they are attached to the motor frame by an adhesive bonding method, usually epoxy. This means the magnets might become dislodged. Another disadvantage with permanent magnets is that they can become *demagnetized* in overcurrent conditions when combined with prolonged overtemperature conditions.

A more expensive type of magnet used is the rare earth version. This magnet is manufactured by a process known as sintering. *Sintering* is the fusion or welding of small particles of metal by applying heat, normally just below the melting point of the metal. This allows the rare earth magnet to be formed into an appropriate shape for later use in mounting to the rotor, which is part of the reason for the added expense. In addition, these magnets are not magnetized until they are installed in the motor frame. They are attached with an adhesive that is somewhat resistant to high-temperature conditions. Because rare earth materials are somewhat expensive, the search is on for an efficient, inexpensive material. One acceptable synthetic magnetic composite is neodymium iron boron. Neodymium (Nd) is one of the rare earth elements (#60 on the periodic chart). Coupled with iron and boron, it has good magnetic characteristics.

Since the flux is basically constant in the stator of a permanent magnet motor, the speed versus torque is a linear relationship even into extended speed range. Figure 4-27 illustrates this relationship. However, with a wound field motor the torque diminishes as it goes into field weakening. (Field weakening is illustrated in figure 4-28.) To increase the speed of the dc machine, the stator's magnetic field intensity is reduced by correspondingly reducing current. When this is done there is simply less available torque-producing current.

Field wound dc motors make up the bulk of the dc motor installations. There are series and parallel wound versions. Figure 4-29 shows a simplified circuit of a series wound dc motor. Also shown is the speed-versus-torque curve of the series wound motor. A series wound motor can deliver a substantial amount of starting torque and running torque at low speeds. It accomplishes this by developing magnetic flux in direct proportion to current, but diminishing as the flux saturates. It is basically a load-dependent motor that keeps generating torque as the load increases. If the load were to suddenly disappear, the series wound motor would keep producing torque current, which could be hazardous. The series wound motor has been used extensively in mill applications, especially on cranes and hoists.

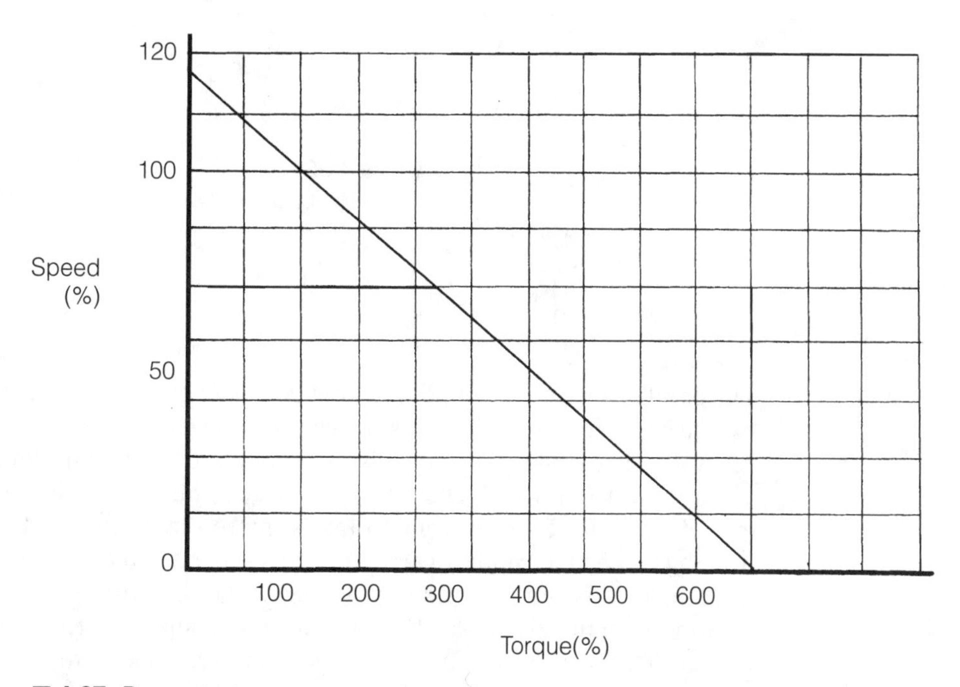

■ **4-27** *Permanent magnet speed-versus-torque curve.*

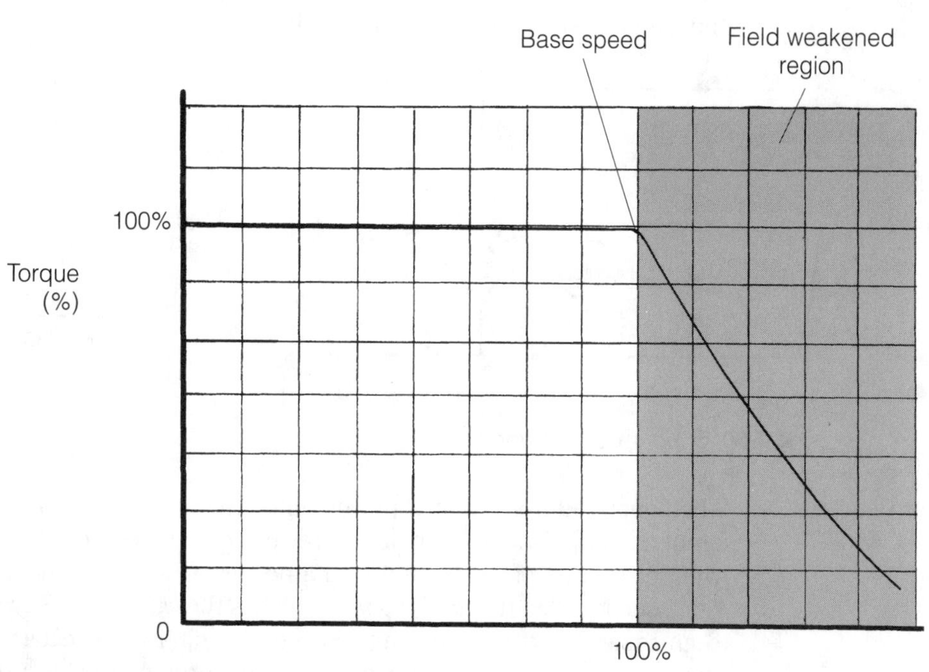

■ **4-28** *Field weakening.*

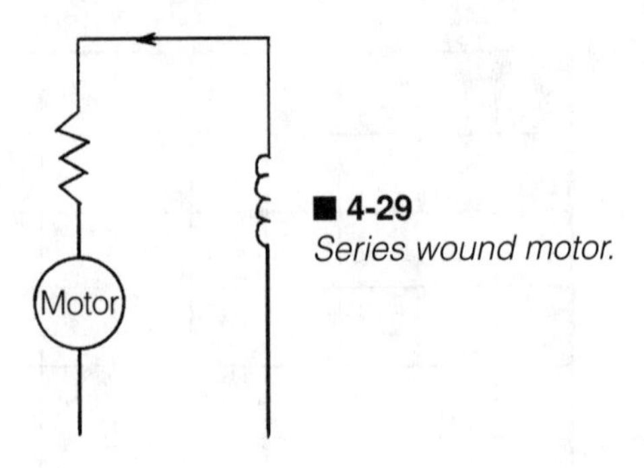

■ **4-29** *Series wound motor.*

The parallel, or shunt, wound dc motor is a more common and less expensive dc motor on the market today. It can be controlled more easily by thyristor-based controls. Typical circuit configurations and curves are shown in figure 4-30. Because a shunt is a short circuit of sorts, it allows current to take a more direct path in the windings. The field windings are supplied a separate voltage, while the armature is provided its own voltage. The armature rotates, creating an induced back EMF that opposes the supplied armature voltage. This back EMF is a predictable quantity; it allows for control by thyristor-based controllers and is self-regulating. Typical field voltage ratings for dc motors are 150/300, 100/200, 120/240. Some are harder to control than others. Typical armature voltage ratings are 180 V, 240 V, and 500 V, although there are plenty of nonstandard armature ratings around.

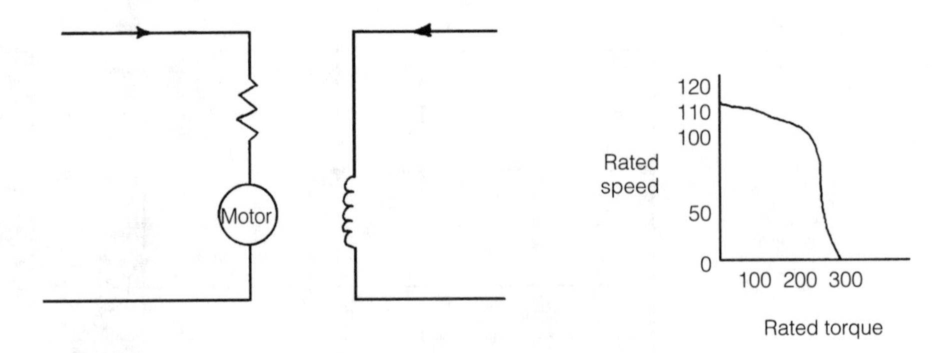

■ **4-30** *Shunt wound motor.*

When operated from an SCR power supply, a dc motor exhibits different characteristics that should be reviewed. This is especially true when placing an older dc motor on a newer SCR-based controller. This issue is called the *form factor*, which is the ratio of RMS (root-mean-square) current to average current. It is an issue because the average current determines the amount of torque available. Also, there can be effects due to excess heating, as well as unwanted sparking and hard-to-commutate conditions. Values have been placed on certain types of rectifiers. These values are selected based on the wave type and amount of SCRs in a given rectifier.

For instance, at base speed and at full load, a three-phase, full-wave, 6-SCR rectifier will exhibit a ripple frequency of 360 Hz and a form factor of 1.01. On the other hand, a semiconverter with 2 SCRs and three diodes, single-phase, will carry a ripple frequency of 120 Hz and a form factor of 1.35. This current ripple is affected by line voltage, motor speed and load, the inductance in the circuit, and most importantly, the type of rectifier. The form factor is used to base the realized current value used to determine heating. The current that we are concerned with for heating is the RMS current. The RMS current takes the form factor times the motor rated current. In this way, predictions as to heating are possible.

Likewise, commutating the motor can be difficult with certain rectifiers. Sometimes no commutation at all occurs; the motor simply arcs, or sparks. Although at times a motor can run adequately with sparking, it should be avoided. Single-phase rectifiers can promote unwanted commutation problems. It should be noted that the lower the form factor the better for heating. This is not necessarily so with commutation issues. Some smaller horsepower-rated motors can commutate with a form factor of 1.35.

A typical dc motor nameplate is shown in figure 4-31. It contains much of the data critical to that motor's design and specific construction. This data is especially necessary whenever applying or replacing a dc motor controller for that particular motor. For the dc motor, information on the nameplate includes field voltage and field current ratings, armature voltage, base speed and speed range, frame, enclosure type, serial number, and special comments about

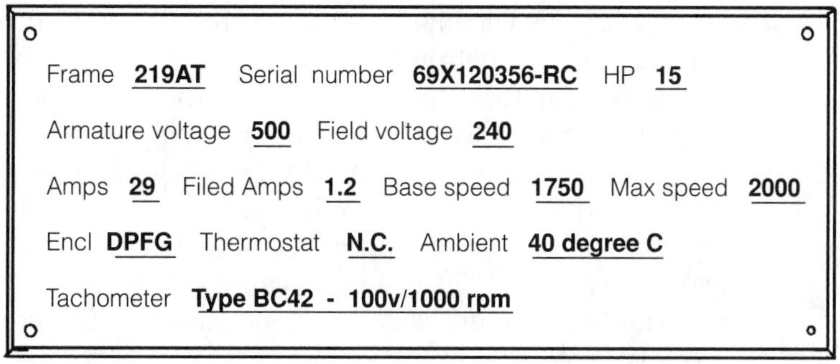

Frame **219AT** Serial number **69X120356-RC** HP **15**

Armature voltage **500** Field voltage **240**

Amps **29** Filed Amps **1.2** Base speed **1750** Max speed **2000**

Encl **DPFG** Thermostat **N.C.** Ambient **40 degree C**

Tachometer **Type BC42 - 100v/1000 rpm**

■ **4-31** *dc motor nameplate.*

the motor. The nameplate shown is common for a standard shunt-wound unit.

Servomotors

Another group of permanent magnet-type motors are servomotors. This type of motor utilizes a scheme of electronic control whereby electric current is changed to the windings, which are embedded in the stator. The servomotor is sometimes called an ac brushless motor or a dc brushless motor. Actually, the terms are basically interchangeable. The motor is, in fact, a brushless motor. It is supplied three-phase power, which is rectified and then furnished as an amplified power signal to the windings. Earlier servomotors did use brushes and a commutator to actuate the rotor, but nowadays, servomotors are electronically commutated, and therefore, brushless systems. Today's servomotors offer less motor maintenance and enhanced motor performance over the brush-type system.

The construction of a servomotor has made it relatively easy to implement from a physical standpoint. Servomotors need to have certain physical features. They should be compact to fit in tight places on a machine, but this compact design should not sacrifice torque output. Like all motors, they need to be able to rid themselves of heat, but usually an auxiliary blower is not practical. They must be capable of being easily retrofitted with a feedback device. Stack length and the mounting face of the servomotor also become issues. Mountings such as C-face and D-flange have become industry standards for mating to machines, gearboxes, and so on.

The external considerations are important, but it's what goes on inside that makes or breaks the application. A view of the inside construction of a typical servomotor is shown in simplified form in figure 4-32. The windings to which current is supplied are located in the stator. On the rotor are located the permanent magnets. A typical servomotor, capable of rotating thousands of revolutions per minute, may have six poles: three N poles and three S poles. Correspondingly, there are six sets of windings for each pole: three positive windings for phases A, B, and C and three for negative, or reverse-motion windings, phases A, B, and C. Sometimes this motor design is called "inside-out" because of the permanent magnet rotor and the wound stator (as compared to the conventional brush-type motor).

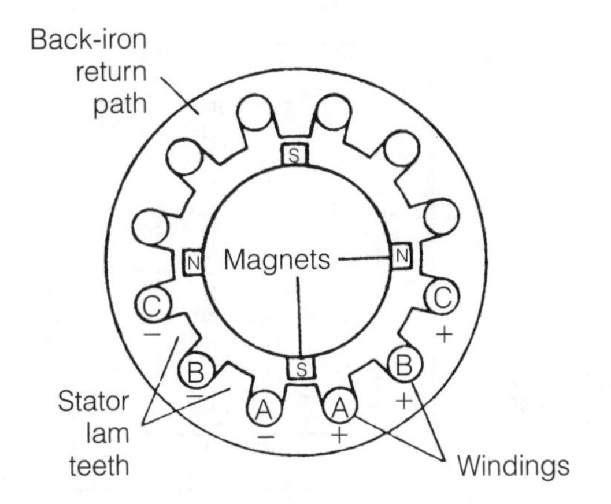

■ **4-32** *Servomotor construction.*

The rotor of the servomotor can be many laminated iron pieces or in some instances can be one solid cylinder. Onto this cylinder is placed the magnets. The magnets are a permanent, rare earth variety. They can be epoxied on or press-fit into place. This composite rotor is then carefully mounted into bearings, which have been permanently lubricated. These bearings can be heavy-duty if side, or radial, loading is expected when the servomotor is mounted. Also, attention must be given to axial loading, which can shorten the life of normal bearings from the standard 20,000 hours.

A typical speed-versus-torque curve for a servomotor is shown in figure 4-33. There are usually two operating zones: an intermittent duty zone and a continuous duty zone. The application will determine which motor is selected based on its peak and continuous current ratings. This curve is also typical in that many servomotors can be matched with higher current output amplifiers, which can yield different performances from the motors. Often there are two modes of continuous/intermittent duty possibilities with a servo. In addition, always know the available voltage because this will affect top motor speed.

The next important component is the feedback device. How it is attached is paramount because inadequate attachment can shut down a factory. The feedback device must be mounted so that it is straight, centered, and free of other interfering wiring. Therefore, the choice of couplings is an important consideration. Will there be shock loading to the motor? Will there be a steady flow of back-and-forth, forward-and-reverse movement at high acceleration

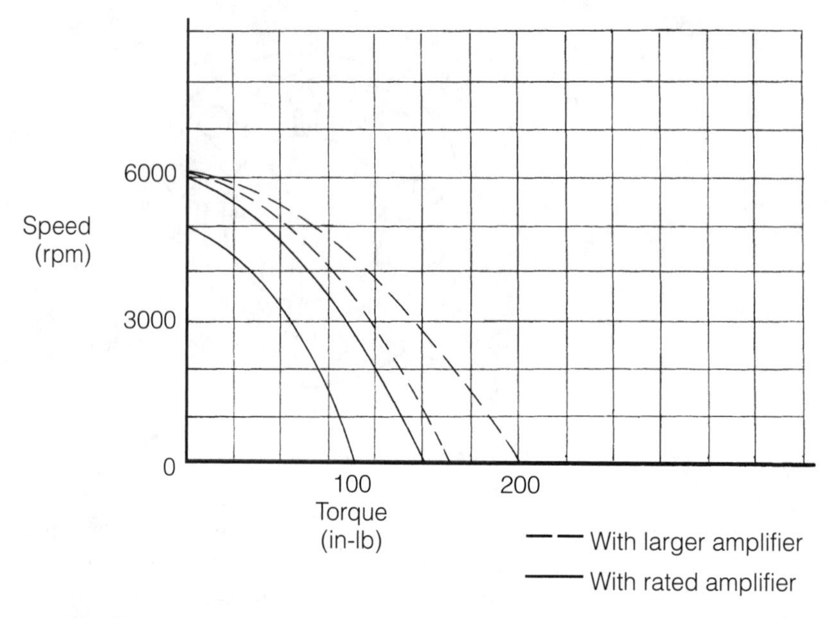

■ **4-33** *Servo speed-versus-torque curves.*

rates? If the coupling is tolerant of some (not much) misalignment, the feedback device will run for a long period of time. The more the misalignment, the shorter the life expectancy of the coupling/shaft attachment.

Often the coupling will fail at a period of high production. This might be because the coupling is at a maximum stress point, or it could just be a coincidence. No matter when a coupling fails, the timing is poor, but when it happens during full production, more individuals take notice. Therefore, once a suitable coupling has been selected, it is important to ensure the integrity of the feedback device's signal. Routing of higher voltage wire away from the feedback wiring and twisting and shielding of the feedback wiring is good practice. Many servomotor manufacturers have developed standard housings for the attachment of the necessary feedback devices.

The connection points for the power feeds to the motor and the feedback wiring are critical in servo applications. The solder joints between the wiring and actual connector have to be proper or else noise can be introduced into the system, especially in the feedback network. Noise is a servo system's worst enemy. Therefore, the cabling and connector should not be treated as insignificant components. Cabling issues are discussed in greater detail in Chapter 8.

One style connector is the MS, or military style connector. The alignment of the pins and the screw-down quality of connection makes for a true and sealed connection. This type of connector also eliminates time-consuming, individual wire connections when running the wire (provided cables are made up ahead of time). No soldering or screw-driven connections are required with this type of quick disconnect.

Other facets of the construction of the servomotor have evolved over time. Motor environment has dictated how the motor is to be constructed. For instance, windings are insulated with a high grade of insulating tape, usually Class F or H. O-rings and shaft seals are also provided to keep the environmental "nasties" out of the servomotor. Purging the servomotor and feedback device with air is a common practice for environments with hazardous gases. Consult local codes, the motor manufacturer, and the electrical code before doing this.

Stepper motors

Another type of motor commonly used to position is the stepper motor. Like its servomotor brother, the stepper requires a drive package in which the electronic control, power supply, and feedback interface can be housed. (The controller for the stepper motor is looked at more closely in Chapter 7.) Figure 4-34 shows a simplified block diagram of the stepper motor operation. Basically, an indexer provides step and direction pulses to the electronic drive amplifier/controller. The current level for each phase to the motor is determined and subsequently output to the motor. An encoder is not necessarily required for stepper motor opera-

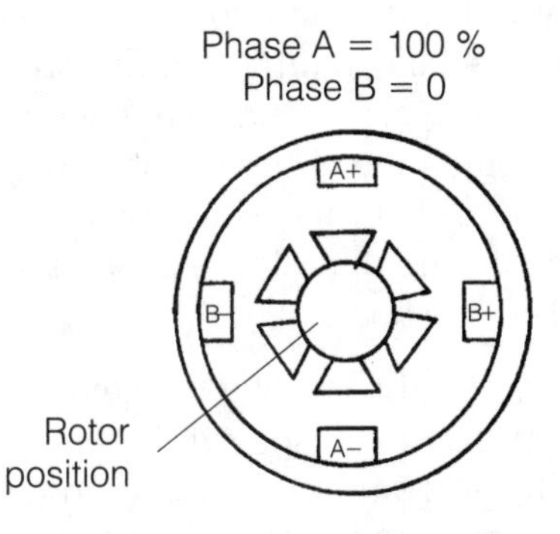

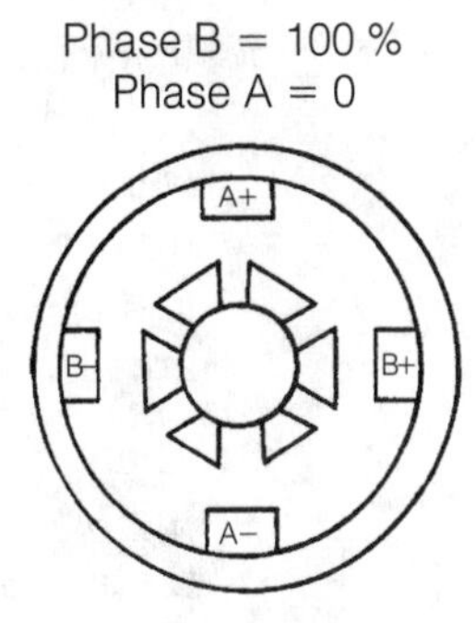

■ **4-34** *Stepper motor simplified.*

tion, although it must be included in high-accuracy positioning applications.

The stepper motor works on the premise that pulses determine the distance for positioning systems, and the frequency of these pulses determines speed. The stepper, or step motor as it is often referred to, is inherently a digital device. It is a brushless motor that can deliver high torque at zero speed and still maintain stability. There is no drift and no errors due to the cumulation. The stepper motor is fully reversible and can withstand numerous shock loads throughout its life. Similar to servomotor construction, the windings are in the stator for quicker heat dissipation. This also allows for low inertia in the rotor. There are two windings in a step motor, and therefore, two phases.

Step motors sometimes exhibit oscillations with every step. This is common with this type of system and is most visible at low speeds. Like the servo system, the goal here is to match motor, amplifier, load, and mechanics so as to minimize the oscillations. Stepper motors tend to run hot due to the PWM waveforms from the controller, high currents at zero or low speed, and duty cycle. When enabled, a stepper motor also tends to jump. This is merely the motor seeking a tooth position. Because of an air gap, the teeth on the rotor and the stator do not come into contact with each other. Therefore, when flux is present, such as in a power-on situation, there can be an initial, internal movement. The stack length or depth will provide the usable torque in both steppers or servomotors.

Closing remarks on electric motors

Electric motors are the workhorses of industry and for most motion in general. Schools, hospitals, and office buildings rely on electric motors for their water, heating and cooling systems, and many other routine tasks that we take for granted. Electric motors are behind the scenes in a lot more places than we think, and they will continue to be for quite some time. Their construction allows for extended use, and even abuse. They have minimal connections and therefore, from the technician's and electrician's standpoint, are not necessarily hard to troubleshoot.

The controllers for the motors tend to be the challenge here for the plant electrician. More wiring, some special, is required with the added effort of having to program a digital electronic drive. Things can get a bit more complicated. Attaching feedback devices and additional motor protection represent the remaining extent of extra wiring that has to be run or traced. Thus, in the basic sense, electric motors are relatively maintenance free, depending on motor type and its use. Sometimes motors have to be rewound because of excessive high-current conditions over a long period of time. Bearings wear, and there are always those unwanted motor short circuits. But relatively speaking, motors can run with little maintenance for many years.

Industry is in agreement that dc motors are more trouble than ac brushless motors because dc motor commutators have to be cleaned and the brushes often get dirty and have to be replaced. Nasty environments accelerate maintenance, and eventually the dc motor is ousted in favor of the ac machine. This trend is happening throughout industry and will continue to happen. Although a time may come when the dc motor is no longer produced, that time is definitely many years away. The ac motor technology must continue to provide all the solutions that dc has for many years. Meanwhile, specialty and other brushless motors will continue to provide solutions for position and other types of control.

Many controller manufacturers argue that the motor is the limiting factor in the technology. Motor manufacturers argue that the controller is the limiting factor. The fact of the matter is that both sides should get together and develop better standards. Some standards exist and there are countless rules of thumb to follow when mating motor controller with motor. The controllers for these motors will continue to change and so, too, will the motors. There are inklings that some users would prefer the motor controller to be mounted integral to the motor housing. This might be achievable someday, but for now there are many other issues to tackle.

The Department of Energy (DOE) has mandated that several minimum standards for electric motors be met by the year 1997. The energy standard is lengthy and many products and issues are discussed. Motor efficiencies and power factor ratings are among the more important issues. These standards will affect not only industrial users but commercial and even some residential markets. The days of wasting electrical energy are gone. The mode is to make currently installed equipment and systems more energy efficient. Installing new energy-efficient motion products today will mean not having to build new power plants and use up available resources in the future.

Direct current (dc) drives

5

THE DC MOTOR HAS BEEN THE WORKHORSE OF INDUSTRY for nearly a century. In order to run that workhorse, a means of changing available three-phase ac power had to be developed. This became the rectifier, which eventually emerged into the adjustable-speed drive for the dc motor. With this basic concept of providing dc power to a plant's machinery and then controlling that power in order to change motor speed, industry grabbed hold of this technology and it spread throughout. Today, the dc drive is still alive, but it is slowly being displaced by ac technology. However, with advances in microprocessor-based systems and power electronics, the dc drive can serve a necessary role in manufacturing and production.

The dc motor theory of operation has already been discussed in the previous chapter. A dc drive can control either a shunt-wound dc motor or a permanent magnet dc motor. The major difference, other than motor construction and performance, is that permanent magnet motors are used most often in applications requiring 5 HP and below. They are not built beyond that size, whereas shunt-wound dc motors are common from fractional to thousands of horsepower. Therefore, this chapter's emphasis will be on shunt-wound applications.

The dc drive utilizes basic power conversion techniques to control both the speed and the torque of the dc motor. These techniques had to factor in a means of controlling the speed of the dc motor under different load conditions. Some loads are hard to start, while others need to have regeneration and/or reversing capability. Other applications might have constant load changes while the motor is running. A dc electronic drive must be able to dynamically change output levels of both voltage and current to a dc motor in order to control speed and torque. It must be able to respond appropriately to all types of load changes. Today's technology utilizes solid-state electronics to accomplish just that.

The dc drive's equivalent circuit is shown in figure 5-1. The two main components that must be controlled are the armature and the field of the dc motor. Beyond maintaining proper field voltage and field current, the dc drive must also control, or regulate, the voltage and current to the armature of the motor, as shown in figures 5-2 and 5-3. The torque T is proportional to armature current, and the speed N is proportional to voltage. Thus, today's dc drives have fairly sophisticated current and voltage regulators. This attention to the armature will make or break a dc drive's overall performance in a given application.

A further look at the dc drive equivalent circuit shows a rectifier circuit. By adjusting the firing angle, either a half-wave (figure

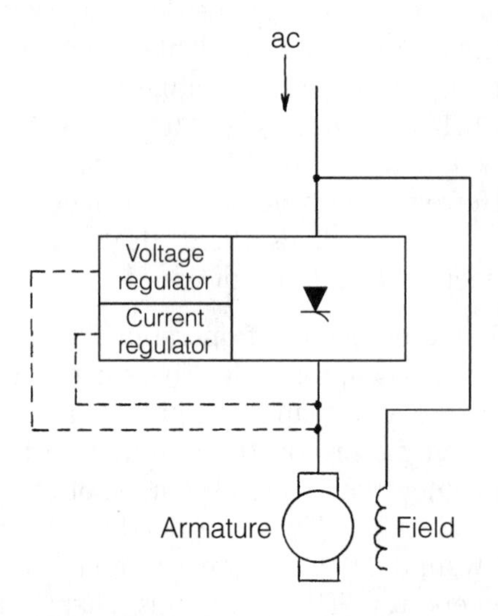

■ **5-1** *Equivalent dc drive circuit.*

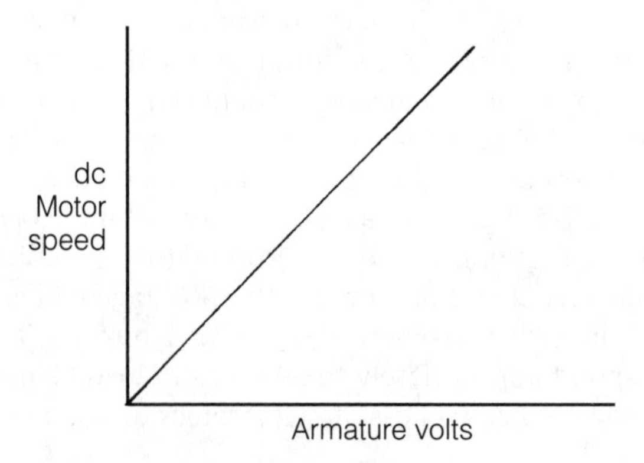

■ **5-2** *Armature voltage curve.*

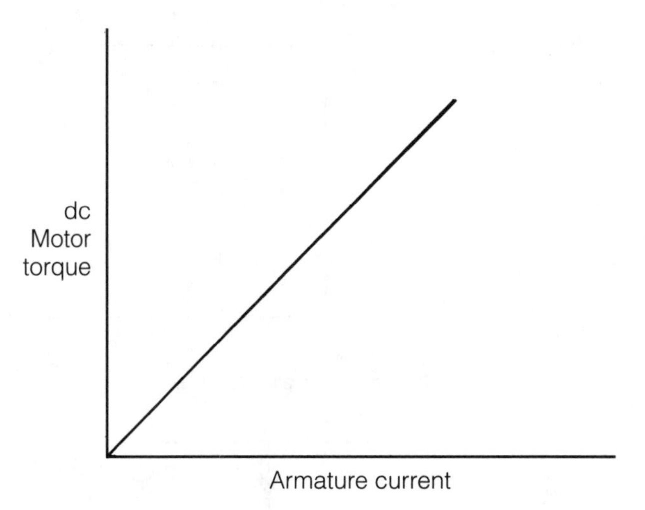

■ **5-3** *Armature current curve.*

5-4), full-wave (figure 5-5), three-phase half-wave (figure 5-6), or a complete three-phase full-wave (figure 5-7) rectifier can be configured. The firing angle can be described as the instant in time when the thyristor is triggered to conduct. Obviously, the immediate differences in these rectifiers are in the single- and three-phase nature of the power, component count, and cost weighed against performance.

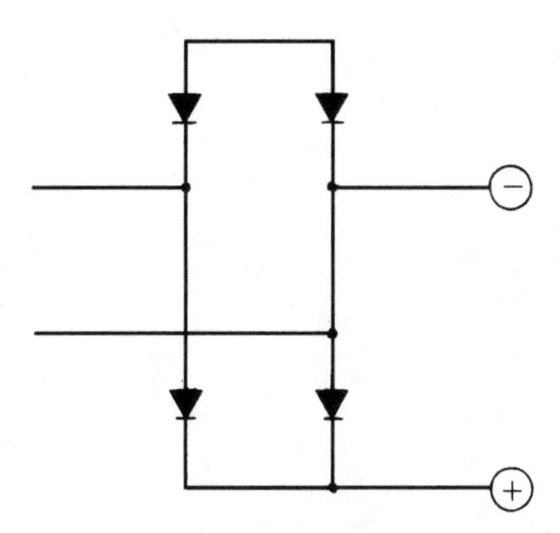

■ **5-4** *Half-wave rectifier.*

In the full-wave rectifier circuit shown in figure 5-7 the alternating current sine wave is converted more efficiently and more completely than the other rectifier schemes. There is a net effect of gaining a positive output from the bridge configuration even when the input goes negative. This is because the output pulses go positive. In the negative half cycle, diodes D-2 and D-4 will be biased for-

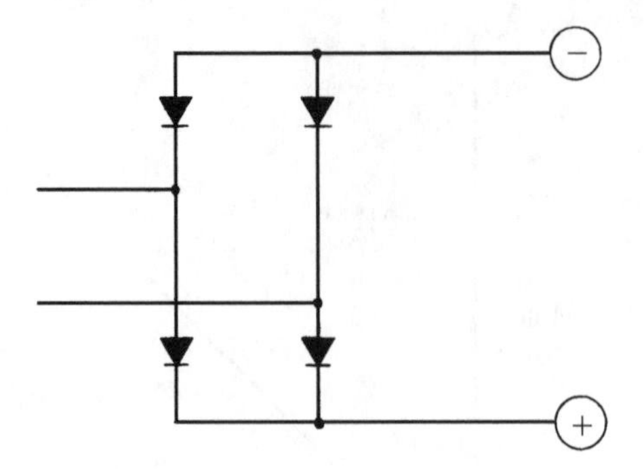

■ **5-5** *Full-wave rectifier.*

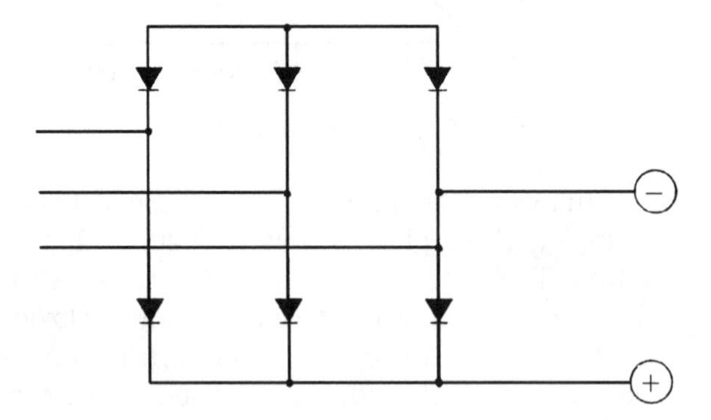

■ **5-6** *Three-phase half-wave rectifier.*

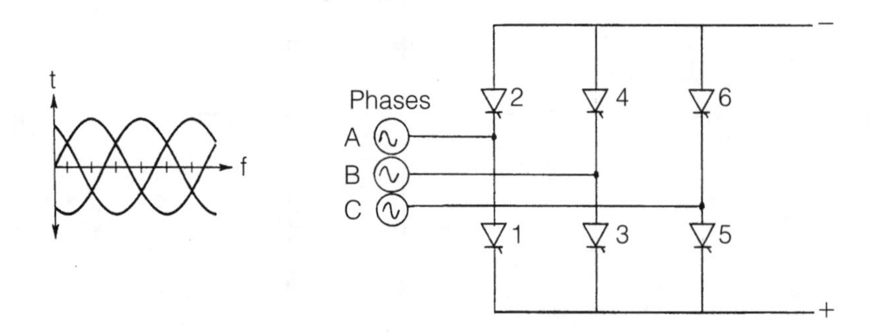

■ **5-7** *Three-phase full-wave rectifier.*

ward, and thus, the polarity will be positive at D-2. Similarly in the positive half cycle, the D-1 and D-3 diodes will be forward biased and thus a resulting positive output is achieved. The thyristor in most of the aforementioned rectifier circuits is the diode; however, there are SCR-based rectifier networks that are extremely common.

The dc drives usually can be classified into two main groups: transistorized dc drives and thyristor, or SCR-based, dc drives. The transistorized dc drive will often be the unit that controls a permanent magnet motor. However, the most common is the thyristor unit because it is available in a much larger power range. This SCR-based dc drive is usually in the form of an SCR phase-controlled circuit. Since the SCR can be given a very small gate voltage compared to the amount of load-carrying capacity it can provide, it is the thyristor of choice in most dc drive designs. Figure 5-8 shows a simplified block diagram of an SCR-based phase control dc drive circuit.

In summary, the speed command and dc motor feedback (both voltage in the form of back EMF and motor current) are compared in the regulator. This updated value is then compared to the zero crossing waveform of the SCRs. Then a new firing command is generated and the cycle is repeated. The amount of comparisons per second and the accuracy of these updates is actually the response of the drive system. This criterion separates good dc drive controllers from bad.

Since the transistorized dc drive usually is applied to specialized dc motor applications such as running a permanent magnet motor, it is not as common as the SCR-based dc drive. In either case the dc drive controls the speed of a dc motor via two major power circuits: the armature bridge and the field bridge. Other necessary components are the control circuitry and the feedback circuitry.

Drive power bridges (dc)

The dc drive's heart and soul is its armature power bridge and its field power circuit. The more complicated of the two normally is the armature power bridge because it is constantly striving to regulate the voltage and the current to the dc motor's armature. In a two-quadrant control scheme, shown in figure 5-9, a dc drive is said to be *nonregenerative*. It can provide torque in the same direction as the motor rotation. The same is true for the opposite direction of motor rotation. Therefore, quadrants 1 and 3 are applicable to this type of operation.

The most common power bridge in a two-quadrant form is the six-thyristor nonregenerative controller for the armature. Usually the six SCRs are provided in three separate modules. This forms a full-wave power bridge. Each thyristor is controlled independently by the gate-firing circuitry. Often there are pulse transformers, which help to isolate voltage. In addition, there can be suppression net-

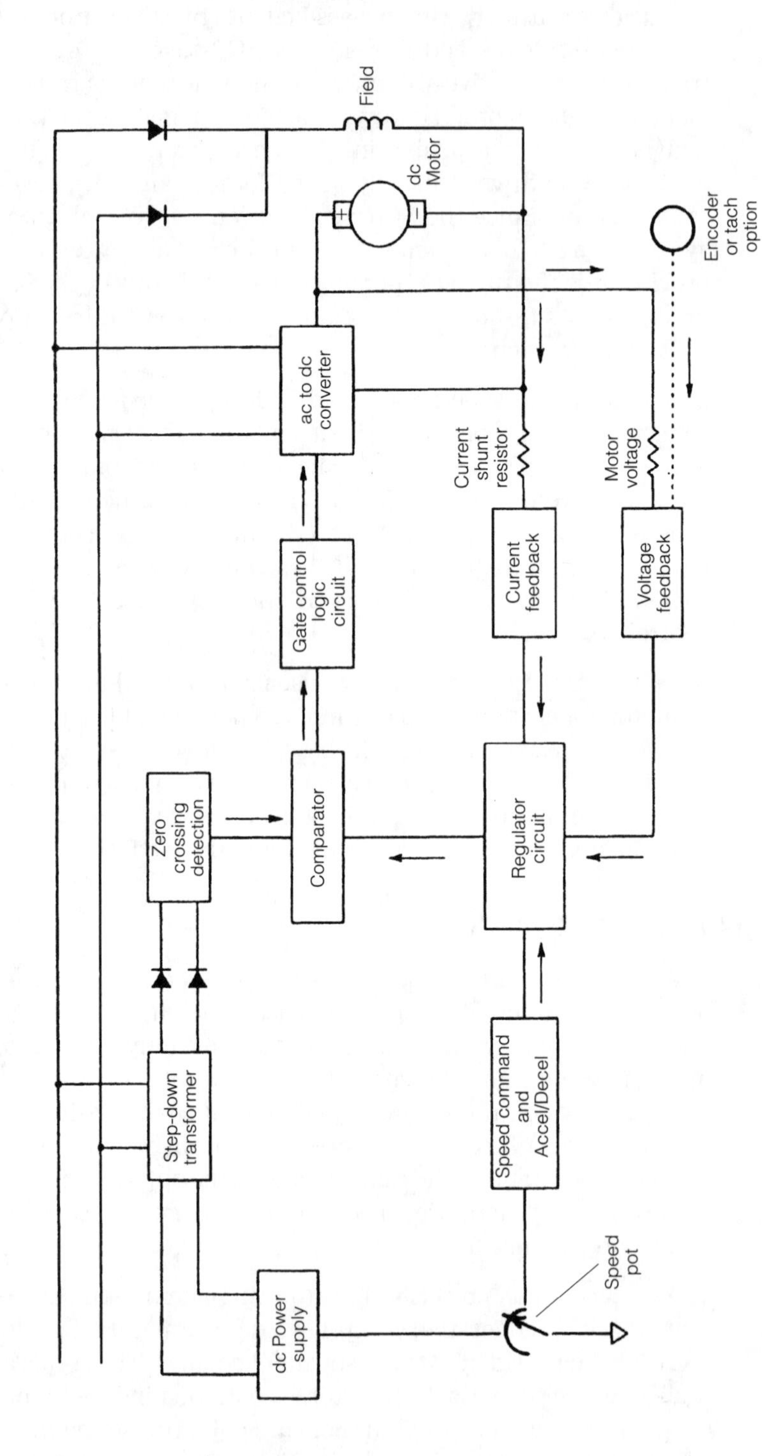

■ **5-8** *SCR phase control block diagram.*

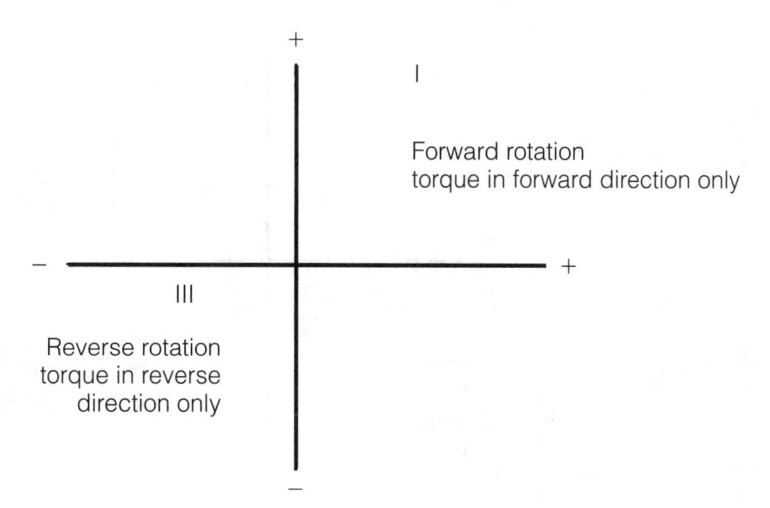

■ **5-9** *Two-quadrant operation.*

works for the power bridge. The RC (resistor/capacitor) type of suppression limits the rate of rise of voltage across the thyristor to minimize the risk of false triggering. Another suppression network often utilized is that of the voltage-dependent resistor, which monitors the circuit to ensure that the voltage does not exceed the maximum limits to which each thyristor is rated.

The four-quadrant power bridge's operational capabilities are shown in figure 5-10. This power bridge is much like the two-quadrant version with the addition of a complete bridge for reversal and regeneration. The addition of another power bridge is commonly called *stacking*. This power scheme allows for motor operation in all four quadrants: motoring and producing torque in the clockwise (forward) direction; motoring and producing torque in the counterclockwise (reverse) direction; regenerating (negative torque) in the clockwise direction; and regenerating in the counterclockwise direction. There is usually not a requirement for additional suppression networks because the two bridges are in reverse parallel. However, it is worth noting that many drive manufacturers suggest fusing as extra protection for the power devices. In most cases the power bridge will protect itself by virtue of electronic overload protection circuitry, but fuses should be incorporated for backup.

Often it is necessary to exact control of the dc motor from the field power bridge. The field is often a product of a thyristor regulator, which controls the field by either simple phase angle control by which the output is a variable voltage source or by

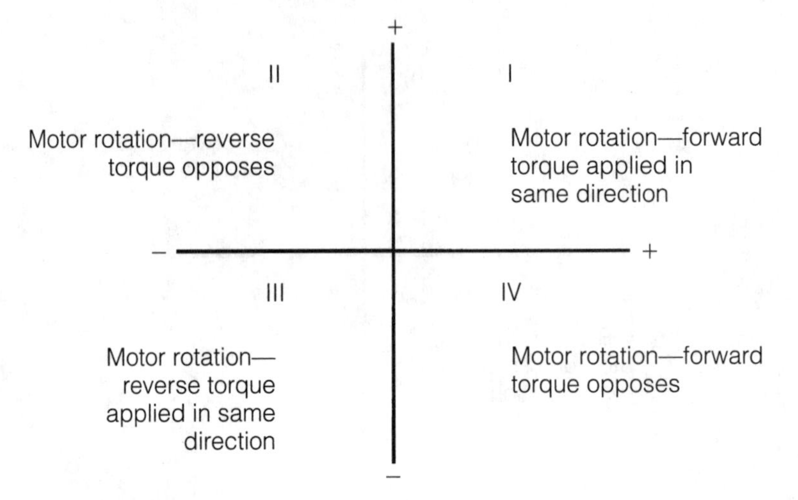

■ **5-10** *Four-quadrant operation.*

actually controlling current in a closed-loop arrangement. Phase angle control is the simplest method to get the field regulator to act like a field rectifier. A full-wave, half-wave, or three-phase half-wave power bridge arrangement can be achieved by fixing the firing angle.

The smoothest output from a dc drive would be that of a full-wave three-phase supplied input to the bridge. Basically, the gating of the SCRs creates pulses of current. The more pulses per second, the smoother the torque output of the motor. Today's dc drive technology allows up to 12 pulses of dc current for every ac cycle (remember that there are 60 cycles to a second). That can be 720 pulses per second, which is not bad for a dc drive system.

By controlling the field current in what is called field current control, the regulator can adjust for variations in the power supply voltage. Field current control is the only choice whenever the dc motor's operation means running into extended speed. This condition is also known as field weakening. If the dc motor is controlled by voltage control, there might be instances where full, maximum speeds cannot be attained, which cause overfluxing of the motor. This type of control is simple, but it can possibly damage the motor if not applied properly. Care and attention should always be given to those applications where extended speed operation is required, especially when using an existing dc motor.

Form factor and ripple

As mentioned in the previous chapter, the ratio of the RMS (root-mean-square) current to the average current is called the form factor. In order to attempt to gain some cohesion and compatibility between dc motors and dc drives, the form factor criteria method was introduced. Form factors are used throughout much of the power electronics industry in order to match equipment. Although no method is absolute, by establishing values of form factor to equipment based on their type of rectification and waveform, we now have a numerical system in place that aids in equipment selection. Ripple, or the disturbance seen in any waveform as various spikes and nonconformities, is the actual culprit. High ripple currents produce additional heating at the motor. This additional heating can be compensated for in reducing the continuous load rating of the motor below base speed. Depending on the motor manufacturer, this derating of load capability can range anywhere from 5 to 15%.

For instance, as mentioned previously, motors have NEMA codes assigned to them, which usually appear on the motor's nameplate. This NEMA rating indicates that the motor is rated for continuous operation off of a rectifier-type controller at its full rated speed and full rated loading. This means that the motor should not incur any extra heating at full load and speed due to running off of the style rectifier in the application.

Typical form factors for a three-phase full-wave converter with 6 or 12 SCRs, nonregenerative or regenerative, is 1.01 with a ripple frequency of 360 Hz and a NEMA code of K. The form factor reference point is 1.00, which represents the form factor of pure dc, as can be found in a dc battery. A three-phase half-wave converter with three SCRs (nonregen) and six SCRs (regen) will have a form factor of 1.10, a 180-Hz ripple, and a NEMA code of E. A semiconductor with two SCRs and three diodes (single-phase) will have the highest form factor at 1.35, 120-Hz ripple, and NEMA K coding. All of these ratings are based on 60-Hz frequencies. Whenever operating from a 50-Hz source, the ripple currents can be up to 20% higher.

Obviously, no system is absolutely right. This is a good attempt at trying to match drives and motors to minimize overheating and failures in the field. Operating conditions, ambient environment, and any abuse in the application cannot usually be predicted by the form factor value. These issues can only be incorporated into the motor and drive selection by the user.

Drive feedback arrangements (dc)

If the dc drive is to perform well in its application, it must provide for and utilize various forms of feedback. In short, the dc drive must, as a minimum, use current and voltage feedback. Current feedback is directly related to motor torque, and voltage feedback is directly related to motor speed. Beyond these, higher performance in both speed and torque regulation can be achieved by implementing pulse generators. As with most control systems, the more elaborate the feedback package, the better the performance and regulation (the cost of the system usually increments accordingly, too).

The torque of the dc motor is controlled by continuously obtaining actual current values relative to the armature. These values are obtained by monitoring the three phases of current to the power bridge via current transformers or current shunt resistors. This gives an indication of what values of current are being sent to the motor. These values are rectified and then assigned a burden in order to achieve the desired current level that the controller can utilize. This is why many drives have similar physical sizes yet different horsepower ratings. The different horsepowers can be attained from the same physical package by merely changing the burden resistor. This derived current value is then used in the regulator circuit to update the output to the motor. Perhaps the most important function of this feedback circuit is to ensure that motor maximum current is limited. Without the protective current limit built into this feedback circuit, the dc motor normally would keep demanding more current until it damaged itself.

Speed feedback gets a little more complicated. Table 5-1 shows three methods for providing speed feedback. The simplest form is voltage feedback as an error signal into the dc drive's regulator. This is actually an indication of the motor's back, or counter, EMF (electromotive force). This voltage signal error is then used to advance or retard the actual firing angle. In this manner the motor speed can be matched to the desired speed reference signal. Typical speed regulation utilizing voltage, sometimes called armature, feedback can be in the 2 to 3% range. A diagram of the voltage feedback method described here is shown in figure 5-11. Many factors enter into this speed regulation equation; frequent load changes at the motor, the severity of the load change, changing of direction under load, and mechanical tightness of the system are just a few.

■ Table 5-1 Various methods for providing speed feedback.

Feedback type	Speed regulation	Extra cost	Hardware required
Armature (voltage)	2–3%	0	Std. w/drive & dc motor
dc tachometer	1%	$	Tach mount to dc motor
Encoder (Digital Tach)	0.01%	$$	Encoder mount to dc motor

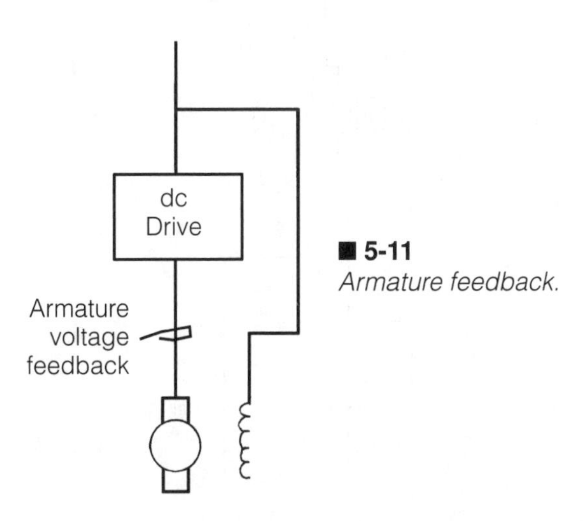

■ 5-11
Armature feedback.

In order to achieve better speed regulation—or more precisely, better control of the speed command versus speed error response loop—other feedback devices can be used. One method that has been in use for many years on dc motors is the dc tachometer (figure 5-12). The tachometer is fitted externally to the dc motor and provides an additional feedback signal to the dc drive. This is an analog device with an output of voltage equivalent to so many rpm's of the motor. This signal is taken into the voltage feedback circuit, scaled and conditioned, and then sent to the speed regulator circuit. Typically, voltage values from the dc tach are 50- or 100-V-per-thousand rpm's. The benefit of using this feedback package is that speed regulation of 1% can be achieved.

The best feedback device in terms of getting the best speed regulation is the digital encoder. However, the dc drive controller must be able to accept this type of signal and consequently utilize that signal to get the desired response. The single-line description for this feedback package is shown in figure 5-13. These encoders, or

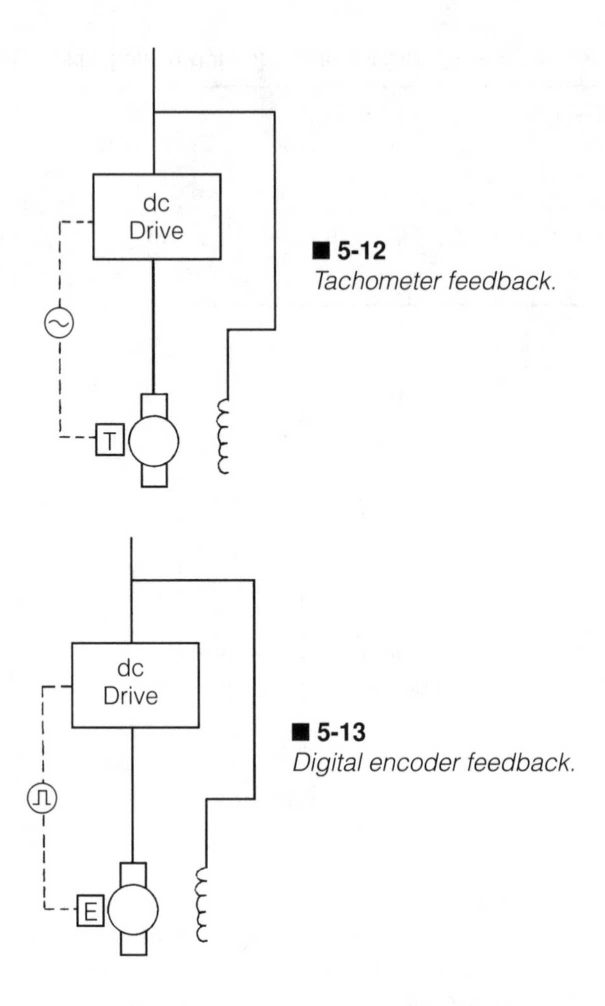

■ **5-12**
Tachometer feedback.

■ **5-13**
Digital encoder feedback.

digital tachs as they are sometimes called, can come in different configurations. The amount of pulses per revolution is the main difference. And as the pulse count increases, so does the cost and complexity of the control scheme. Yet the speed regulation with this type of feedback device is dramatic. Speed regulations of 0.01% can be achieved in ideal conditions.

As discussed previously, there are many forms of feedback for speed regulation. The first question to consider is whether to affix another component to the system (always with potential for failure), as with the add-on tach or encoder. If speed regulation utilizing armature voltage feedback is adequate, it is probably best not to introduce another physical component that could shut the motor and drive system down. Often it is possible to use IR compensation to improve speed regulation without adding an external feedback device. Frequently referred to as IR COMP, this circuit is a positive-feedback circuit that readjusts the firing angle to get more armature voltage or to boost that voltage whenever the load at the dc motor changes dramatically from light to full. This adjustment, usually available with most dc drive controllers, can provide a betterment in speed regulation over standard armature voltage feedback.

Regenerative versus nonregenerative dc drives

A dc drive with an SCR bridge is referred to as a nonregenerative drive. It has no inherent braking ability. If braking is required for the application, the addition of a dynamic braking circuit is the next-best alternative. The dynamic braking circuit will dissipate the motor's braking energy as heat to a resistor bank. These braking circuits are rated only for stopping and should not really be considered for continuous holdback in an application. They should also not be used as a holding brake. The braking effort is extremely high with the initial larger torque and reduces to no torque at zero speed. Nonregen drives are not a good selection in these types of applications where high inertia loads have to be repeatedly started and stopped.

Additionally, the nonregenerative dc drive has no inherent reversing ability. If the application requires the motor to reverse, the method to use is a reversing contactor. This contactor or switch approach reverses the polarity of the dc voltage out to the motor. Again, the nonregen drive is not well suited for applications that have to incorporate a reversal scheme frequently. Occasional reversals are practical with the addition of the reversing contactor, but the better choice is to apply a full regenerative, dual SCR bridge drive.

The regenerative dc drive is built so that it can inherently perform both full holdback braking and frequent reversals. It electronically regenerates by allowing the kinetic energy produced by the motor and its driven load to pass through the SCR bridge and return to the power supply. This can be seen in the simplified diagram in figure 5-14. The nonregenerative version simply does not have the second bridge to accomplish this. With this arrangement the regenerative drive can regulate braking torque from full-speed, high-torque conditions all the way down to zero speed, and perform the function anywhere in between. That is why it is the best choice for those applications requiring continuous holdback.

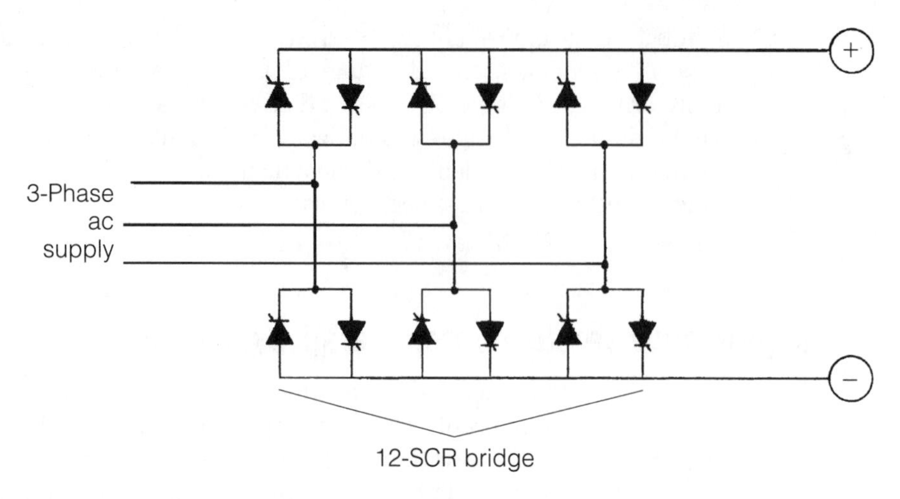

■ **5-14** *Regenerative bridge scheme.*

In the converting industry, regeneration is essential for exact control of the tension zones in the web-fed machine. And because this type of application requires tension control, speed accuracy is not the primary function. When tension feedback is introduced at any section, speed is "put on the back burner." However, with today's microprocessor-based dc drives, speed regulation is maintained much more easily and quickly.

Figure 5-15 shows various sections of a tension-control, hold-back-type machine. The in-feed and out-feed sections must be nipped and will require regeneration. Additionally, the unwind will require holdback, or regen, capability. The winder will also have to be capable of regeneration in order to ensure that no runaway condition can exist. Regeneration is also the means by which the entire drive system will stop. However, the dc drive must have control of the stops and the regenerative energy in order to perform the stop function. This is why many installations have dynamic braking systems in place; If the power is lost to the control portion of the dc drive, emergency stops and safe stops can still be achieved.

Regenerative braking should not be confused with dc injection braking and plugging. Regenerative braking can take place virtually anywhere in the operating speed range of the dc drive. Plugging, on the other hand, is a function of the ac drive controller and motor, which reverses the ac motor line voltage polarity so that the ac motor develops a retarding force of countertorque to pro-

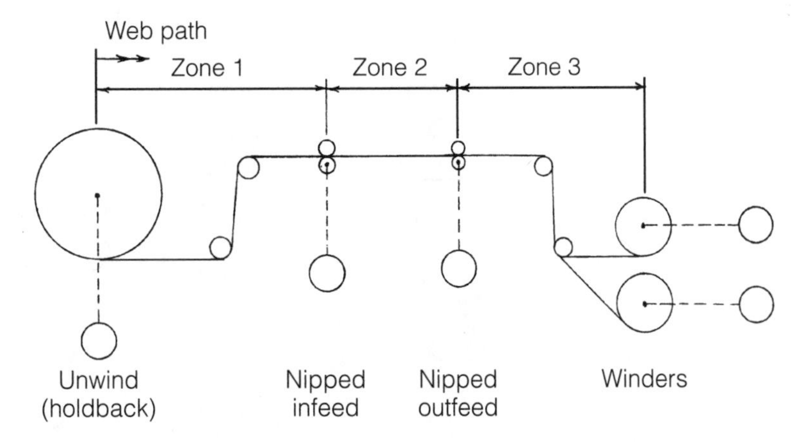

■ **5-15** *A web converting line.*

vide a form of braking. Because this cannot happen at high speeds, antiplug protection is usually built into the controller. This function is sometimes referred to as dc injection braking and is discussed further in Chapter 6.

Reversing with a regen drive is another built-in capability. The motor polarity is electronically reversed by the drive. Thus, no reversing contactor is required with possible contacts to burn or wear out. For applications requiring frequent reversals, a regenerative drive must be considered.

M, or loop, contactor

In order to ensure correct power-up sequencing, a power contactor must be placed in series with the main power path, hence the name main or M contactor. Also called the loop contactor, the main contactor is shown in figure 5-16. This contactor should not be confused with a reversing contactor, which is not always used with a dc drive controller, nor does it perform the same function. The main power contactor can be either ac or dc and should be initiated by the controller by way of an isolating relay. This isolating relay usually drives the contactor coil with the same voltage as that of the auxiliary power supply. The auxiliary power supply must be powered at all times whenever control action is needed. Therefore, no additional switches should be incorporated between the incoming ac and the auxiliary power supply terminals. Additionally, no other contacts should be connected in series with the contactor coil because this might cause control sequencing problems.

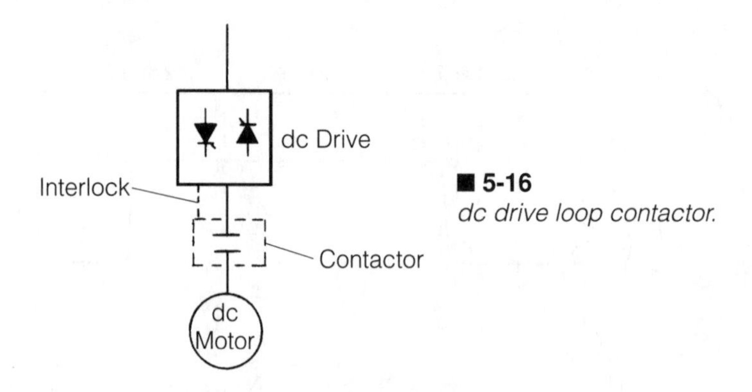

■ **5-16**
dc drive loop contactor.

Speed scaling

Typically, the speed of the dc motor is in direct relationship to the voltage supplied. Today's higher voltage-rated shunt-wound dc motors have 500-V armatures and 300-V fields. Lower ratings are also available in 240-V armatures and 150-V fields. Therefore, if 100 V is supplied (20%), the motor will run at 20% speed or 350 rpm (when the base speed of the motor is 1,750). At 75% voltage the motor will run at ¾ speed. And at full voltage, or 500 V, it is running at full speed, or 1,750.

When setting up the speed scaling circuit for the motor/gear reducer/conveyor system shown in figure 5-17, it is important to take the tachometer feedback into account. This application will use an analog 100-V-per-thousand rpm tach (obviously a pulse generator would have to be treated differently). The other assumptions are that the motor has an armature voltage of 500 V maximum, the maximum motor speed at full armature voltage is 1,800 rpm, and the gear reduction is 10 to 1. The desired maximum speed at the main conveyor roll is 100 rpm. Therefore, the steps to get the proper speed scaling for this application are as follows:

1. Calculate maximum motor speed:

   100 (conveyor roll rpm) × 10 (gear ratio) = 1,000 rpm

2. Calculate desired tachometer output:

$$100\text{ V} \times \frac{1{,}000}{1{,}800} = 55.5\text{ V}$$

3. The armature voltage should be 500 V × 1,000/1,800 = 277.8 V at the desired maximum speed.

At this point the maximum speed parameters should be entered with the calculated values, and the system is now set up for the desired results. Often the tachometer is placed elsewhere in the

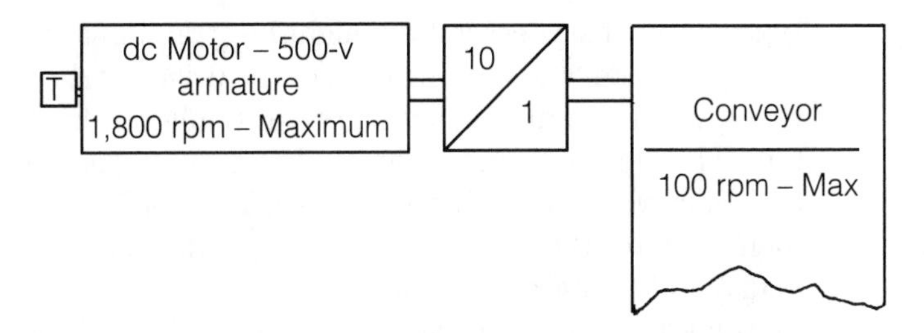

■ **5-17** *dc motor with tach mounted.*

drivetrain. As is seen in figure 5-18, the tachometer could be placed away from the motor and on a main roll of a conveyor. The calculations have to be done in this case to reflect a scaled speed and tach voltage from the desired conveyor speed. This type of dc drive installation is sometimes referred to as a *tach follower*.

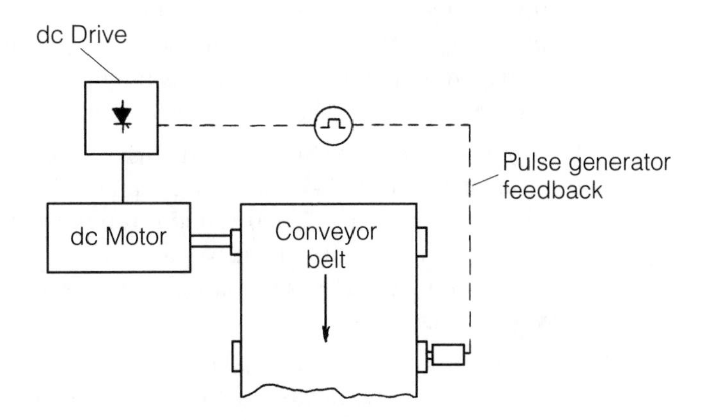

■ **5-18** *Tach follower.*

The speed feedback loop is just one of several loops that need tuned. Follow similar steps to set up the field voltages and then move on to the current loop/current regulator setup functions. These tune-up procedures are touched upon within this chapter but actually have to be dealt with on an individual basis. The manufacturer of a particular dc drive usually has specific instructions to set up the drive's control loops. Once these loops are set up and tuned for the actual loads and speeds, the dc drive is operational.

Traction drives

The term "traction drives" is somewhat of a misnomer. Because the drive, power conversion, and control are traditional dc drive types, the better description might be "traction applications." It is the application and modification of the drive for the traction usage that gives it a name. "Traction drive" implies a drive for any hauling or locomotive application, mainly seen in the mining industry and usually thought of as a dc drive. The application and environment dictates that these drives be rugged, capable of severe overloads and duty cycles, and packaged compactly. There is a movement toward ac drives, but the majority of these applications are still done with dc drives.

With respect to the power circuit and rectifier, traction drives are the same traditional dc bridge but with some modifications. Several issues must be taken into account when applying dc drives (or ac for that matter) to traction applications:

1. *Cooling.* In most other applications, cooling is normally done by way of traditional heatsinks and/or cooling fans. In mining operations there must be a means of transferring the generated heat to a water cooling plate, since water is usually available on mining equipment. This might involve a repositioning of the SCRs in the drive package.
2. *Voltages.* Incoming power to the dc drive is not usually the traditional 480- or 230-V three-phase power. Instead, it is in odder voltages such as 1,000 V, and often noise or disturbance is associated with it. Robust drive networks are therefore required, along with proper voltage and current ratings for the power devices.
3. *Ambient.* Another major consideration is that of ambient temperature rating. In case of a failure in the water cooling system, the drive must be able to continue working properly and safely in a much higher than normal temperature range. Often in an underground situation the temperatures can reach 80°C for a drive ambient rating.
4. *Overloading.* This requirement comes with the territory. These applications are harsh, rugged, unforgiving, and they run 24 hours per day. Along the way, there will be instances where the drive will be expected to deliver 200% or more current for short periods of time. The drive must be built accordingly.
5. *Size.* Often the drive has to ride along with a machine that is already streamlined to snake through underground tunnels. Therefore, the drive must also be small and compact.
6. *MSHA (Mine Safety and Health Act) approvals.* In the mining industry most equipment for underground use has to

comply with MSHA rules and regulations. This might mean explosion-proof systems, among other things. Usually third-party certification is required on the equipment.

Chopper drives

The chopper drive is actually well-named. It could be called a "valve control" drive, but that mechanical affiliation might cause more confusion among electronic drive users. Actually, the chopper drive's name comes from its power bridge construction, rather than its application. The rectifier of the chopper drive is affixed with an additional leg to its circuit. This leg has an extra power semiconductor circuit that functions as a valve.

The simplified chopper rectifier is shown in figure 5-19. An SCR is incorporated ahead of the dc bus and after the rectifier circuit. It allows the dc voltage to rise to a predefined level. Once at this level, the chopper switch, or valve, opens to deenergize, and the voltage decreases. In effect, the chopper acts like a bleed-off valve, "chopping" the voltage rise. This design allows for better control of highs and lows of the dc voltage during operation.

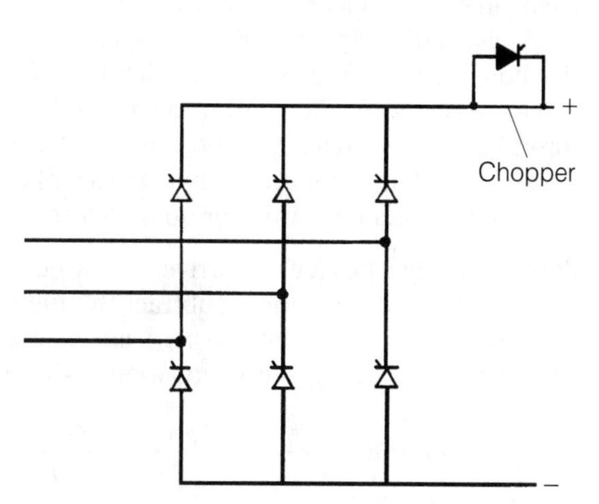

■ **5-19** *dc chopper circuit.*

Drive features (dc)

Field weakening This is a feature that has to be implemented in order for top speeds, above base speed, to be achieved. The typical sequence in setting up a drive for field weakening is as follows (assume that speed control is by varying armature voltage and that the field is constant up to base speed): First, run the drive up to base speed and check to see that the motor voltage is correct as to the conditions previously mentioned. Then increase the speed above base speed and check that the armature voltage now remains constant while the field starts to fall in value. This procedure is then followed until maximum desired speed is attained. At this time, the maximum speed and the minimum field setting should be trimmed.

Field economy Some dc drives do not have this circuitry, especially those with no built-in field regulator. Essentially, this feature allows for the field current to be lowered whenever the motor's armature is disconnected. This lower value is preset and represents a means of saving some energy in the normal everyday operation of the dc drive.

Field loss If the field current is weakened to a very low level, a fault should occur. This low-level value is usually predetermined by the installer of the dc drive and is proportional to the motor field current nameplate rating.

Diagnostics With the available microprocessors and data storage media in present-day dc drive controllers, information relative to a particular fault, trip, or alarm is now easily retrieved. Many times this information can be stored by the dc drive for a period of time and even added to when new occurrences happen. Most manufacturers of dc drives have their own nomenclature and arrays for the information that is important enough to gather and store. Consult the dc drive manual always for quick reference.

Drive speed regulation (dc) To arrive at that percentage value that is the actual speed regulation, subtract the full-load speed from the no-load speed, divide by the no-load speed, and multiply that value by 100. This regulation equation is shown below:

$$\textit{Speed regulation} = \frac{\textit{(No-load speed)} - \textit{(Full-load speed)}}{\textit{No-load speed}} \times 100$$

Example:

No-load speed = 1,800 rpm, full-load speed = 1,750

$$1{,}800 - 1{,}750/1{,}800 = 0.02777 \times 100 = 2.77\%$$

To achieve the optimum in speed performance, the speed loop must be tuned as well as it can for the given load. The speed loop's proportional and integral gains must be adjusted in real time. The gain of the speed loop system is defined as the ratio of the system output signal to the system input signal. Adjustments can be made to this ratio to "gain" optimum performance. This is done by entering discrete values in the appropriate parameters in a digital dc drive (accomplished by tweaking the correct potentiometer in an analog dc drive). These values are determined by observing the tach feedback signal as it relates to overshoot (see figure 5-20). The key here is to keep the overshoot to a bare minimum under loaded conditions. This is observed on an oscilloscope by directly monitoring the tach feedback. Most dc drive manuals can direct you to the proper terminal from which this observation can be made.

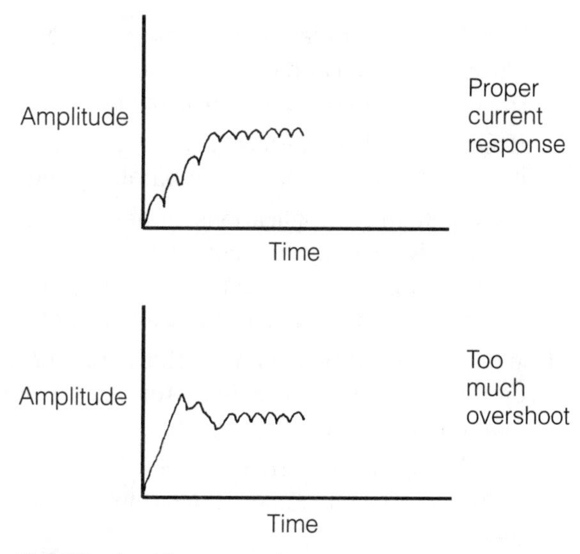

■ **5-20** *dc drive current response.*

Wide speed range Capable of 1,000 to 1 if motor is cooled by auxiliary means. With a 1,800-rpm base speed motor and a desired speed range of 40 to 1, the motor should be capable of running down to 1,800/40 or 45 rpm. The real question then becomes, "how long must the motor stay at that speed and is there a large amount of current going to the motor at that speed?"

Built-in field regulation Today's dc drives have the ability to provide constant field current for armature feedback. They can also be configured to automatically field weaken when speeds are well beyond base speed.

Current limit The current limit is usually a preset value of 110%, but it can be set higher. When the dc motor runs into an overload condition, the controller will reduce the current to 110% in order to protect the motor. Motor thermal protection should always be integrated into the circuit for added protection.

Drive selection (dc)

When should a dc drive be used and why? This is the million-dollar question. Sometimes the dilemma is simply choosing dc over ac or vice versa. Other times it is a matter of selecting some other speed-changing device. Following are some points to ponder that might help to justify using a dc drive in a particular application:

1. The dc drives are less complicated than ac drives—in both hardware design and control scheme. The dc drives have just the one power conversion from ac to dc.
2. Because the dc drive has only one power conversion bridge, it is more efficient than a comparably sized ac drive.
3. For regenerative applications, either overhauling or hold back, the dc drive is much less complex and much less expensive than a comparable ac regen drive. Regen dc drives are common and can run continuously in a regenerative state.
4. When properly loaded and with the brushes set appropriately on the commutator, the dc motor can provide years of low-maintenance service.
5. The dc motors can provide excellent ride-through capability in frequent load change applications due to their higher rotor inertias.
6. The dc drive and motor can deliver tremendous amounts of torque for short periods of time and at very low speeds. In fact, the dc system can deliver at least 100% of rated torque throughout its speed range.
7. The dc drives and motors have been in use for decades, therefore, the user can usually feel comfortable with the application.
8. The dc drive by itself is less expensive than a comparable horsepower-rated ac drive.
9. Speed ranges of dc drives today can exceed 1,000 to 1 and speed regulation can be better than 0.1% when necessary.
10. Perhaps the biggest reason of all: plant personnel know this technology and can work on these dc systems. Also, engineers and management might just prefer the dc over the ac.

Maintenance of dc drives

Like most electronic devices used in industry today, dc drives are built to last for many years. However, it cannot be overemphasized: If the drive is to run trouble-free for a long time, some routine maintenance should be performed. And the irony is that there is no real mystery to much of the maintenance; it just involves using some common sense and monitoring the installation, especially in a dirty, dusty environment. With the necessary maintenance, the drive equipment will definitely last. Also, it must be stated again: **All input power must be removed from any dc drive before doing any maintenance.** If there is a lock-out mechanism in place on the input power line, it should be used before doing any maintenance.

A routine inspection should be done on a dc drive every 3 to 4 months—more frequently if the environment around the drive is very dirty. The inspection and subsequent maintenance should include the following:

1. Visually check the drive for any dust buildup, corroded components, and any loose connections.
2. Using a vacuum cleaner with a plastic nozzle to minimize damage to components while cleaning, brush dusty and dirty devices. Particular attention should be given to the rectifier heatsink areas. Excessive accumulation of dirt and dust will eventually lead to an adverse overheating condition (probably at a time of peak production).
3. Clean and retighten any loose electrical connections.
   a. All fuse ends and bus bar connections should be cleaned according to the manufacturer's instructions.
   b. In the case of screw pressure connectors, the power connections should be checked often and retightened if necessary during the first few weeks after installation.
   c. If any components are unfastened during the maintenance procedure, the mating surfaces should be treated with the proper joint compound (per the manufacturer's recommendation) before replacing. After replacing any power modules subject to certain torque values to their heatsinks, it is suggested that another check be made on the tightness of the screws.
4. Touch up any exposed or corroded components with paint.
5. If the dc drive is mounted within an enclosure, check the fans and filters. If necessary, clean or replace the filters and replace the fan if the shaft does not spin freely.

Troubleshooting and repairing dc drives

Each manufacturer of dc drives makes their drive in some ways unique. Basic power conversion, available devices used, and standard wiring and protection techniques can be assumed to be similar from manufacturer to manufacturer. However, control schemes, gate firing circuits, packaging, and so on will definitely vary. Therefore, consulting the manufacturer's installation, maintenance, and repair manuals is mandatory when performing any work on dc drives. This section will discuss standard and common techniques, along with practical applications and problem-solving. Again, the danger of working with electrical equipment cannot be overstated: **Electrical shock can cause serious injury or even death. Remove all incoming power before attempting to do any work to any electrical device!** This is another reason to have the manufacturer's manuals on hand; they usually can take the repair person safely step-by-step through the procedure.

Many times when the drive has quit running and the factory is at a standstill, it is much faster to just replace the suspected bad component with a brand-new component from the spare parts inventory. Hopefully, there will be modules, boards, and plenty of fuses on hand. If not, then shame on the personnel who decided that this issue was not important. Parts can be usually sent via overnight mail, but nothing is better than disconnecting the bad device, going to the stockroom, pulling the spare device, and reinstalling it. This is particularly gratifying at 3:00 A.M. during third shift!

Every manufacturer of dc drives has a recommended spares list of those parts that should be kept on hand in the event of a drive failure. The best time to purchase these parts is at the time of drive purchase because this is when the buyer has his or her best buying power. After that, the price of the spare parts becomes two to three times what they were initially.

For a typical dc drive, various fuses at the ratings throughout the drive should always be on hand. Input fuses and branch fuses are a must, since they protect the bulk of the drive's components from incoming power surges. These fuses do not have to be purchased through the drive supplier as long as the current ratings and sizes for the recommended fuses are matched. Power modules, thyristors, or SCRs should be stocked as spares in groups of three, six, or twelve (for three-phase drives—regen and nonregen), obviously in the exact ratings for the particular drive. Initially, these should be purchased through the manufacturer because there might be some characteristics inherent to the devices being used.

For instance, the devices might have certain turn on/turn off values that might cause problems later if not matched accordingly. When an SCR fails, it is common practice to replace the other SCRs in the leg of the circuit because they might have become stressed and might be prone to quick failure.

Other components that should be stocked as spares are the printed circuit boards for the drive. These boards must be ordered from the drive manufacturer. The control boards and gate firing boards are the most common to stock, but consult with the drive manufacturer for any others. Also, consider electrostatic discharge (ESD) when handling any printed circuit boards. A grounding wrist strap is ideal but rarely practical when servicing an out-of-service drive. The next-best precaution is to have an ESD-protected bag available. Minimize handling of the boards, and don't touch small components. Use one hand whenever possible to prevent an arc of voltage from forming. Place the board in the bag as quickly as possible.

Suppression components might be in place to minimize further component damage in the event of surges. These can be stocked as spares, but can be found in many electrical hardware supply stores. They are the capacitors and resistors in the snubber network. Power supplies are other items that sometimes fail; these, too, can be purchased from supply stores. Determine whether or not keeping these items in stock is worthwhile.

Finally, any fans, tachometers, special boards (for feedback, etc.), and any temperature sensors should be kept on hand if the drive runs 24 hours per day, seven days per week. As a further precautionary measure, continue to train various technicians in the plant on how to troubleshoot, replace, and repair drive components. This training must be specific to the drive in place and can usually be performed by the drive supplier.

Whenever the drive manufacturer can provide the field service, training, and additional required support it is always practical to take advantage of these services. However, they are usually very expensive, and not everyone in the plant can become an expert on every drive. That is one reason why plants try to standardize on one or two particular drive manufacturers. Keep in mind, however, that drive manufacturers change their designs over time, and the user must often relearn the product anyhow. The following sections will attempt to cover many of the common problems, faults, and diagnostics that are inherent to most dc drives. Many drives have on-board diagnostics that can help greatly in tracking the problem. The drive manual, which should be stored in a pocket on the drive door and not in someone's office, should be consulted.

When a drive is down the first thing to do is to kill incoming power and lock it out. Be wary of drives that contain capacitors, which can carry charges for a period of time. Many drives carry an LED that indicates, when it's on, that the capacitor charge has not been fully bled off. Be careful! Upon opening the drive cover or door, look for any apparent signs of internal component damage. The smell of fresh electrically produced smoke is a good indicator. Next look for any burnt wiring or components. Some components might even swell, especially the capacitors. Any loose or disconnected wires should be noted, as should any marks on the drive enclosure walls from apparent electrical arcs. Once the visual inspection is done, continue troubleshooting and tracing from the suspect point.

Check all input fuses to see if any are blown. This might be visually apparent, or a physical check with an ohmmeter might be required. Either way, you must determine whether the problem lies before, at, or after the fuse section. Many times the fuses are blown, but there is still a second or third problem with the drive. This is where the fun of troubleshooting begins. Again, some drives have the capability of saving multiple trips and faults. Use this tool if possible.

Since the heart and soul of the dc drive is its thyristors, let's start with an in-depth look at troubleshooting this circuitry. SCRs can be tested for short or open circuits. To do this the SCR power section must be disassembled. After removing the appropriate nuts and bolts to get at the SCR, a test reading can be made. The SCR does not have to be removed from the heatsink in order to be checked. An ohmmeter is used at this point to gather representative readings from the SCRs. Use the same ohmmeter throughout all the testing and for each of the SCRs in the drive.

Depending on whether the SCRs are of the hockey-puck type or the modular type, there will be different resistance values expected. Consult the drive manual or the power semiconductor manufacturer for these expected values. The first reading of the first SCR, or the one suspected of being bad, should be duly noted and compared to acceptable values. Do not replace the SCR if the readings fall dramatically outside of the acceptable values until a similar check is made of the other SCRs. Don't do any extra work than is necessary.

The SCR meter reading values for resistance can be mid-range, which normally makes the SCR's condition questionable; high or greater-than range, which usually means that the SCR is good; or low range, which means that the SCR is shorted. To perform these meter readings, place the positive probe on the SCR anode and the negative probe on the cathode, in the case of the modular type SCR (see figure 5-21). Because the hockey-puck version has no anode lead, it must be tested in the heatsink.

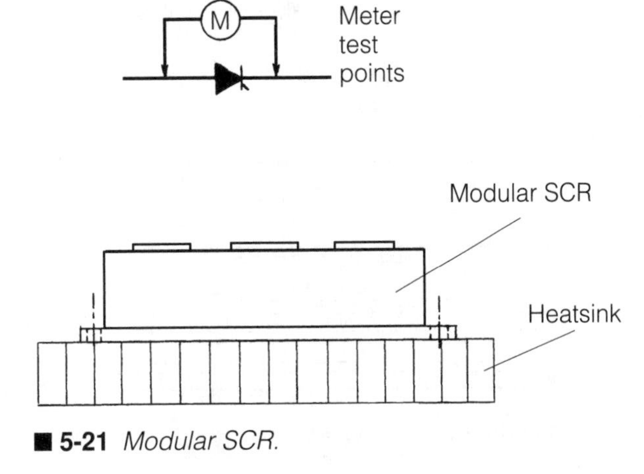

■ **5-21** *Modular SCR.*

The second part of the test involves taking a meter reading to check the SCR's gate resistance. Here again will be high, low, and midpoint ranges of resistance. Additionally, this reading might indicate a shorted or even an open circuit, indicating that the SCR should be replaced.

At this point the bad SCR must be removed from the heatsink and replaced with a new one. This is when a module-type power package is nice because it is easily installed and removed (but more expensive). The screws and washers are now removed, and the SCR is pulled off the heatsink. Select the new SCR, apply a thin coat of thermal joint compound, and mount with exactly the same washers and hardware as used previously. Some drives have SCR packages that require a torque wrench while others do not. This is a critical stage of the replacement process. Note if a Belleville washer was used. Once flattened, this washer indicates that the proper torque has been applied. Clean the assembly and then reassemble the remaining bridge components.

Hockey-puck SCRs have two heatsinks, the ac and the dc (figure 5-22). The hockey-puck style of power semiconductor is required for drives of higher current ratings. This version is a bit more troublesome to work with, but is very common. The principles of testing and replacing are similar to those of the modular-type SCR, but the manufacturer's manual and instructions should be followed.

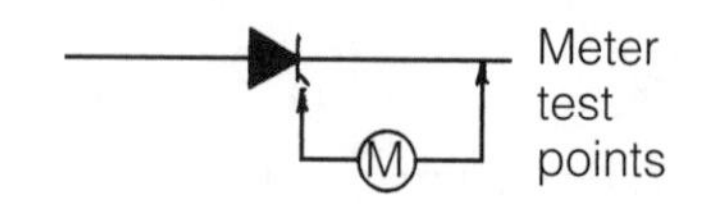

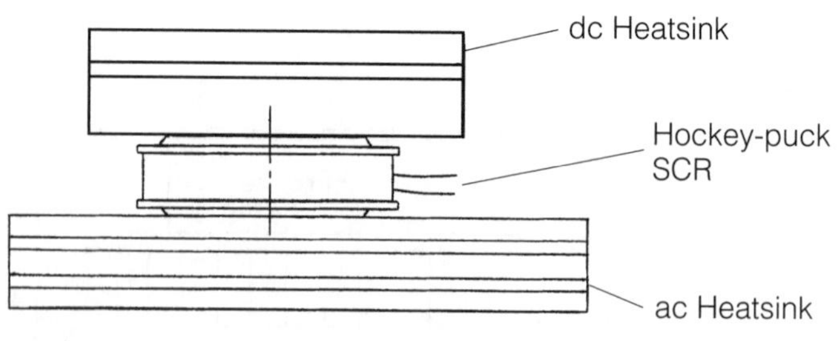

■ **5-22** *Hockey-puck SCR.*

Common drive faults: Causes and solutions

Today's dc drives tell us much more than older drives. When the drive trips, we can usually get a diagnosis from the unit itself. Some even provide text on a terminal screen that tells the troubleshooter where to troubleshoot. Many drives trip, or fault, for the same reason, but the solution might mean looking at a certain terminal or adjusting a certain parameter for that particular drive. For this reason, refer to the drive manual. Common drive trips (sometimes called alarms or errors) are as follows:

Low line input, undervoltage, or power dip

Possible cause: Input fuse(s) blown; low input voltage; power outage

Solutions: Check and replace fuse if blown; check to see that input voltage is within drive controller's allowable high and low range; check for possible problems with utility; sometimes lightning can cause this fault.

Overtemperature

Possible cause: Heatsink temperature sensor has caused trip; cooling fan has failed; drive ambient temperature has risen to extreme levels.

Solutions: Check thermistor with ohmmeter. Replace if bad. Check whether heatsinks are dusty and dirty. If drive is in an enclosure, check whether filters are clean. If not, clean or replace them. If there is a cooling fan, check whether it is operational. If not, repair it.

Overcurrent or overload

Possible cause: A badly tuned current loop; motor can't turn

Solutions: Tune current regulator so that high current transients don't exist. Load might be too large for motor; a larger motor might be required or the gearing might need to be changed. If there's a jam in the drivetrain, clear it. Some dc drives fault on either armature overcurrent, field overcurrent, or both, so check motor.

Loss of tachometer signal

Possible causes: Broken or loose coupling or connector; feedback wires not connected or have been cut; tachometer input board faulty.

Solutions: Fix or tighten coupling or connector; ensure feedback wire integrity and ensure tach is wired properly; replace faulty tachometer input board.

Short circuit

Possible causes: Motor might have a short circuit; power supply might have failed.

Solutions: Check motor armature and have motor repaired; replace power supply.

Field loss

Possible causes: Motor field has an open circuit; armature voltage is changing too rapidly in field weakened state; field economy setting might be too low.

Solutions: Check and repair motor; limit acceleration to slow change in armature voltage; raise field economy value.

Main fuse(s) have blown

Possible causes: Bad SCR(s); short or bad connection in power circuitry.

Solutions: Check SCRs and replace if defective; trace and check connections in power circuitry.

Speed at motor is not correct; speed is fluctuating

Possible causes: Speed reference is not correct; tach feedback line might be carrying interference; speed scaling is set up incorrectly.

Solutions: Ensure that speed reference signal is correct, isolated, and free of any noise; filter noise on tach feedback line and/or eliminate noise; recalibrate speed scaling circuit.

Contactor will not enable

Possible causes: Supply power logic for on/off improperly sequencing; contactor coil or wire malfunction; drive is faulted.

Solutions: Check on/off logic and reconfigure if necessary; check coil fusing and voltage, replace; clear fault.

Motor stall trip

Possible causes: Not enough current to drive the load; too-low field current.

Solutions: Check calibration of current for drive and motor; recalibrate field current scaling.

There are certainly other types of faults for any given dc drive and motor installation. Digital dc drives can always contribute the ever-deadly CPU failure or show no response at all when the main microprocessor board is not functional. Also with digital drives, if the drive can automatically tune itself with the motor and an operational error is detected, the auto-tune sequence is aborted and an alarm is indicated. Additionally, if any drive controllers are set up to accept other external feedback and it is not present, the controllers might trip when powered up. Also, motors can overspeed or run away, but usually the drive controller can be set up to trip at a maximum speed for protection.

Closing remarks on dc drives

The dc drive can provide 200% of full-load current for short periods (and if sized accordingly, can produce even higher amounts for starting) and even 150% for longer periods. It has a simple yet rugged design that has kept it around for many years and will keep it around for many more. Contrary to the prediction that all dc applications will eventually be replaced with ac drives and motors, there is evidence that this is happening much more slowly than anticipated. Granted, there are several reasons to switch from dc to ac; however, some applications still require the performance of dc. And since dc drives themselves have kept pace with digital and solid-state electronics, many extra functions are now possible with the dc drive.

Digital dc drives can also perform many logic functions. This is not to say that the newer dc drives are going to displace the programmable controller, but many low-level inputs and outputs are now part of the drive package. Start and stop logic in the form of low-voltage discrete inputs as well as the ability to accept 0–10-V analog signals have become a standard to dc drives. Some dc drives can also carry a software program that is application-specific and can communicate with other computerized devices within the factory. With microprocessor technology, high-speed data transmission to other smart drives and host controllers is now possible. We are now on the leading edge of automation with the dc drives of today.

Digital dc drives can also provide diagnostics for faults when they occur. The drive can store and log the faults, recording when they occurred and what they were. Some can go as far as directing the repair person to the "scene of the crime"—right to the suspect circuit or component. Many drive systems can now send much of this specific diagnostic data to a host computer in the plant so that the plant personnel can be made aware of pending motor or drive problems ahead of time and take preventive measures.

As mentioned previously in this chapter, a dc motor is really a dumb animal that will keep accepting electrical current until it burns itself up. The drive acts as the protective portion of the drive and motor circuit. The dc drive will trip if output current remains at a harmful level for too long. Thermostats in the motor are always worth the extra few dollars. They act as an insurance policy to stop operations if the motor gets too hot. The thermostats should be wired into the circuit or into the dc drive so that power to the motor is ceased accordingly. Thus, the dc drive protects the motor in more than one way.

To date, dc drives and dc motors are used quite extensively in industry. The ac drives are displacing dc drives in some areas but are not doing a complete change-out. The dc drives are good—and sometimes necessary—choices for many applications such as web-fed tensioning systems, extruders, machines with severe or rapid load changes, and others. The dc motors have solid benefits and definite features. As long as the technology continues to enhance them, they will be used for years to come.

Alternating current (ac) drives

6

DURING THE 1970S THE OIL EMBARGO CHANGED THE LIVES of all Americans. It actually changed the entire world. The world economy was kicked into high gear and so too was technology. We were in the midst of an energy crisis. Gasoline had to be rationed. Prices of oil-related products went out of sight virtually overnight. Something had to be done. No longer could we waste energy. Energy was to be conserved in all forms, most importantly, with regard to automobiles, heating, and electrical equipment. Efficiency of electrical equipment had to be increased, and new ways of saving energy had to quickly be developed.

One available but seldom used technology involved converting alternating current energy into direct current, then inverting that energy back into alternating current and varying the frequency to an electric motor. This technology saved energy, but the cost to implement such a device was extremely expensive. Therefore, it sat in the shadows until the electric utility companies nearly panicked. Existing oil-burning power plants had to be converted to coal, and new plants had to be coal or nuclear powered. No more oil-burning plants! And the price of electricity went up and kept going up. Now it was time for the industrial users of electrical equipment to panic.

What should be done in the factory to reduce energy consumption? One possibility was to make every piece of equipment as efficient as possible. Another was to install premium-efficiency ac motors whenever practical. But what could be done with all the ac motors existing and running in the plant? Of particular concern were those motors that always ran full speed but incorporated mechanical means of slowing or reducing the flows of the liquids or gases. These were the motors running all the fans and pumps, constantly running at their full speed whether they had to or not. They were good motors, too, so replacing by an energy-efficient motor wasn't justified because that wouldn't solve the main problem.

The problem was that many hours per day these existing motors did not have to run full speed because the loads and demands were not that severe. In fact, over the period of a year the motor's energy costs were exorbitant. Fan systems had mechanical dampers in place that would be adjusted to restrict the flow of air as necessary to get the desired flow; all the while as the motor was running full speed.

The same was true with the centrifugal pumps in the plant. The traditional method of getting the desired reduced flow was to close a valve somewhere in the pipe system. The motor still was running full speed! In fact, these mechanical methods of reducing flow actually were hard on the system. The analogy is that of constantly fighting the full power of the electrical motor. Dampers and valves would eventually fail and wear out and need to be replaced. Weak locations in the piping system and the ductwork often ruptured, causing more maintenance. There was a better way, and the energy crisis made its implementation cost-effective and practical. Since then, the technology of ac drives has never looked back.

The use of ac drives has proliferated—not only because of energy conservation but for better process control, soft-starting capability, and motor protection. With this proliferation there have been major strides in cost reduction (isn't competition nice?), packaging, and most of all, in the features and technology of the drives themselves. Different types of drives have emerged, all useful in some application.

Interestingly, ac drives got their main application thrust from industry and now have become commonplace in the commercial marketplace. As costs come down, they will even become prolific in the home. This chapter will define what these ac drives are, and describe the different types, how to apply them, how to troubleshoot and repair them, and how to "speak the language" of drives.

The word *drive* can mean something different to many people. Some interpret it as the device in your computer where you store a program or place a floppy disk. Others might consider a drive to be all the mechanisms required to move part of a machine. Even when referring to it as an ac drive, many still don't have a clear understanding of what it is. Someone in the factory might know that the drivetrain is being powered by an ac motor and simply assume that this is the "ac drive." To clarify, a drive that electrically changes the electrical input to a motor should be called an *electronic drive* or, better yet, an *electronic ac drive*. *Electronic* because today's drive combines electrical power elements and microprocessor technology to perform its tasks, and *ac* because it runs an alternating current motor. But the drive is more than that, and this chapter will attempt to confirm that.

Electronic ac drive basics

The main components of an electronic ac drive are the power bridges and the control section. Figure 6-1 shows in simplified block form the two main power sections: the dc link and the control scheme. The power bridges, the way the drive derives electrical feedback from the motor, and the drive's output waveform all define the type of drive being used. Like their dc counterpart, all ac drives must have a power section that converts ac power into dc power. This is called the *converter bridge* (figure 6-2). Sometimes called the "front end" of the ac drive, the converter is commonly a three-phase full-wave diode bridge. Compared to the phase-controlled converters of older ac drives, today's converter provides for improved power factor, better harmonic distortion back to the mains, and less sensitivity to the incoming phase sequencing.

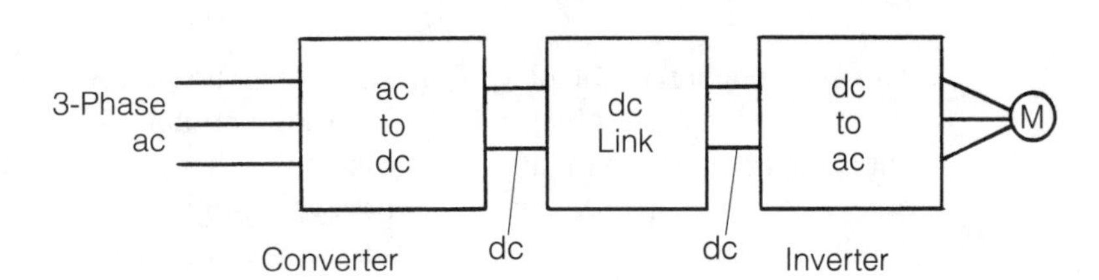

■ **6-1** *Electronic ac drive in simplified block form.*

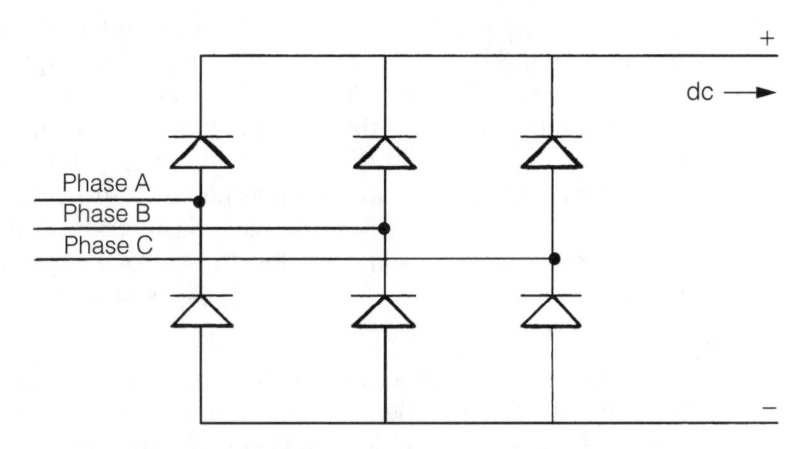

■ **6-2** *Converter section.*

The next component is the *dc bus*, or *filter*. This section is shown in figure 6-3 and is common to all electronic ac drives in some form. This is the section of the drive's circuitry where many drive manufacturers filter the dc bus voltage. Capacitors or chokes are used to ensure that the desired dc voltage or dc current is supplied to the inverter section. Here at the dc bus, valuable protective functions take place. The dc voltage is monitored for surges and compared to a maximum limit to protect devices from overvoltages. Also, the dc bus can provide the "quick outlet" for braking energy to a bank of resistors whenever a motor becomes a generator. Often the dc bus is called the intermediate circuit, bus, or link.

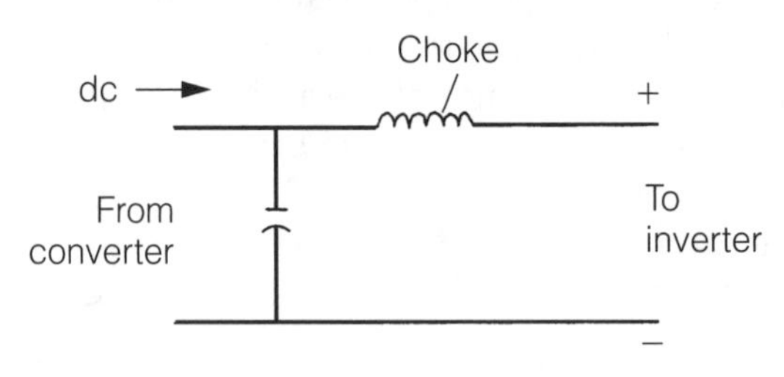

■ **6-3** *dc bus.*

The main portion of the electronic ac drive is the *inverter section*. From a basic standpoint, this power bridge actually is the differentiating component in drives. This is where that constant-voltage dc energy is inverted back to ac energy through the power semiconductor network. The dc drives don't have an inverter section, so from the outset, ac drives are more complex and expensive. However, an inverter section is a good component to have. Without it we cannot vary the frequency of an ac motor.

The inverter takes on many shapes and sizes. Its design is so important to the ac drive that many individuals have simply started calling ac drives "inverters." This is not a proper name because the drive has functions other than inverting dc to ac, including converting ac to dc, filtering the dc, and controlling all these functions. The inverter section is also where most of the drive differences among different manufacturers' designs will lie. Some designs utilize thyristors, while most modern inverters use some type of transistor. However, the inverter's principle of operation remains the same—changing dc energy into ac.

Inverters are classified as voltage sourcing, current sourcing, or variable-voltage types, depending on the form of dc that the inverter receives from the dc bus. It also is a function of how the drive has been designed to "correct" its own electrical feedback loop. This loop is actually part of a comparison with inverter output to the motor and the motor's load. In order to continue driving the motor at the desired speed, the drive must constantly be correcting the motor's flux.

If the drive receives a constant dc voltage, it is said to be a *voltage source inverter (VSI)*. As shown in figure 6-4, in this condition the inverter has to pay attention to both the frequency and the amplitude of that constant dc voltage. If it receives a dc voltage that varies, it is called a *variable-voltage inverter (VVI)*, as shown in figure 6-5. In this case, because the voltage is variable, the inverter is mainly concerned with frequency to maintain control. The last type of inverter, called a *current source inverter (CSI)*, sources a dc current from the dc bus (figure 6-6). This current might be variable and, as with the VVI, the inverter has to control mainly the frequency for proper operation. Note that the current sourcing inverter must have many extra components over the voltage regulating inverter and is therefore more complex.

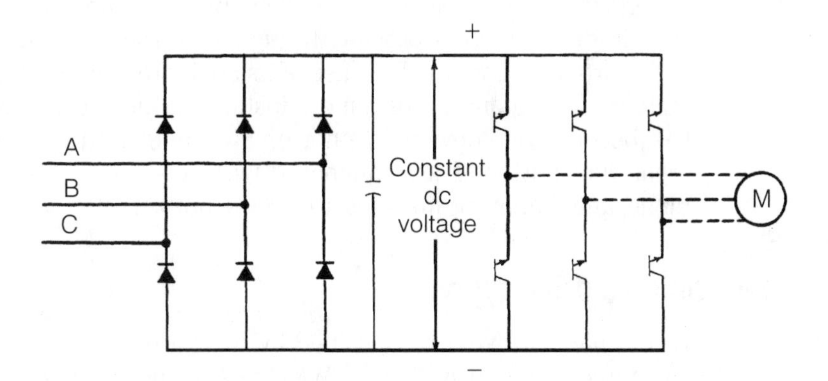

■ **6-4** *Voltage source inverter (VSI).*

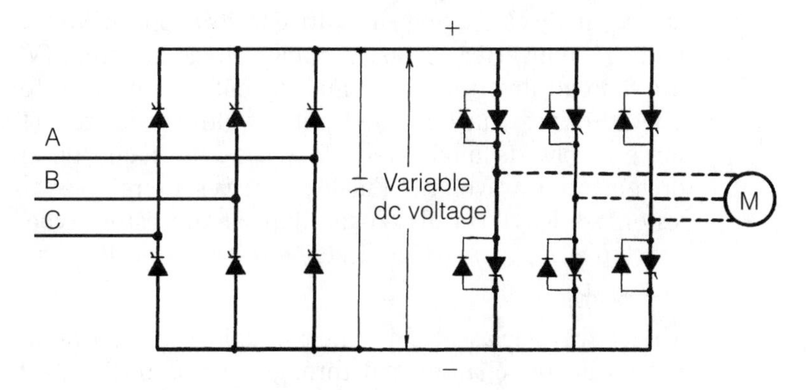

■ **6-5** *Variable-voltage inverter (VVI).*

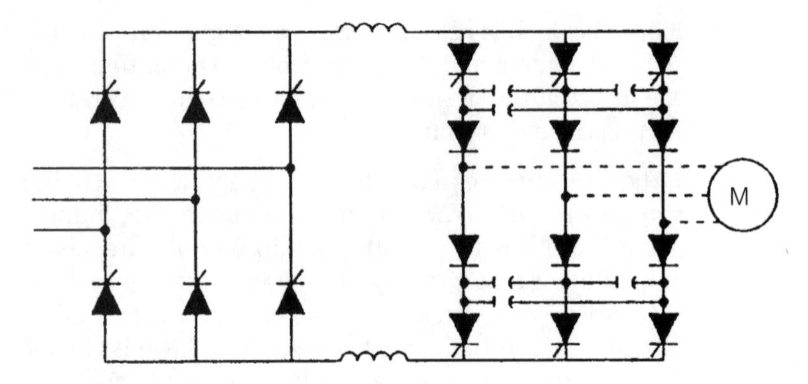

■ **6-6** *Current source inverter (CSI).*

Early drive manufacturers used various designs of inverters: thyristor-type semiconductors, different methods of cooling, different style heatsinks, and so on. But the basic concept of the inverter was the same, and eventually the thyristor gave way to the faster switching transistor technology. Because the thyristor has to wait for the current passing through it to reach zero, it has become virtually obsolete. Because of its ability to change from conductive to nonconductive almost instantaneously, the transistor has become the device of choice. Switching frequencies (the bread and butter of a frequency drive) now are in the 15-kHz range, and higher frequencies are on the horizon.

Electronic ac drive types

Electronic ac drives are classified by their use, dc bus voltage or current sourcing, waveform (PWM or PAM), or the type of power device used in their inverter section. By use, there are ac traction drives, vector drives, load-commutating inverters, spindle drives, and so on. By classifying an ac drive by its supply of voltage or current to the inverter, we get variable voltage inverters (VVI), voltage sourcing inverters (VSI), and current sourcing inverters (CSI). The ac drives can be called pulse width modulated (PWM) or pulse amplitude modulated (PAM), as the name describes the drive's output waveform. Finally, ac drives are referred to as transistorized, IGBT (insulated-gate bipolar transistor) type, or even six-step SCR types, which describe the inverter devices being implemented.

The ac drives classified by their function or application are discussed in this chapter and throughout this book. Electronic ac drives are usually classified by their output and the shape of that output's waveform. The main objective of the ac drive is to vary the speed of the motor while providing the closest approximation to a sine wave. After all, when an ac motor runs directly off of 60-Hz power, the signal to the motor is a sine wave (as clean as the local utility can provide). Put a variable-speed drive in the circuit and vary the frequency to get the desired speed. Sounds simple enough, but the industry is continually striving to address all the side effects and to provide a pure system.

The *pulse width modulated (PWM) ac drive* is the most common today. It will most often integrate transistors into the inverter section to accomplish the switching pattern. The switching pattern controls the width of the pulses out to the motor. The current and voltage waveforms are shown in figure 6-7. The output frequency of a PWM drive is controlled by applying positive pulses in one half-period and negative pulses in the next half-period. The dc voltage is provided by an uncontrolled diode rectifier. Thus, by switching the inverter transistor devices on and off many times per half-cycle, a pseudo-sinusoidal current waveform is approximated. As was shown in figure 6-7, a six-pulse PWM inverter produces some harmonic content. An 18-pulse modulator provides a dramatically cleaner waveform (figure 6-8).

The *pulse amplitude modulated (PAM) ac drive* is more concerned with the amplitude of the pulse than with the frequency of it (figure 6-9). Whereas a PWM drive and its high switching frequency might affect audible motor noise, the PAM drive can also

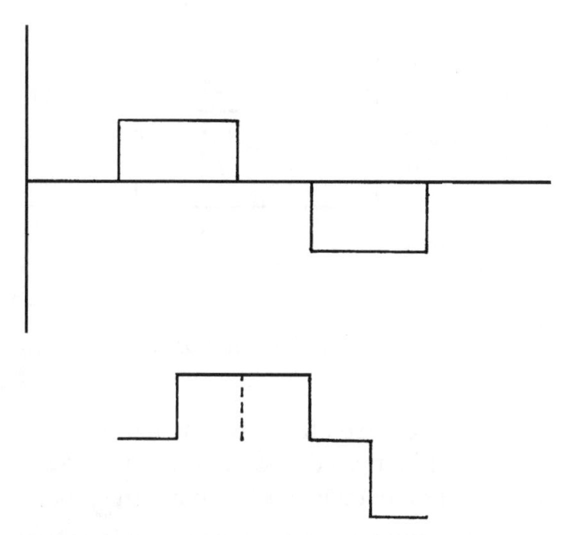

■ **6-7** *Pulse width modulated (PWM) six-pulse waveform.*

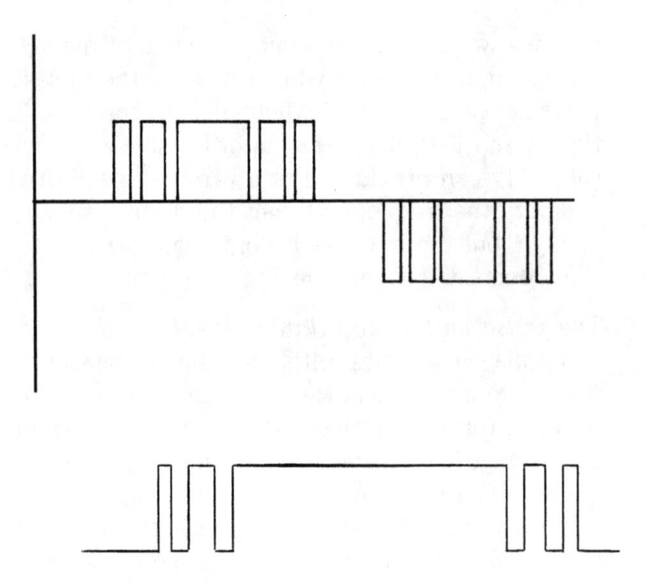

■ **6-8** *Pulse width modulated (PWM) 18-pulse waveform.*

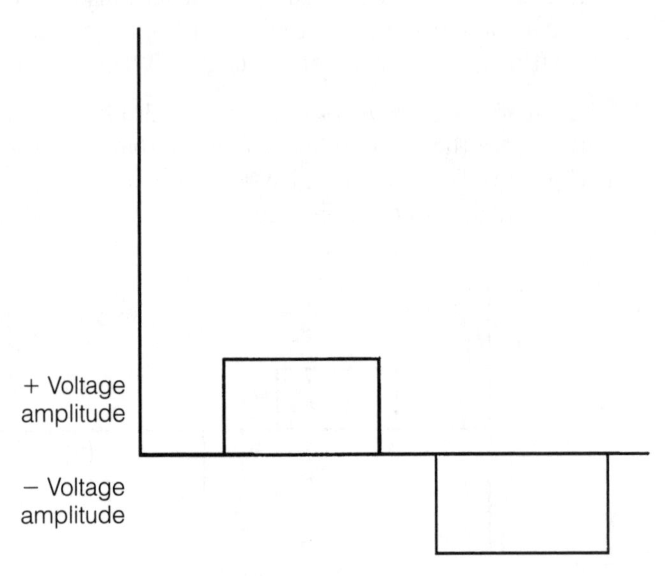

■ **6-9** *Pulse amplitude modulated (PAM) waveform.*

have some adverse effects on the motor, including cogging at low speeds and increased heating due to voltage spikes in the waveform. PAM drive controllers are not that common at this time.

As discussed earlier, the electronic ac drive is made up of a converter section, dc bus link, and an inverter section. Therefore, there are many converter designs and the same is true for the inverter section. Generally speaking, the converter section, or the front end of the variable-frequency ac drive, is the dc drive for a dc motor, with some modifications and a field regulator. However, with the exception of a specialized cycloconverter, all ac drives have to rectify ac power through a converter section.

The more common designs for the converter utilize diodes or thyristors to rectify the ac incoming voltage into dc voltage. One is called the *diode rectifier*. In figure 6-10, the three-phase incoming 60-Hz power is channeled into three legs of the converter circuit, each leg with two diodes. This creates a constant dc voltage. From here, this constant dc voltage, through the dc link, goes to the inverter circuit to be changed into variable frequency ac to the motor.

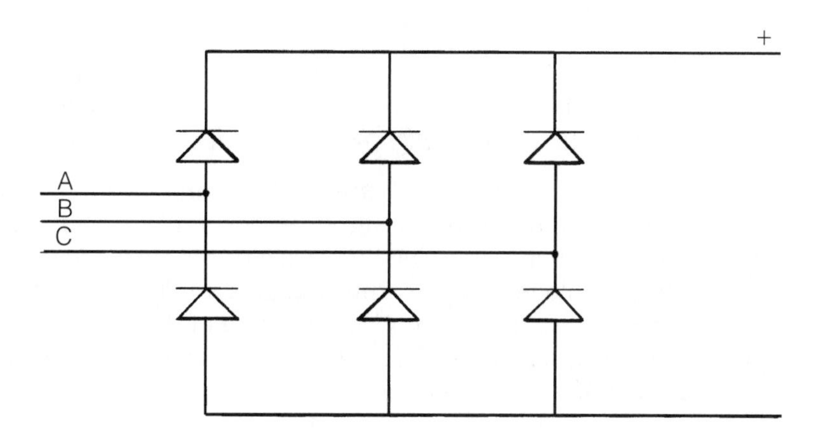

■ **6-10** *Diode rectifier.*

The diode rectifier is the most popular design because it is simple and the least expensive. Advantages of this type of rectifier include unity power factor, less distortion backfed to the supply, and resilience to noise in the converter itself. The biggest drawback to using diodes is that no voltage can return to the source, thereby allowing no regeneration. Another separate power bridge must be added to the converter section if regeneration is required, which can be expensive.

The diode rectifier with a chopper is basically similar to the aforementioned simple diode rectifier. The difference is shown in figure 6-11. A thyristor is incorporated, usually an SCR, to act as a valve to allow the dc voltage to rise. Once the voltage reaches a predefined level, the valve, or chopper switch, opens to deenergize and the voltage decreases. This process controls the high and low lev-

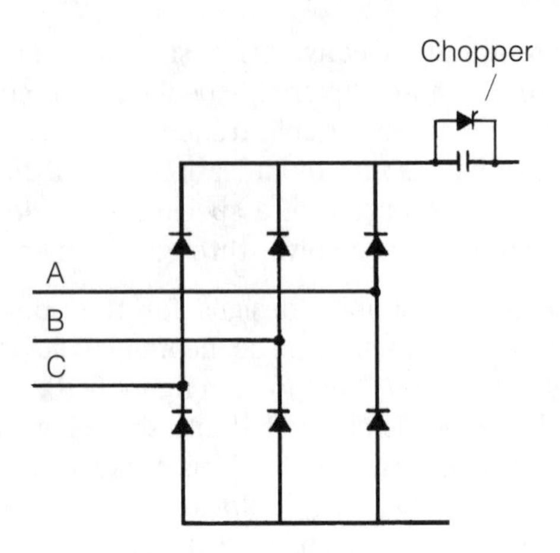

■ **6-11** *Diode rectifier with chopper network.*

els of dc output voltage to the inverter. This design allows for better control of a constant volts-per-hertz ratio, although the cost to produce this design is slightly higher than a straight diode rectifier.

The *silicon controlled rectifier (SCR) converter* has more capabilities than the diode converter. Although more robust and flexible, SCRs cost more. The SCR converter is shown in figure 6-12. Like the diode rectifiers, this design is a full-wave-type rectifier. Six SCRs are used to control the gating of the device. (*Gating* is the term given to controlling the SCR's time of conduction by turning the SCR on and off.)

The SCR cannot be turned on until it has deenergized after being commanded to turn off. This is sometimes referred to as the *zero-crossing* of the current. Therefore, many SCRS have different

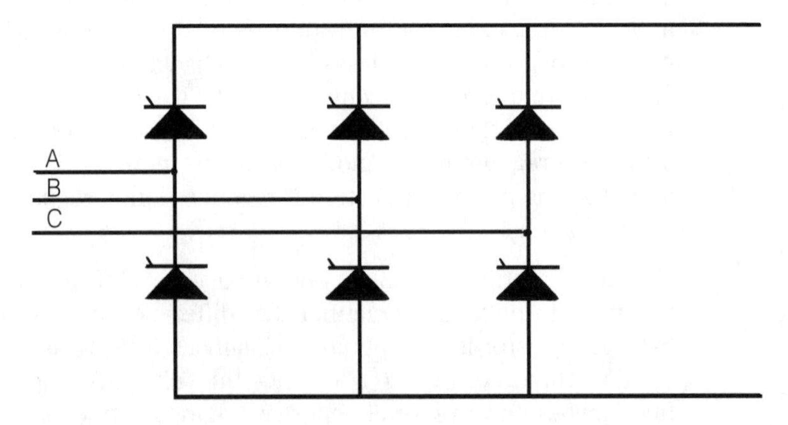

■ **6-12** *SCR converter.*

turn-off times, and it is sometimes necessary to get all six SCRs to match in one drive's circuit in order to ensure proper, smooth gating. The drive's logic circuitry provides the means of control for this gating sequence, thereby controlling the output voltage to the inverter.

Unlike the diode rectifier, the SCR, or thyristor-type converter has some distinct disadvantages and advantages. As speed is decreased, so too will the power factor of the system. Distortion fed back to the supply is another major concern. Many times a choke or otherwise special circuit must be added to minimize disturbances. Additionally, these SCRs are more susceptible to line disturbances, which can result in nuisance drive tripping. As long as manufacturers take measures to protect against these problems, this type of rectifier is attractive, especially because it has the ability to regenerate power back to the ac supply simply by gating the SCRs in reverse sequence.

Many designs of inverters are available, and there has been much discussion on which one is best. The answer is that they all are, when selected for the right application. Some designs will cost more than others, while other designs might have horsepower limitations or address harmonic distortion. Basically, most lower horsepower inverters incorporate high-frequency transistors, while some higher horsepower inverters do not. Individual power device costs and the paralleling of devices to get the higher currents can be expensive.

The latest technological design method for inverters is based around the *insulated gate bipolar transistor (IGBT)*. This transistor is a combination of features provided by the MOSFET transistor and the bipolar transistor. It has good current conductance with lower losses. It possesses very high switching frequency and is easy to control. Figure 6-13 shows its simplified circuit. This technology has gained much momentum because the IGBT can be used up to several hundred horsepower.

Other inverters utilizing high switching frequency transistors will incorporate either MOSFET (metal-oxide semiconductor field-effect transistor), bipolar transistors, or Darlington transistors. The transistor's basic advantage is that it can be switched from conducting to nonconducting at will. It does not have to wait for a zero-crossing condition much like its diode and SCR counterparts.

Also, since higher current ratings of transistors are now available, higher resulting horsepower drives are being built, sometimes with transistors in parallel to get the desired output to the motor.

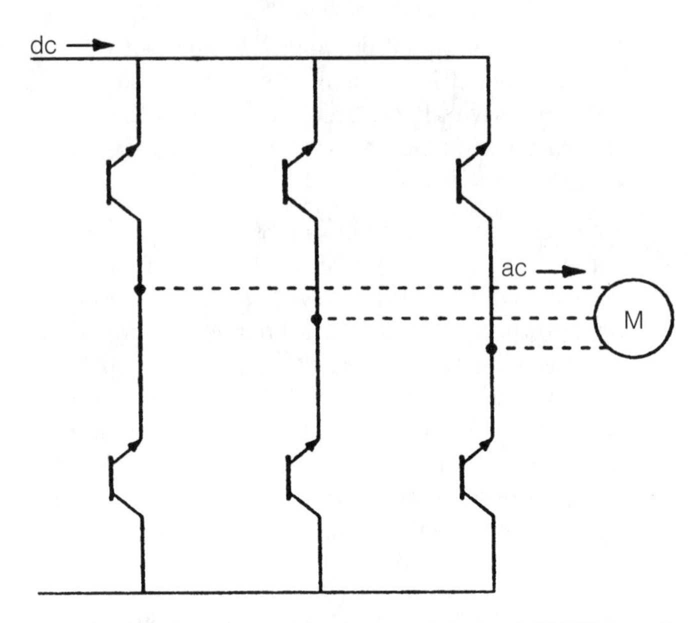

■ **6-13** *Insulated gate bipolar transistorized (IGBT) inverter.*

These transistors have the ability to switch at several kilohertz. This virtually eliminates audible noise at the motor—an earlier objection to using high-switching frequency transistors. This also means that the current waveform to the motor is close to sinusoidal, while the voltage can be modulated much easier.

Other types of inverters utilize diodes and SCRs. The six-step or variable-voltage inverter needs a chopper in front of the diode bridge or individual SCRs to switch on and off to produce the desired six-step voltage waveform. Each power-switching SCR's conduction time is controlled to get the desired increase or decrease of each individual step, thus changing the frequency output to the motor. Commutation, or the act of turning on and turning off semiconductors, is accomplished with the aid of an extra circuit, usually of capacitors, to provide power to switch off devices. Transistors and GTO devices do not need this commutation circuitry.

Another type is the *current source inverter (CSI)*. The equivalent circuit for this inverter is shown in figure 6-14. This inverter will normally utilize SCRs as switches to gain a six-step current waveform output. Here, the conducting time is changed up or down for each individual step, resulting in a longer or shorter cycle time.

All in all, electronic ac drives have to simply function in this manner: take three-phase ac line frequency power, convert it to dc,

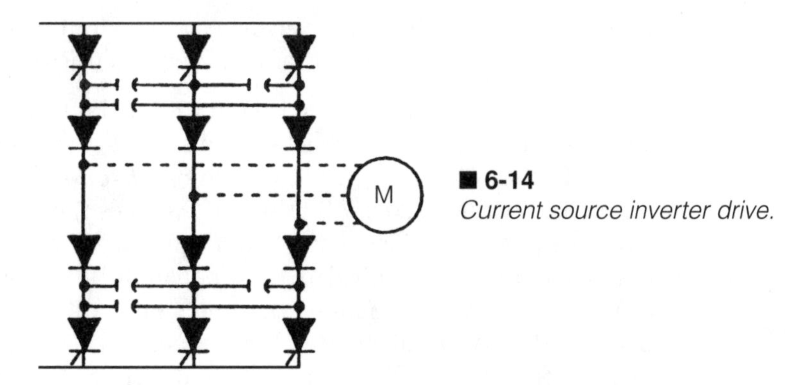

■ 6-14 *Current source inverter drive.*

and invert it back to variable frequency ac. Sounds fairly straightforward and simple, right? Not so, as we have seen. Higher voltages and currents, transients, misapplications, bad motors, and so on make for interesting problems and challenges. However, the technology keeps evolving and improving. Time will tell!

How electronic ac drives work

Understanding how the variable-frequency drive converts and inverts energy is valuable, but that is only part of the equation. Not only does the electronic drive control the ac motor, it also becomes part of the motor's electrical circuit. The drive needs the motor and the motor needs the drive. The proper operation of the motor/drive system is critical to each supplying the other with voltage, current, or both, and vice versa.

In order to really understand the application and operation of an electronic ac drive, we must first understand how an ac motor functions. An earlier chapter has already discussed this. However, certain motor characteristics have to be programmed in when commissioning the drive and motor combination. Motor speed concepts and speed-versus-torque concepts are critical. In the following formula for an ac motor

$$\text{synchronous speed} = 120 \times \frac{\text{frequency}}{\text{number of poles}}$$

the value of 120 is a constant and cannot be changed. It is derived from the electrical relationship given to synchronous machines with fixed poles, location of those poles in a given half-cycle, and the frequency (cycles per second and seconds per minute). Thus, for a given motor the number of poles has to be constant; they are physically in place on the motor's rotor. Therefore, in order to change the speed, all that can be changed in the formula is the frequency, which is exactly what a variable-frequency ac drive does.

In the equation, if the frequency is 60 for normal 60-Hz supplied power, then the motor speed will be the maximum for an equating number of poles. This is referred to as the *synchronous speed*. As an example, 60 times 120 is 7,200, which becomes the numerator. A two-pole motor has just that—two magnetic poles, north and south. Therefore, 7,200 divided by 2 equals 3,600. This is the speed, in revolutions per minute, that the motor will run at if it is applied 60-Hz power. Likewise, if the frequency is 30 Hz, the resulting speed is half, or 1,800 rpm. Table 6-1 shows synchronous speeds for various motors with different pole configurations. Those shown represent the most common configurations.

■ Table 6-1 An ac induction motor speeds/poles chart.

Number of poles	Synchronous speed (60 Hz – rpm)	Synchronous speed (50 Hz – rpm)
2	3,600	3,000
4	1,800	1,500
6	1,200	1,000
8	900	750
10	720	600
12	600	500

The number of poles in the ac motor is fixed. These magnetic regions on the motor's rotating element, the rotor, have a permanent polarity, plus or minus. As the stator windings receive electrical energy from the ac drive, a change in magnetism takes place, and thus motion of the rotor takes place. How often this current flows through the windings is the frequency. Figure 6-15 is a simple diagram showing this relationship. It should be noted that when an ac drive first applies power to the ac motor, a certain amount of current is needed to get magnetic flux built-up enough to even begin motion. This is called *magnetizing current* and must be accounted for by the drive before any torque-producing current can be utilized.

When given the task of driving an actual load, the actual shaft speed of an ac squirrel cage motor will be slower than the synchronous speed. This is called the *full-load speed*; it is a function of a motor characteristic called *slip* and is typical of all ac induction motors. Because the motor is always dynamically correcting to maintain speed, when loaded, it lags behind in actual motor rev-

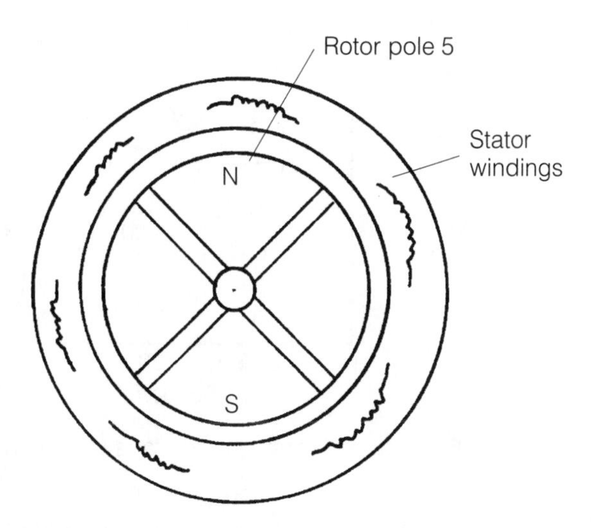

■ **6-15** *ac motor poles and stator relationship.*

olutions per minute. The percentage of slip can be found by using the following formula:

$$\textit{slip percentage} = \frac{\textit{synchronous motor rpm} - \textit{full load rpm}}{\textit{synchronous motor rpm}} \times 100$$

The amount of slip in the typical NEMA B squirrel cage motor is usually 2 to 3% of the motor synchronous speed. For example, when supplied three-phase power at 60 Hz and with no loading, a 1,800-rpm synchronous speed motor will run at 1,800 actual rpm's. However, when the motor shaft is applied a load, the actual speed is now 1,746 rpm's. The speed at which a motor can run fully loaded is called *base speed*. A motor can run above or below its base speed by increasing or decreasing the frequency.

As for loading and slip, these items must be factored into the application whenever running dramatically higher than or lower than base speed. Also, as was seen in a previous chapter, NEMA A, C, and D motors carry different values of slip and therefore will have varying needs for current, particularly magnetizing current. Their slip percentage can range from 5 to 8%. This must be a consideration when applying the variable-frequency drive.

The motor slip will directly relate to the ability to drive any given load at a given speed. Torque can be said to equate to load, which equates to current. Therefore, there are given relationships to speed and torque. A typical speed/torque curve for a NEMA B motor is shown in figure 6-16. Slip will remain constant anywhere on the curve as the frequency is reduced to achieve the desired

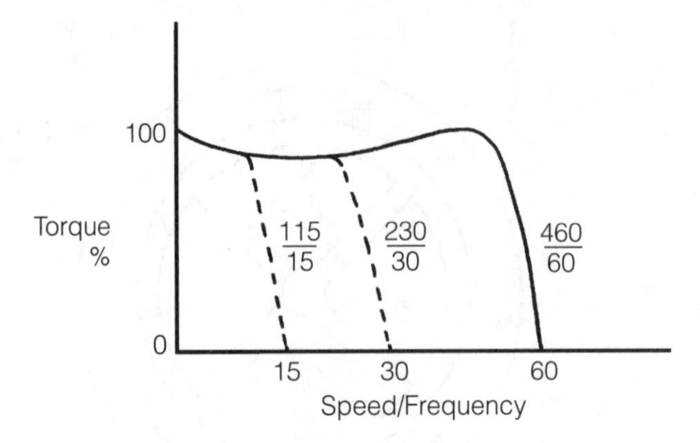

■ **6-16** *Speed versus torque for a NEMA B motor.*

speed. Slip is a crucial element in controlling an ac motor, especially at low speeds. In essence, controlling slip means the motor is under control. The flux vector drive, which will be discussed later in this chapter, gains the best control of this slip function with the ac motor. However, electronic ac drives with volts-per-hertz capability can control slip very well down to low speeds. The biggest factor is the loading. Light, or even centrifugal loads are much easier to control at slow speeds.

As the induction motor generates flux in its rotating field, torque is produced. This flux must remain constant in order to produce full-load torque. This is most important when running the motor less than full speed. And since ac drives are used to provide slower running speeds, there must be a means of maintaining a constant flux in the air gap of the motor. This method of flux control is called the *volts-per-hertz ratio*. When changing the frequency for speed control so, too, must the voltage change, proportionally, to maintain good torque production at the motor. The volts-per-hertz ratio nominally is 7.6 to 1 for a 460-V, 60-Hz system (460 ÷ 60 = 7.6). Thus, at half-speed on a 460-V supplied system, the frequency is 30 Hz and the corresponding voltage is 230 V to the motor from the drive.

This pattern saves energy to the motor, but it is also very critical to performance. The variable-frequency drive tries to maintain this ratio because if the ratio increases or decreases as motor speed changes, motor current can become unstable and torque can diminish. This is the reason that variable-frequency drives start to have control troubles below 20 Hz.

Again, the flux vector design variable-frequency drive is one solution for maintaining better control of the volts-per-hertz pattern at very low speeds. Another method of increasing the voltage at low speeds to produce adequate torque is to incorporate a voltage boost function available on most drives. However, if the motor is lightly loaded and voltage boost is enabled at low speeds, an unstable, growling motor might be the result. Voltage boost should be used when loading is high and the motor must run at low speeds or start with a high load.

Our electronic ac variable-frequency drive also allows for motor operation into an extended speed range, sometimes called *overspeeding* or *overfrequency operation*. Sometimes the application requires that the motor run beyond 60 Hz. Frequencies of 120 (twice base speed), 300, 600 Hz, and even beyond are possible with faster switching inverters. Higher speeds can be achieved, but torque diminishes rapidly as the speed increases. The volts-per-hertz curve for this type of extended speed operation is shown in figure 6-17. Trying to maintain a constant ratio between the voltage and the frequency from zero to full speed is desirable in these kinds of applications.

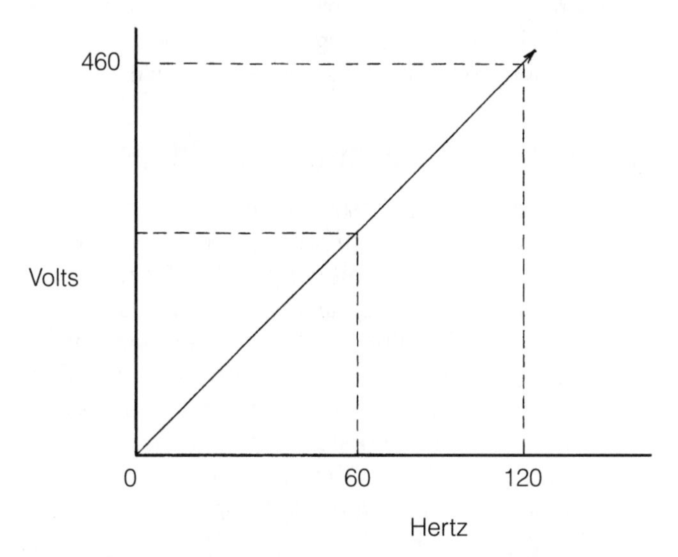

■ **6-17** *Extended speed range—volts-per-hertz curve.*

This capability is a real benefit in those pumping applications where getting a little more speed out of the motor can increase the flow just enough to satisfy the demand. When there is capacity in the motor and the ac drive can be programmed to do this, a new, larger motor does not have to be purchased and installed. The old motor does not have to be taken out. In a flow-and-demand type of system, often the ac drive can run 10 to 20% higher (66 to 72 Hz) in speed and make up for lost capacity.

There is presently a need for this type of operation for many refrigeration systems still using old-style—and outlawed—refrigerants. These CFC-containing gases are compressed by a motor. The new refrigerants, those allowed in place of the old, have different physical properties, and the motor has to speed up in order to get the desired flow and net capacity of the refrigeration system. The electronic ac drives might come to the rescue in many of these instances.

Beyond the minor speed increase here and there, there are very high-speed applications, some needing 10 times base speed and faster. These include test stands and dynamometers. While speed is an attractive feature of the variable-frequency drive, care must be applied when utilizing this function. Many applications cannot tolerate going extremely fast. Physical components such as gearboxes, couplings, fan blades, and so on can actually explode at too high a speed. This can be a dangerous situation. However, many test stands are actually rotating an object as fast as possible to see when it does break apart. For safety's sake, investigate the mechanical system accordingly before experimenting with high-speed operation from a variable-frequency drive.

Energy savings with electronic ac drives

The reality of energy savings attributed to electronic ac drives is evidenced by the rebate programs in place around North America. Many electric utility companies are offering substantial rebates to end users in hopes that ac drives will be considered more often as an energy-saving alternative. The mere fact that the electric utilities are offering monetary incentives to implement ac drives is a sound testimonial that these devices actually save energy. And the results are usually outstanding! Of course, the utility hopes that the end user will be so happy with the energy savings that subsequent ac drives will be funded elsewhere.

The rebates are offered regionally and periodically. They can range anywhere from $20 per horsepower to completely covering the purchase and installation of a drive system. Electric utilities have to offer these rebates. The biggest reason is because the prospects of bringing a new power plant online will take years of lobbying (once the EPA, DER, and other environmental groups give their approval) with the locality for the site, then add on the years of construction and phase-in, and of course the costs: millions if not billions of dollars to get the new electricity down the wire! The better scenario is to entice users to conserve. Heating, ventilating, and air-conditioning markets as well as pump applications are perfect candidates for energy-saving electronic ac drives and even premium-efficiency motors.

So why all the fuss over ac drives? Not only can they provide exceptional cost savings while operating, they can also offer a power factor improvement, which further makes the utility happy. The ac drives limit inrush current to motors. They also provide a buffer to the utility's supply so that line fluctuations caused by leading and lagging elements on the supply line don't appear as often and as severe. No more penalties from the utility to the user of the ac drive. This alone would justify the cost of installing an ac drive. However, the real savings is in actual operating costs saved over time. These costs are more dramatic in areas where electrical costs are very high.

An electronic ac drive can be used in place of several mechanical restrictors in most centrifugal fan or pump systems. With the fan system, an ac motor traditionally runs full-speed all the time, and the only way to slow the airflow is to open and close dampers or use inlet guide vanes. Surprisingly, this method is still common in many commercial buildings even today. With pump systems, the same reduction in flow was achieved by mechanical means. A pump motor runs at its full-load speed. Discharge or bypass valves are in place in the piping system to be opened and closed either manually (most often) or automatically (when the user can afford it) to reduce the flow of liquid. These dampers and valves can get stuck, wear out, or corrode. In addition to being maintenance items, they waste energy. That is where the ac drive comes in.

The ac drive is installed ahead of the ac motor in the electrical system. It now reduces the flow electronically. It slows the ac motor down and puts out only the voltage and current required to drive the load. And with centrifugal applications the load is dramatically lowered as speed is lowered. These fans and pumps follow certain principles of physics called the *affinity laws*, discussed further in Chapter 11. Figure 6-18 shows a curve that indicates how the ac drive compares in line-to-shaft efficiency with various traditional methods of flow control. This curve shows that the ac drive can be a major factor in saving energy over mechanical means even when running close to full speed. It will supply full voltage and current if required at full speed, but a centrifugal load doesn't usually require 100%+ torque at full speed, as illustrated in figure 6-19.

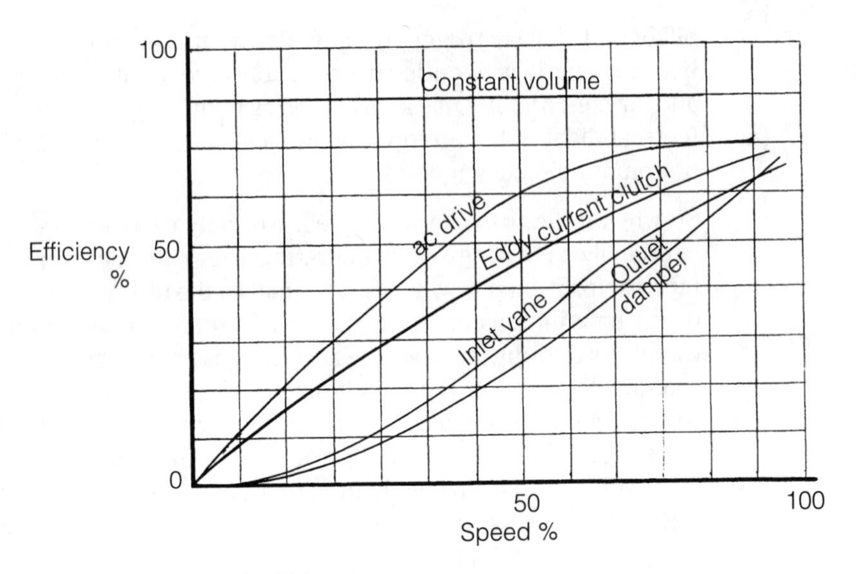

■ **6-18** *Line to shaft efficiencies.*

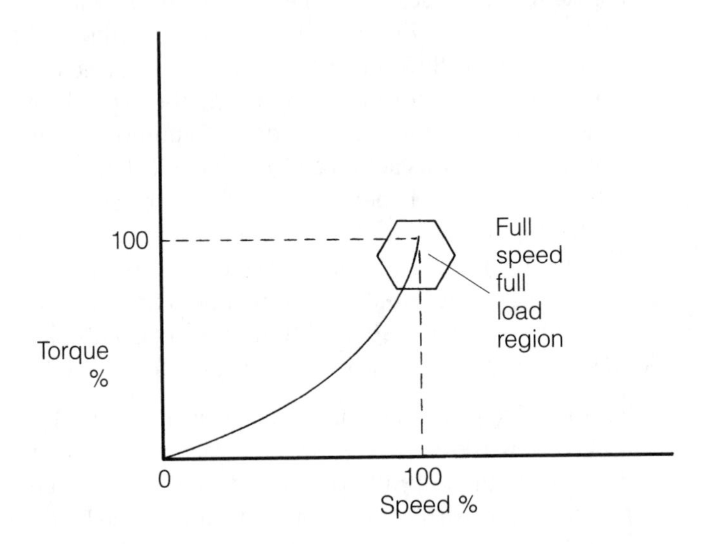

■ **6-19** *Centrifugal load curve.*

The affinity laws, along with other basic formulas for fans and pumps, should be understood by the user, engineer, or installer of the drive. This knowledge is useful in getting the full capability out of the variable-frequency drive. Even before applying the drive, it is sometimes necessary to completely justify the drive from an energy-savings viewpoint. An example of calculated savings is shown in figure 6-20. Minimal data required are the cost per kilowatt-hour,

$ VFD $

Centrifugal pump using a variable frequency drive cost savings example:

Givens:

- The pump runs 24 hours per day, 7 days/week, 52 weeks/year (8736 Hrs total).
- Motor is 75 HP (HP times 0.746 = kW, thus 75 × 0.746 = 55.95 kW)
- Electricity cost is 6 cents per kW-Hr.
- The pump flow is controlled by a discharge valve.
- A VFD will save approximately 50% in kW operating the pump in a centrifugal manner.

Therefore:

(55.95 times 0.5) × (8736 Hrs) × (0.06 cents/KW-Hr) = $14,663 savings/year

■ **6-20** *Example of calculated energy savings with variable frequency drive.*

motor size, hours of operation, and present means of flow restriction. Once this data is entered into the formula, the calculation is fairly straightforward. Sometimes the exercise is not even necessary if the facility where the drive is being considered has a very high cost per kilowatt-hour. These are regions where there is high power consumption, and production capacity of that power is limited. Thus, the cost for electricity is high, and rebates are probably available.

Other formulas for calculating the brake horsepower (BHP) are shown below. These formulas are necessary for calculating the horsepower in a new installation for fans or pumps. Otherwise, in existing installations, motor nameplate data is all that is needed. However, checking the existing BHP to a recalculated BHP will verify that the motor is not oversized (another waste of energy). The formulas are as follows:

Fan application:

$$BHP = \frac{CFM \times psf}{33{,}000 \times fan\ efficiency}$$

or if pounds per square inch are known:

$$BHP = \frac{CFM \times psi}{229 \times fan\ efficiency}$$

Pump application formulas:

$$BHP = \frac{GPM \times feet \times specific\ gravity}{3{,}957 \times pump\ efficiency}$$

or if pounds per square inch are known:

$$BHP = \frac{GPM \times psi \times specific\ gravity}{1{,}713 \times pump\ efficiency}$$

BHP = brake horsepower
CFM = cubic feet per minute
GPM = gallons per minute
PSF = pounds per square foot
PSI = pounds per square inch

where head in feet is equal to 2.31 pounds per square inch gauge (psig) and the specific gravity of water is equal to 1.0.

Thus, if we were to predict what the energy savings would be for a motor whose brake horsepower was determined to be 5 BHP, the first step would be to convert that horsepower value into real power, or watts. Therefore, 5 × 0.746 (746 W equals 1 HP) = 3.73 kW (kilowatts). A variable-frequency drive is assumed to be at least twice as efficient as a pump with a discharge valve, and therefore, we apply a ratio to each. Pump systems notoriously operate at 65 to 75% of their maximum flow rates; by using pump curves, which can be supplied by any pump manufacturer, we can establish typical efficiencies and other data useful to confirm our ratios.

Next, we calculate the power consumed by the pump with the discharge valve and at a ratio of 0.98 to the drive at 0.5. We come up with 3.73 kW × 0.98 = 3.655 kW and 3.73 kW × 0.5 = 1.865 kW, respectively. Subtracting the smaller value from the larger, we now have the difference in energy between the two types of pump motor control. That difference is 3.655 – 1.865 = 1.79 kW. This means that the actual power that will be saved every hour is 1.79 kW.

By applying two more factors, time and money (actual hours of operation and cost per kilowatt-hour), we can put a dollar value on the savings. If the pump runs 8 hours per day, 5 days per week, then in one year, it runs 2,080 hours. If the utility charges 9 cents per kilowatt-hour, we simply take 0.09 × 2,080 × 1.79 kW, and we arrive at a savings of $335 per year. Considering that a 5-HP drive might cost $500 (using the $100-per-horsepower cost estimate), saving $335 each year means that the drive will pay for itself in 18 months.

Obviously, the previous example makes some assumptions. However, if the fan or pump runs 24 hours per day and/or the electrical costs are higher, then the actual savings will be much greater. Likewise, if the motor is much higher in horsepower rating, the savings are more quickly realized. The bottom line is that a variable-frequency drive will save money in a centrifugal application. How much it will save and how long it will take until the investment of the drive is returned can be answered for each individual application and depends on hours of actual motor running, the cost per kilowatt-hour, and how good a deal you get on the drive purchase. When the rebate is a possibility, implementing the ac drive should be done as quickly as possible for two important reasons: the rebate period will expire and every hour the motor runs without an ac drive controlling it is more money not saved!

Variable torque and constant torque

The concepts of variable torque and constant torque are important to understanding ac drives. An earlier chapter defined torque as the force that moves the objects and loads in any machine. Horsepower doesn't really do the work. And, logically, one might assume that if there is any work to do, torque will be needed. If we are continuously running and doing work, then we are always in need of torque. This could be construed as constant torque, right? Not exactly. In fact, this misconception has led to more misapplications with ac drives than we can imagine. Using the curve in figure 6-21, we'll attempt to segregate the two types of torque.

As can be seen in the curve, variable-torque applications require very little torque at starting or low speeds. These are often centrifugal loads which don't develop the need for high torque until their speed has increased dramatically. These are the "easier" applications for ac drives to operate and where true, exceptional energy savings are justified. While it is true that some torque is required anywhere on the speed curve, the full, high-load torque seen in constant torque applications is not required. The safest approach to applying the ac drive is to know the load requirements at all possible operating speeds. Select the drive based on the worst case. Some engineers feel that if they cannot turn the motor shaft by hand (indicating a light load to start), then the ac drive might have trouble. Many fans and pumps are variable-torque, centrifugal applications and are good candidates for ac variable-frequency drives.

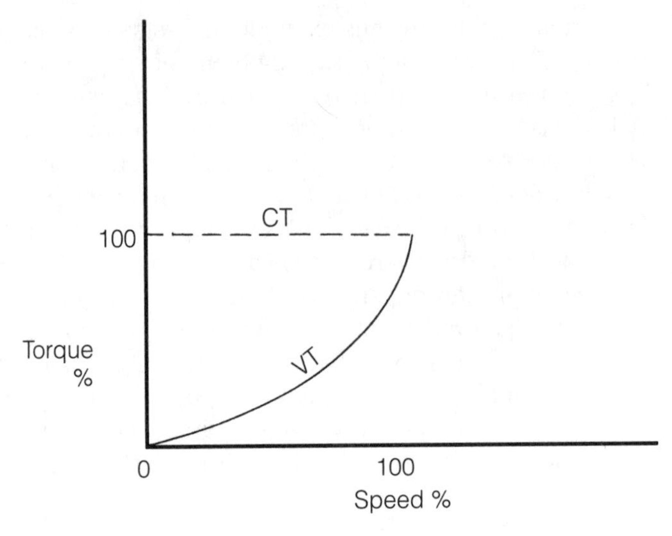

■ **6-21** *Constant torque versus variable torque comparison.*

Constant-torque applications, on the other hand, have to be analyzed closely. How much torque is required and when? How low in speed will the drive have to operate with that load? These are questions that must be answered first in constant, or continuous high-torque applications. These applications will usually have high starting torque requirements as well. The ac drive can be sized to handle these types of loads, but upfront investigation can save embarrassment and hassles later. These applications are not justified for energy-saving reasons, but rather for better process control or soft starting. The ac drives today with higher carrier frequencies and with vector capabilities can handle most constant-torque applications. Figure 6-22 shows a curve that relates horsepower to torque in a constant-torque speed range. Note what happens to both at the point where 100% speed is reached.

Drive selection (ac)

Why use a particular electronic ac drive over another? This is a hard question, and usually there is more than one answer. Each application dictates what is required. The first issue is deciding whether or not a dc motor or an ac motor will be used. Once that issue has been decided, the application requirements take precedence. Will there need to be braking, or regeneration? What is the horsepower? What is the supply voltage? What speed regulation will the application require? What torque regulation? Is cost a factor (isn't it always)? Is plant floor or wall space at a premium? Is a

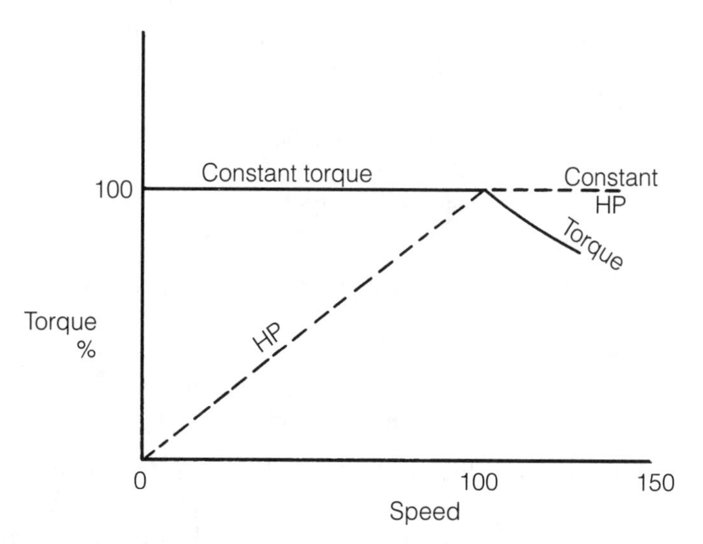

■ **6-22** *Constant torque and constant horsepower.*

digital or analog drive preferred? What will the maintenance person find when there is a problem? What kind of duty cycle, or loading is predicted? Does the plant have a "clean" voltage supply? Do we need an efficient drive? The questions seem to be endless. Let's look at the issues.

Once it has been decided that an ac motor will be used for the application, we must determine the horsepower, voltage, and enclosure. Most importantly, we must determine whether the motor selected can run properly with any ac drive's output. In other words, is there a good match between motor and drive?

Misapplying drives is, was, and will be a problem. Engineers applying drives today have probably lived through a bad application or two and want to avoid another. Some individuals stick with a certain ac drive technology or a certain type of drive. Others are willing to push the technology and try something different. Regardless, the application really dictates what to do. Be wary of manufacturers of drives who might have biased reasons to support or degrade a particular design. Listen and learn. Ask the same question of many. The best suggestion to be given is: Take the time and perform an in-depth analysis of the current equipment available. Also, completely understand the application from a mechanical, electrical, and connectivity vantage point.

A manufacturer's variable-frequency drive might be well suited for one application but not for another. There are hundreds of considerations to ponder when selecting a VFD. Section 15, Electrical, of *Architectural Specifications* usually contain written descriptions of drives to be used on building and construction projects. Try to secure a copy before purchasing a VFD. Talk to as many technical vendors as possible because the industry is constantly changing. They can advise if your application is right for their drive. Some important considerations:

Constant torque or variable torque How much current is required on a continuous basis for the particular application? What does the speed-versus-torque curve look like? Perhaps the application only needs starting or peak torque for a few seconds. If so, one frame size of an ac drive might be more suited than another.

Complexity of the ac drive's circuitry The more components, the more risk of component failure. Also, consider how long a particular complex drive type has been in production. Where are similar applications in service with the same design? Reliability is crucial, and in an ever-changing industry, how long will the chosen design be built and supported?

Digital or analog ac drive control While most manufacturers promote digital designs with hundreds of functions and features, analog types are usually simpler and less expensive. Digital drives offer much more capability in diagnostics, protection, and communications.

Speed range What is the speed range, or turn-down ratio, and what is the loading like at various speeds? Turn-down ratios of 10:1 are common for variable-torque applications. Low speeds while fully loaded are the most difficult for ac drives to control. Similarly, hard-to-start loads might require high breakaway torque. Typically, a drive's slip compensation function can provide adequate adjustment.

Encoder feedback Check to see if a drive design can be retrofitted with a feedback device to gain better speed regulation. If a module can be added to the drive control circuitry to accept pulses from an encoder, this information can be used by the drive to correct the speed error.

Auxiliary blower at motor When a motor runs at low speeds, fully loaded for long periods of time, it must be fitted with an external blower with small motor. Also, the starter for this blower motor, often just a single-phase device, must be included within the drive package. This auxiliary blower will run constantly regardless of the main motor's speed and will blow air continuously over the skin of the main motor.

Braking and regeneration Is braking required or can the motor coast to a rest via the friction in the drivetrain? What will happen in an E-stop situation? The drive has no control when it is not being supplied power, and thus cannot commutate. If the drive is not commutating, it is not able to regenerate. Therefore, it is common to use dynamic braking even if a drive can regenerate.

Adequate power supply Some drives are more sensitive than others. Some are also phase sequence sensitive. Determine prior to drive installation if the power supply will be stable. If the supply has dips, frequent outages, and surges, take necessary precautions at the drive.

Cable sizing Motor voltage is directly proportional to speed in an ac drive variable-speed application. Therefore, any voltage drop is very noticeable at low speeds. When the motor is running near full speed, a voltage drop in the wire doesn't really affect the speed.

Location of drive with respect to motor Running wire and cable is often preferable to locating a drive in a nasty or outdoor environment. Even though special enclosures can be built for these environments, they can be expensive. Also, it is recommended that the drive be as close as possible to the motor.

Proper earth grounding The ac drive and motor applications requiring three-phase power need to use four-conductor cable. All equipment in the motor/drive circuit must be tied to earth ground at one location (by *earth ground* we mean the central ground point for all ac power and electrical equipment within a factory). The input power to the drive should be from a WYE configured source. All current will be contained within these four wires and minimize interference on the input. The output wiring should allow for the fourth wire, the ground conductor, to be used as the fixed ground connection between the ac drive's enclosure and the motor itself.

Inrush currents The ac drives generally limit the amount of inrush current to the motor to no more than 100%. This lengthens the life of the motor while providing softer, smoother starting. When line-starting an ac motor, typical inrush currents can approach 600% of motor rating.

Replacing a dc drive with an ac drive Check the horsepower and torque requirements. An identical-horsepower ac motor and drive might not be enough to do the job. For example, a center-winder with a buildup ratio might require a much larger ac motor than the previous dc motor. Many times a dc motor can be run into the field-weakened speed range, thereby keeping its horsepower size down. Again, an ac drive and motor might need to be upscaled in size to get the proper torque at all operating speeds.

Multiple motor operation Does the application require that several motors be run from one ac drive? If so, because it is load-dependent, a current source drive will have limitations. Only a few, if any, motors can be taken off line when running the entire set. A voltage source drive is more suitable here. One important precaution is to apply individual motor thermal overload protectors. The ac drive could inadvertently supply too much current to a smaller motor because the drive doesn't necessarily know it is running smaller motors. Its electronic overload protection is basically for a motor sized for the drive's rating.

Ground fault and short circuit protection Does the drive manufacturer require an isolation transformer? How will the drive handle a ground fault or a motor short circuit? Conversely, what happens to the inverter in an open circuit situation?

Power factor Most ac drives today have a unity power factor in the 0.98 range. Power factor must be taken into account because of utility penalties, where the utility imposes a penalty on demand charges for a low power factor. A drive with a constant power factor throughout its speed range is attractive, especially when it will replace a drive with a poor power factor. Power factor will be analyzed further in Chapter 12.

Single-phase input to a three-phase drive It is not recommended that a three-phase ac drive run a single-phase motor. On the other hand, a single-phase input to a drive running a three-phase motor is feasible; however, the drive will have to be derated for the desired output at the selected motor. Consult the drive manufacturer for sizing.

Harmonic distortion, or content Variable-frequency drives create disturbances back to their supply and out to the motor. These are input and output harmonics and are prevalent in devices that convert ac and dc power. Determine what your particular system can tolerate. Usually, the ac drive will have little trouble operating in the harmonic distortion it has created. Harmonics is discussed in detail in Chapter 12.

Higher horsepower and medium voltage ac drives Motor horsepowers can sometimes dictate the type of inverter required. Some transistors are not available in higher current ratings. This means that power semiconductors might have to be placed in modules and/or connected in parallel in the power circuit. For medium voltage ac drives having to directly run a 2,300-V or 4,160-V motor, this capability raises numerous issues. Costs and sizes of equipment in this class of voltage are very high. The drive trips that can stop a low horsepower drive/motor application are magnified with larger motors and drives. These units often are the heart and soul of a facility, and if out of service, can cost thousands of dollars in lost production. Explore the variable-frequency drive manufacturer's experience, testing capability, and complete design when entering into the high horsepower arena.

Drive efficiency (ac) The ac variable-frequency drive efficiency is just as important as the cost. Make the analysis of each with both the motor and the actual drive. Consider efficiencies at all speeds and mainly for the predicted speeds to be operated. Remember, the VFD is supposed to be an energy-saving device.

Cooling VFDs are usually air-cooled, although some are water-cooled or air-conditioned. Because of the cooling fans, air-cooled drives tend to be noisier than others. Make sure the plant can tolerate higher decibel levels. Water-cooled drive systems involve a pump, piping, and heat exchanger system in addition to the drive's converter and inverter package. These water-cooled systems, although very efficient when maintained, have the potential for leaks, must be cleaned, and have higher initial costs.

Ventilation A ventilated, cool ac drive is a drive that will function well and for a long time. When physically installing any drive, whether ac or dc, special attention should be given to the heat generated by the drive. The drive has current running through it. This current will produce heat, and this heat has to go somewhere. It can naturally dissipate if there is a light-duty cycle. However, if loading is heavy, provisions for cooling or ventilating are necessary.

First, take a look at the ambient environment around the drive. Is the room in which the drive will be located warm or hot naturally? What is the temperature on the hottest summer day? Next, is the drive going into an enclosure? Is this enclosure going to be completely sealed? Determine if the drive is going to be heavily loaded. Will it run 24 hours a day or intermittently? Most of the time, ventilation fans, pulling ambient air in to the enclosure and up across the drive, will suffice. Figure 6-23 shows typical drive cabinets with different ventilation schemes.

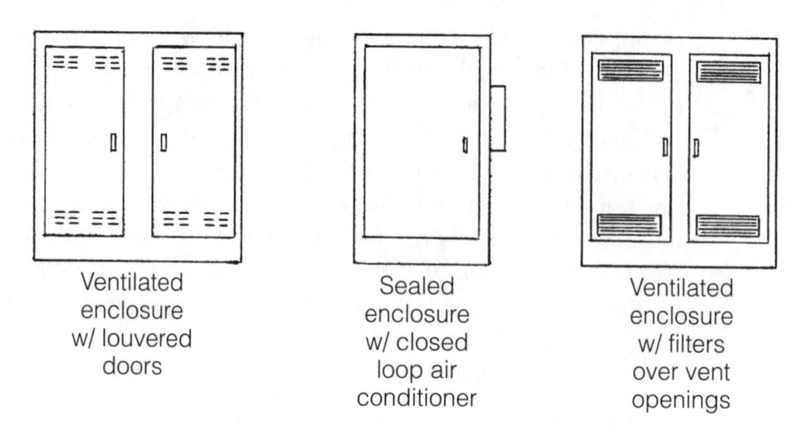

■ **6-23** *Drive enclosure ventilation schemes.*

Another question to be answered is how clean is the ambient air? If dirty air is brought into the enclosure, a new potential problem can emerge. Dust and dirt will collect on the drive and virtually suffocate it. Eventually no heat will be able to escape from the drive, and it will overheat. Most newer drives will trip on an overtemperature fault. This protects the drive but is a nuisance because the drive will have to cool off before starting it again.

Another means of handling drive enclosure heat is by air-conditioning the cabinet. This is the more expensive approach but sometimes the only answer. When the ambient air is too warm or too dirty, air-conditioning makes sense. There are many manufacturers that specialize in small, compact air conditioners that attach directly to the wall of a drive enclosure. These are self-contained, closed-loop units that keep the inside of the drive enclosure completely cooled.

Keep in mind that if the ambient area is dirty, the air conditioner will have to be kept clean in order to operate efficiently. Also, if the air conditioner stops running for whatever reason, the drives will trip quickly because they are now in a completely sealed enclosure with no way for heat to escape. Many air conditioners today are made to work without any CFCs. However, they will need cooling water supplied to them and will require a drain. They are worth looking into.

High altitude High-altitude locations can be a problem for drive installation. Knowing the elevation conditions ahead of time can head off drive-related problems. As the altitude increases, so too does that air's inability to dissipate heat. Thin air, which one might get above 3,300 feet above sea level (or approximately 1,000 meters), can't hold as much heat as an equivalent amount of air at sea level. Therefore, transferring the heat from our heat source, the drive, is more difficult at higher altitudes. Drives often have derating values for equivalent horsepower, or continuous current output, for a given level of elevation.

High humidity and excessive moisture Typically, this is a problem for a drive if the excessive moisture in the air condenses on the drive's components. Water will conduct electricity and if enough water forms on the drive, short circuits can occur. The maximum level of moisture in the air is a relative humidity of 95%. This again causes nuisance tripping, and likewise downtime will occur until the drive is dried. One means of dealing with condensation is to install space heaters within the drive enclosure. These should come on when the drive is off; when the drive is running, enough heat is generated by the drive itself to keep moisture from condensing. Some installations with smaller drives can even make use of a standard lightbulb as the "space heater."

Another moisture-related problem is harmful: corrosive gases in the air such as acid mist, chlorine, saltwater mist, hydrogen sulfide, and others. Problems can range from slow deterioration of the printed circuit boards to actual corrosion of the bolts that hold the drive together. Protective coatings to the boards can help, but the better solution is to keep the contaminated air from getting to the drive at all.

Long distances between motor and drive This can be a factor depending on the type of drive being used. A voltage source drive tries to maintain a constant voltage in the motor/drive circuit. Long runs of cable can create voltage drops, which can affect the motor's ability to maintain speed under heavier loading. One solution is to increase the gauge of the cable to minimize drops in voltage. This might have to be done if the actual supply voltage is lower than nominal. For instance, if the supply is supposed to be 460 V, most drives are able to handle a 10% range above and below nominal. However, if the supply is lower than the projected 460 V, a long run of cable can further reduce that value to the motor in less-available volts per hertz. This might limit speed and torque capabilities. Typical distances, though, are less than 300 feet and usually present no problems.

On occasion, it is necessary to locate the actual electronic drive farther away from the motor than 300 feet. If the drive is a voltage source drive with very high switching transistorized output, a different phenomenon can exist. This is sometimes called the *stand-*

ing wave condition. This condition can make for high peak voltages and possibly damage motor windings. The standing wave condition is more prevalent with these types of drives and very long runs of cable (usually over 300 feet). One solution is to install output reactors between the motor and the drive. The reactors smooth out the voltage but add impedance to the system. This can add to the overall voltage drop and must be considered when evaluating this option as a solution. Again, if the voltage supply is steady and higher than nominal, then all should be fine at most speeds and load conditions.

Two other considerations relative to this standing wave condition: 1. The longer the distance between motor and drive, the higher the impedance output reactor be used; and 2. Check the class of insulation in the motor as it might be better able to withstand voltage peaks without substantial degradation.

Input and output contactors Often a motor is located very far from the variable-speed drive, even completely out of sight. If a maintenance person wants to work on the motor, he or she must ensure that there is no electricity going out to the motor. One common practice is to install a contactor or disconnect near the motor, in the circuit between the output of the drive and the motor. This is fine from a safety standpoint, but it can be potentially harmful to the drive.

For example, suppose a maintenance person goes out to the roof, where a motor is driving a fan. The variable-speed drive is located two floors below in a mechanical room. The fan is running along at full speed and full load, and the maintenance person decides to "kill" power to the motor by opening the contactor at the motor. This possibly could cause a high energy spike back to the drive and could blow an output device.

Rather than assume that all maintenance personnel are trained to never open an output contactor under load, it is more practical to interlock a contact that faults the drive first and then opens the output contactor. This will ensure that there will be no power flowing through the drive, thus eliminating the possibility of a spike. This contact must be an early auxiliary contact.

Likewise, an input contactor located on the supply side of an ac drive can and should be interlocked to open in certain situations; for instance, the drive should not be able to receive any input voltage or current when a fatal fault has occurred or the drive is attempting to dynamically brake. No new energy is wanted at the time, so the input contactor acts as the shut-off valve and as the disconnect. Make sure all interlocking is proper.

Input reactors Do variable-speed drives require isolation transformers? First determine what you are trying to accomplish. Are you trying to minimize noise in and out of the drive at the supply point? Do you need ground fault protection? The isolation transformer can provide all of this, but the user should understand the type of drive being supplied before requesting input line reactors.

Output reactors Sometimes it is necessary with certain types of drives to reduce ringing at the output to the motor. However, two major reasons to use output reactors are to protect the inverter power devices from a short circuit on the drive's output wiring and to overcome long cable runs from the drive to the motor by smoothing the voltage and adding impedance to the system.

Other ac drive functions and features

Many electronic ac drives today have many built-in features. The power of the ac drive is not only in the converting and inverting sections, but also in the software that controls the drive. Early versions of ac drives had to include extra printed circuit boards or plug-in modules to accomplish many specialized tasks. Now, with digital technology, high-speed microprocessors, and ample memory, all you have to do is call up the parameter on the drive's display and make the necessary change. Setting up an ac drive is 10 times easier today than it was when dip switches and potentiometers had to be tweaked. Following is a list of many common ac drive features and functions.

Acceleration This function allows the user to select the amount of time desired to reach full speed. Often referred to as *ramping*, this function actually provides the soft-start capability that limits inrush current to an ac motor. High-inertia loads might require several seconds, even minutes for some large fans, to accelerate to full speed. If this were not adjustable, the drive would constantly trip off line under these conditions. Some drives even can provide two acceleration settings, or a two-stage ramp. As the drive begins to accelerate the motor, a contact closure can signal that it is time to "shift gears" and go into a faster ramp up to speed. Any acceleration setting will be operational even when any speed change is desired. Going from low to high speed will utilize the set acceleration rate. Special acceleration and deceleration curves can be programmed with many drives. The ever-popular S-ramp for very soft starts is common.

Deceleration Deceleration is similar to acceleration in setup, but most often with a different value. Deceleration is a controlled stop provided by the ac drive. It has limits and is load dependent. If a high-inertia load is to be brought to a fast stop, the drive must have some method of handling the motor-generated energy. Regeneration or a dynamic braking circuit will provide an outlet for the brake energy. If not, the drive can provide some braking power by allowing its dc bus voltage to rise slowly. If this voltage rises to the maximum set level (set there to protect devices), the drive will trip on an overvoltage fault. The solution here is to try a longer deceleration rate.

Automatic restart Some ac drives come equipped with the ability to automatically attempt to restart themselves when conditions permit. For instance, a drive can be programmed to automatically attempt three times, at two-minute intervals, to restart. As long as power is available to the control circuitry, the drive's logic unit will allow restart attempts. If after the select number of attempts the drive cannot restart, it remains in a faulted condition until someone manually resets it. Many times a manual reset is preferred by plant personnel for safety reasons. A typical automatic restart function is appropriate whenever supply power is known to dip below acceptable levels of voltage (undervoltage).

Loss of automatic signal Many drives can be programmed to run manually from a potentiometer or keypad setting right at the drive. Other times it is desirable to receive a 0- to 10-V or 4- to 20-mA (milliamp) signal to scale as the speed range. Four milliamps will equal minimum speed and 20 mA will equal maximum speed. These minimum and maximum values are also programmable for the particular application.

When running in the automatic mode, it is necessary to have a safety built in, which handles those conditions where the automatic reference is zero (absent) or loaded with electrical noise and outside the min/max settings. Most often the drive will go to a preset safe, slow speed under these circumstances, rather than allowing the motor to run away. Some drives can be programmed to also shut down and fault, while other applications might need to have the motor keep running while a fault is announced. Likewise, many ac drives that might be accepting an encoder or tachometer feedback signal, might need to fall back to a safe, slow speed if that signal is lost. Look into this ahead of time.

Skip critical frequencies In many drive applications, especially fan and pump applications, there is always the possibility that resonant frequencies can exist. And many times these resonant frequencies can cause severe vibration in the mechanical drivetrain of the drive and motor system. If the drive were set at this frequency and run continuously there, possible premature mechanical failure could occur. The ac drive can be programmed to avoid certain frequencies. It will select a frequency above or below a set bandwidth in order to skip the known resonant frequency. This is illustrated in figure 6-24.

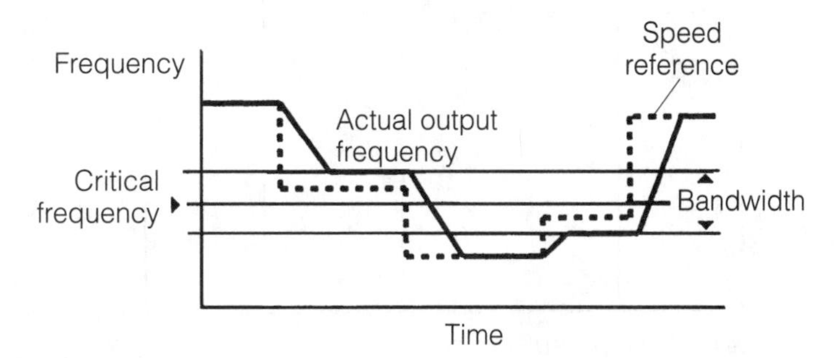

■ **6-24** *Critical frequency rejection circuit bandwidth.*

Fault logging Many digital drives are kind enough to display the fault that has taken them out of service. Many are not. Some, which automatically restart, keep running and the plant personnel are not aware of any problems. Many ac drives have enough memory to store faults as they occur and thus have a record of when a fault happened, what it was, and how the drive was reset (automatically or manually). Figure 6-25 shows a typical fault log. Usually these faults are stored in a first in, first out manner.

Traverse or P jump This function can be seen in figure 6-26. It is commonly used in textile and fiber winding applications where the traverse function is better done electronically than mechanically. This triangular, or sawtooth, waveform is accomplished by programming the traverse period, maximum traverse, and the P-jump value. The drive will move the thread back and forth in a diamond-like pattern. This distributes the thread evenly across the tube surface. The pattern is altered in order to prevent a buildup of thread at the same location on the tube. This alteration is accomplished by varying the speed of the traverse in an on/off, cyclical manner throughout the speed range.

Power loss ride-through An ac drive has the ability to ride through short power interruptions. The drive can continue to run off of

Fault history log for 100 HP, 460 V, 3 phase, 60 HZ VFD

Fault #	Fault type	Reset method	Input volts	Output amps	Time	Day
1	Fuse failure	Manual	545	98	1355	7/23
2	Fuse failure	Manual	567	95	1320	7/24
3	Undervoltage	Auto	390	66	0906	8/4
4	Overvoltage	Auto	512	92	0650	8/9
5	Fuse failure	Manual	660	99	1205	9/30
6	Overtemperature	Manual	455	97	1215	10/5

■ **6-25** *Drive fault log and history.*

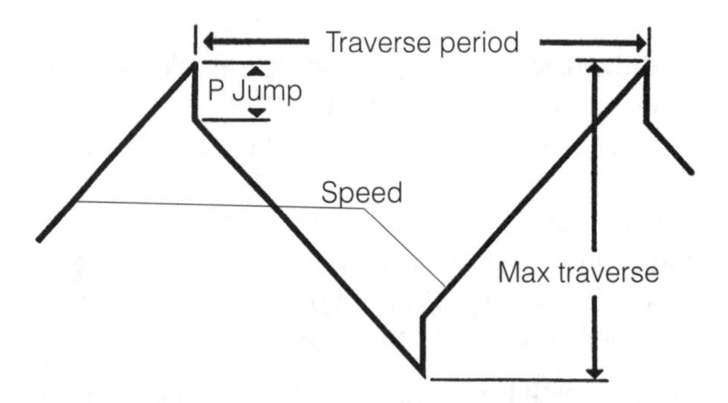

■ **6-26** *Traversing, or P-jump, function used in textile winding applications.*

stored dc bus energy until the bus voltage reaches a level where it can no longer supply enough voltage to keep running and it then faults. Most drives can only ride through 1 to 2 seconds.

Slip compensation The amount of slip in an ac motor system is proportional to load. With increased slip comes an increase of necessary torque to continue driving the load. However, many times the speed of the rotor actually slows in order to continue driving the load. In applications where this speed shedding is undesirable, slip compensation provides for a solution. As the load increases, the ac drive can automatically increase the output frequency to continue providing motor slip without a decrease in speed. The actual amount of slip compensation will be proportional to the increase in load. Usually one setting of slip compensation will cover the entire operating range with speed regulations of 0.5%.

Pick up a spinning load Many ac drives are expected to catch a motor spinning with load at some speed and direction, and take it to the actual commanded speed and direction. This means that the drive has to be able to determine the present motor speed instantaneously, determine the direction, and then begin its output at that speed. The drive then has to reaccelerate or decelerate the motor to the desired, reference speed. Applications with large, high-inertia fans are good applications for this function. A large fan blade can take several minutes to come to a rest after running full speed. The ac drive can start right into this coasting motor.

Motor stall prevention There are some instances where a motor can get into a temporary overload condition. The drive's normal operation is that it wants to protect the motor and shut down. This can be a nuisance fault and might not be acceptable. Many fan systems that begin moving cold air experience these overload conditions. But once the air warms, the motor becomes less loaded and will continue to operate as required. This function allows the drive to lower the output frequency until the output current levels begin to decrease. In this way the drive will ride through the overload condition without tripping.

Maintenance of ac drives

NOTE: All input power must be removed from any ac drive before doing any maintenance. Also, live voltages can exist in drive capacitors. Check to see that all energy has dissipated. Also, if a lock-out mechanism is in place on the input power line, it should be used before doing any maintenance.

A routine inspection should be done on an ac drive every 3 to 4 months, and more frequently if the environment around the drive is very dirty. The inspection and subsequent maintenance should include the following:

1. Visually check the drive for any dust buildup, corroded components, and any loose connections.
2. Using a vacuum cleaner with a plastic nozzle to minimize damage to components while cleaning, brush dusty and dirty devices. Particular attention should be given to the rectifier

heatsink areas. Excessive accumulation of dirt and dust will eventually lead to an adverse overheating condition (probably at a time of peak production). Cleaning the drive will prevent these types of problems.

3. Clean and retighten any loose electrical connections.
   a. All fuse ends and bus bar connections should be cleaned according to the manufacturer's instructions.
   b. In the case of screw pressure connectors, the power connections should be checked often and retightened if necessary during the first few weeks after installation.
   c. If any components are unfastened during the maintenance procedure, the mating surfaces should be treated with the proper joint compound (per the manufacturer's recommendation) before replacing. After replacing any power modules subject to certain torque values to their heatsinks, it is suggested that another check be made on the tightness of the screws.
4. Touch up any exposed or corroded components with paint.
5. If the ac drive is mounted within an enclosure, check the fans and filters. If necessary, clean or replace the filters and replace the fan if the shaft does not spin freely.

Troubleshooting and repair of ac drives

Many ac drives are manufactured in a similar fashion to one another. Surface mount device (SMD) technology has helped make drive circuit board integrity very good, and many drive manufacturers are using this technique. Basic power conversion, available devices used, and standard wiring and protection techniques can also be assumed to be similar from manufacturer to manufacturer. However, control schemes, gate firing circuits, packaging, and so on will definitely vary from drive to drive. Therefore, consulting the manufacturer's installation, maintenance, and repair manuals is mandatory when performing any work on ac drives. Especially with the numerous parameters that can be set on digital drives and the extensive diagnostics provided, the manual will be the only place that will define many of the displays.

This section will discuss standard and common techniques along with practical applications and problem solving. Again, a warning about working on electrical equipment: **Electrical shock can cause serious injury or even death. Remove all incoming power before attempting to do any work to any electrical device!** This is another reason to have the manufacturer's manuals on hand; they usually can take the repair person step-by-step, safely, through the procedure. Whenever possible, work with another individual. Also, when using an oscilloscope be aware that it, too, must be properly grounded.

Many times in the heat of the moment, when the drive has quit running and the factory is at a standstill, it is usually much faster to just replace the suspected bad component with a brand-new component from the spare parts inventory. Hopefully, there will be modules, boards, and plenty of fuses on hand. Parts can be usually ordered and sent via overnight mail, but nothing is better than disconnecting the bad device (work that has to be done even in a repair situation), going to the stockroom, pulling the spare device, and reinstalling it.

Every manufacturer of ac drives will have a recommended spares list of those parts that should be kept on hand in the event of a drive failure. The best time to purchase these parts is at the time of drive purchase, when buying power is at its peak. After that, the price of the spare parts becomes two to three times what they should have been. Besides, not having them when they are needed most results in lost production, not to mention the costs of having parts overnighted.

For a typical ac drive, various fuses at the ratings seen throughout the drive should always be kept on hand. Because they protect the bulk of the drive's components from incoming power surges, input fuses and branch fuses are particularly important. These fuses do not have to be purchased through the drive supplier as long as the current ratings and sizes for the recommended fuses are matched. Power modules, transistors, or SCRs should be stocked as spares in groups of three, six, or twelve (for three-phase drives), obviously in the exact ratings for the particular drive. Initially, these should be purchased through the manufacturer because there might be some characteristics inherent to the devices being used. The devices might have certain turn on/turn off values that might cause problems later if not matched accordingly. When an SCR or transistor fails, it is common practice to replace the other devices in the leg of the circuit because they might have become stressed and prone to quick failure.

Other components to stock as spares are the printed circuit boards for the drive. Unfortunately, you can only purchase these boards from the drive manufacturer. The control boards and gate firing boards are the most common to stock, but consult with the drive manufacturer for any others. Also, consider electrostatic discharge

(ESD) when handling any printed circuit boards. A grounding wrist strap is ideal but rarely practical when servicing an out-of-service drive. The next-best precaution is to minimize handling of the boards and to have an ESD-protected bag available. Don't touch small components on the board. Use one hand whenever possible to keep a voltage arc from forming, and get the board into the bag as quickly as possible.

There might be some suppression components in place to minimize further component damage in the event of surges and these can be stocked as spares but can be found in many electrical hardware supply stores. They are the capacitors and resistors in the snubber network. Power supplies are other items that sometimes fail; these are usually not manufactured by the drive supplier. Determine whether or not keeping these items in stock is worthwhile.

Finally, to round out the necessary spare parts in the drive's inventory, any fans, feedback devices, special boards (for feedback, etc.), and any temperature sensors probably should be kept on hand, particularly if the drive runs 24 hours per day, seven days per week. As another precautionary measure, continue to train various technicians in the plant on how to troubleshoot, replace, and repair the drive components. This training must be specific to the drive in place and can usually be performed by the drive supplier.

Whenever the drive manufacturer can provide the field service, training, and additional required support, it is always practical to take advantage of these services. However, they are usually very expensive, and not everyone in the plant can become an expert on every drive. That is one reason why plants try to standardize on one or two particular drive manufacturers. Keep in mind, however, that drive manufacturers change their designs over time, and the user must often relearn the product anyhow. The following sections will attempt to cover many of the common problems, faults, and diagnostics that are inherent to most ac drives. Many drives have on-board diagnostics that can help greatly in tracking the problem. The drive manual, which should be stored in a pocket on the drive door and not in someone's office, should be consulted.

When a drive is down, the first thing to do is to kill incoming power and lock it out. Be wary of drives with capacitors, which can carry charges for a period of time. Many drives carry an LED that indicates, when it's on, that the capacitor charge has not been fully bled off. Be careful! Upon opening the drive cover or door, look for any apparent signs of internal component damage.

That smell of fresh electrically produced smoke is a telltale sign. Next, look for any burnt wiring or components. Some components might even swell, especially the capacitors. Any loose or disconnected wires should be noted, as should any marks on the drive enclosure walls from apparent electrical arcs. Once the visual inspection is done, continue troubleshooting and tracing from the suspect point.

Check all input fuses to see if any have blown. Double-check with an ohmmeter if the blown fuses are not visually apparent. Either way, you must determine whether the problem lies before, at, or after the fuse section. Many times the fuses are blown, but there is still a second or third problem with the drive. This is where the fun of troubleshooting begins. Again, some drives have the capability of saving multiple trips and faults. If possible, use this tool.

Since most ac drives contain diodes and transistors, it is common to suspect these devices whenever there is an apparent problem. Many drives have the capability to display which transistor has failed to turn on or has been forced off abnormally by excess current. One probable cause might be that a high-inertia load has been accelerated too quickly. In this case, merely setting a longer acceleration time will solve the problem. Another cause could be the voltage boost. While a handy tool for hard-to-start loads, it is also sensitive. Reducing its value will overcome motor losses and keep the transistor from failing. The last probable cause, a failed transistor, is certainly not the easiest to correct.

In order to determine if a transistor module is faulty, we have to measure the resistance across its terminals with an ohmmeter. Measuring normally with the one times range, we can get readings that, when compared to the manufacturer's acceptable and nonacceptable levels, will indicate whether the transistor needs replacing. Transistors commonly come in modular form and are fairly simple to remove and replace. The same procedures should be followed for any of the diodes.

Common drive faults, possible causes, and solutions

Electronic ac drives today tell us much more than older drives. When the drive trips we can usually get a diagnosis from the unit itself. Some even provide the actual text on a terminal screen that tells the troubleshooter where to look. Many drives trip for the same reason, but the solution might mean looking at a certain terminal or adjusting a certain parameter for that particular drive.

Therefore, as always, consult the drive manual. The common drive trips (sometimes called alarms or errors) are as follows:

The motor will not run

Possible causes: No line supply power; output voltage from ac drive too low; stop command present; no run or enable command; some other permissive not allowing drive to run; drive could be faulty.

Solutions: Check circuit breakers and contactors on input power side.

Check to see if stop command is absent when trying to run motor. Check for run, enable, or other permissives being present. Try another ac drive if the one in place does not respond.

Low line input, undervoltage, or power dip

Possible cause: Input fuse(s) blown; low input voltage; power outage.

Solutions: Check and replace fuse if blown; check to see that input voltage is within drive controller's allowable high and low range; check for possible problems with utility; sometimes lightning can cause this fault.

Overtemperature

Possible cause: Heatsink temperature sensor has caused trip; cooling fan has failed; drive ambient temperature has risen to extreme levels.

Solutions: Check thermistor with ohmmeter. Replace if bad. Check whether heatsinks are dirty. If so, clean. If drive is in an enclosure, check whether filters are clean. If not, clean or replace. If there is a cooling fan, check whether it is operational. If not, repair.

Overcurrent or sustained overload

Possible cause: Incorrect overload setting; motor can't turn or is overloaded.

Solutions: Check overload parameters and change to fit motor loading. Load might be too large for motor; a larger motor might be required or the gearing might need to be changed. If there's a jam in the drivetrain, clear it.

Motor stalls or transistor trip occurs

Possible causes: Acceleration time might be too short. High-inertia load. Special motor.

Solutions: Lengthen acceleration time. Readjust volts-per-hertz pattern for application.

Short circuit or earth leakage fault

Possible causes: Motor might have a short circuit; power supply might have failed. Excess moisture in inverter or motor.

Solutions: Check motor and cables for short. Have motor repaired. Replace power supply.

Overvoltage

Possible causes: The dc bus voltage has reached too high a level.

Solutions: Deceleration time is too short for load inertia; supply voltage too high; motor might be overhauled by load.

Main fuse(s) have blown

Possible causes: Bad transistor(s) or diode(s); short or bad connection in power circuitry.

Solutions: Check devices and replace if defective; trace and check connections in power circuitry.

Speed at motor is not correct; speed is fluctuating

Possible causes: Speed reference is not correct; speed reference might be carrying interference; speed scaling is set up incorrectly.

Solutions: Ensure that speed reference signal is correct, isolated, and free of any noise. Filter noise online and/or eliminate noise. Recalibrate speed scaling circuit.

Peripheral relays or control communication circuits tripping

Possible causes: Drive carrier frequency too high; other electrical noise present.

Solutions: Lower carrier frequency. Eliminate electrical noise.

Certainly other types of faults could exist for any given ac drive and motor installation. Digital ac drives can always contribute the ever-deadly CPU failure or show no response at all when the main microprocessor board is not functional. This is sometimes called a watchdog timer fault with the main control board. Also with digital drives, if the drive can automatically tune itself with the motor and an operational error is detected, the auto-tune sequence is aborted and an alarm is indicated. Additionally, if any drive controllers are set up to accept other external feedback and it is not present, the controllers might trip upon powering up. Also, motors can overspeed or run away, but usually the drive controller can be set up to trip at a maximum speed for protection.

Flux vector drives (ac)

The precise, accurate control of motor torque is necessary for good transient and dynamic performance in any electronic ac drive system. The performance of a dc drive, that which would have a dual converter, can be obtained with a standard ac induction motor and a drive with the proper modifications. In the dc motor and drive system, magnetic flux is produced by the field winding and controlled independently from the armature, which is busy producing torque. With the ac induction motor, these same qualities can only be controlled through the windings in the motor's stator. The interaction of motor CEMF (counter electromotive force) and flux produces torque in the motor. CEMF is produced as the primary current flows through the torque-producing and flux-producing branches of the ac motor. The flux-producing current is often referred to as the magnetizing current.

The vector control principle states that the magnitude and phase of the ac motor's stator current vector is to be controlled by producing constant magnetic flux while generating the necessary component of torque-producing current. This electronic ac drive is called the *flux vector drive*. It takes complete control of the motor. Today's flux vector drive does this by using special algorithms, high-speed microprocessing, and digital feedback from the motor itself. Often an ac flux vector system (motor and drive controller) can outperform an equivalent-horsepower dc system from a speed and torque regulation standpoint.

Figure 6-27 shows a simple ac motor diagram. This diagram shows the vector relationship to the torque angle. The torque angle mathematically is the arc-tangent of the torque vector divided by the magnetizing vector. It is shown in figure 6-28 as it relates to the magnetizing current, I_m. Keep in mind that these events are taking place within the air gap of the induction motor. Physically within the motor, the torque-producing current is trying, through all load changes, to remain 90 degrees from the magnetizing current. The magnetizing current vector is continuous and completely independent of the torque vector. Often in dc drives there can be discontinuous current. This makes the continuous current feature of the ac flux vector drive attractive.

The flux vector drive incorporates the basic ac power bridge, converter and inverter sections, and some extra control algorithms.

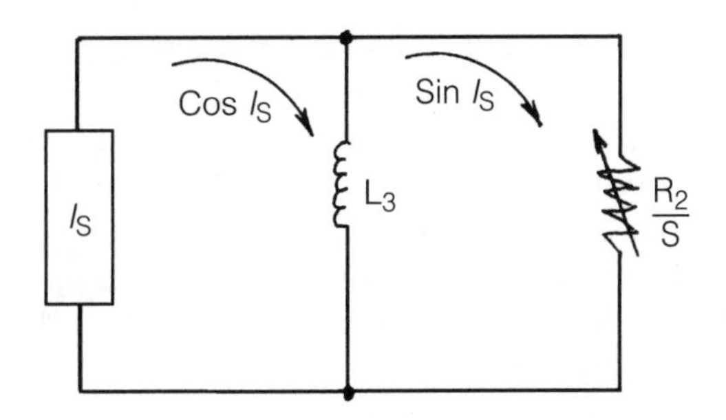

■ **6-27** *Simple ac motor circuit.*

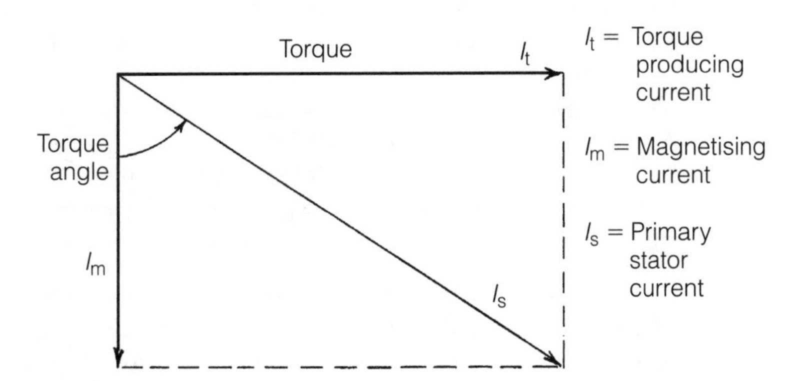

■ **6-28** *The torque vector.*

Therefore, any design of ac drive could be modified into a flux vector version. Pulse width modulated drives, current source drives, and others have been made into vector packages. The power and performance of the individual vector design is in the control circuitry, software, and integrity of the feedback device. The direct access to the torque-producing vector even allows the vector drive to function as a torque reference follower, or torque helper. A torque reference signal can be sent from the drive, or the drive can receive a torque command.

By attaching a pulse generator to the motor's rotor and feeding this signal back to the drive, motor slip information and rotor position can be determined. This feedback signal is crucial to the flux vector drive's ability to maintain tightly regulated speed and torque. If this signal is lost (i.e., if the encoder fails at the motor), then the drive might fault. Make sure the encoder is securely and properly mounted to the motor. This encoder signal is compared to the commanded speed. By knowing what the motor slip is, a precise speed correction can be made in the form of new voltage output to the motor. This is made possible by high throughput microprocessors that crunch the data and create a newly corrected output continuously. Thus, our flux vector drive is constantly correcting as the motor runs, much like an open-loop drive, but on a higher level. This feedback now makes the flux vector drive a closed-loop drive.

In this chapter much importance has been placed on the flux vector drive's control algorithms for torque and speed control. Also of crucial importance is the feedback device. The last, but perhaps most important, component of the flux vector ac package is the induction motor. Attention must be given to the ac motor, which is being controlled by the flux vector drive. Its nameplate and test data should be available so the correct parameters can be entered into the drive. Also, the motor, physically, must be equipped with a pulse generator and most likely an auxiliary blower for cooling at low operating speeds. Many times it is better to procure the motor from the same supplier as the ac flux vector drive to ensure compatibility.

However, many times the motor needing vector control is already existing in the field, and test data has long since disappeared. Sometimes even the nameplate has been destroyed. Now the task of getting pertinent motor information into the vector drive is more of a challenge. This situation is not as bad as it seems, however. Actual current measurements to show no-load current (which is actually magnetizing current) and full-load currents are going to be accurate values. The motor has to be modified to accept a pulse generator anyway, and the motor's inductance and reactance values will have to be predicted.

All this information now gets entered into the flux vector drive's memory, and the experimentation can begin. In short order, the drive and motor will have become matched. Running the motor/drive system over a wide range of speeds and loads will satisfy the complete range of operating parameters. Therefore, in many instances a standard ac drive can be physically modified in the field to become a flux vector drive. This usually can be avoided if the upfront application evaluation is done.

Thus, the world awaits the next-generation flux vector drive. This drive can perform as well as traditional dc drive and motor systems—with certain limitations. One major limitation is that most ac drive power bridges are nonregenerative. Therefore, if the application requires continuous regen capabilities from the drive, a dc SCR-based package might be in order. However, ac drives can be fitted with dynamic braking resistors whenever braking is an intermittent need.

Another application requirement is to have a flux vector drive without having to attach a pulse generator to the motor. Today, ac drives attempt to do this. Called voltage vector drives, they require no additional feedback. These drives make predictions of what is going on at the motor under load changes. At the least, they must have as much information about the motor in order to make these predictions. The slip of the motor must be furnished into the drive's memory so that these predictions can work when the new output from the drive is tempered.

Following are a few additional remarks concerning flux vector drives:

- ☐ *Speed regulation.* A flux vector system can usually provide speed regulation to 0.01%.
- ☐ *Torque at zero speed.* A flux vector drive and motor can provide full holding torque with the shaft at standstill. This means that full current is being provided to the motor and there had better be some auxiliary means of cooling the motor!
- ☐ *Torque linearity.* Some applications require smooth torque throughout the speed range even during acceleration and deceleration. Torque pulsations can actually leave marks on material. The flux vector drive exhibits smooth, linear torque regulations continuously, regardless of loading.
- ☐ *Speed and torque requirements.* Determine, beforehand, the speed range and the actual torque requirements, especially at the low speeds. Also, does the load change frequently? The answers to these issues will help you select the proper size and type of controller to use.
- ☐ *Forward and reverse deadband.* The ac flux vector drive has no deadband and can reverse direction quickly. This is done electronically and requires no contactor. The better question is whether the load can be stopped quickly enough to reverse.
- ☐ *Troubleshooting the flux vector drive.* The flux vector drive is built from the same blueprint as the standard open-loop ac drive. Therefore, many issues and concerns are similar. Troubleshooting the flux vector drive will be very similar to the ac drive with some subtle differences.

Common misconceptions about flux vector drives are as follows:

- ☐ *We can get more torque out of the flux vector drive.* Not necessarily so. A flux vector drive provides exceptional torque control but will not create more current (needed to produce more torque) for a given horsepower sized drive. In other words, a 25-A rated ac drive will produce 25 A continuously to a motor (and sometimes 150% for short periods, such as a minute), whether or not it is a flux vector drive.
- ☐ *The ac flux vector drive systems cost more than standard ac drives.* This is true, so be sure the application really justifies the extra cost. The major contributors to the cost increase are the motor with auxiliary blower, the added control and feedback modules, and the pulse generator.
- ☐ *The encoder is a minor component in the system.* Not so. In fact, it can shut down the entire process. Affixing the feedback device concentrically with no misalignment and with a good, flexible coupling is most critical. Take special care in mounting these devices.

Troubleshooting and diagnostics of flux vector drives

The ac flux vector drives can be diagnosed and repaired in much the same matter as the base electronic drive from which it is built. However, critical attention must be given to the other elements in the flux vector package. For instance, the motor running from the vector drive's output might have other features that could cause a problem in the circuit (encoder mounted, auxiliary blower, special bearings, etc.). In addition, the flux vector drive might have more interaction with host computers or PLCs, causing troubleshooting to become more extensive. Begin troubleshooting with the basics and, as always, have the manufacturer's manual on hand for the particular vector drive in question.

Flux vector ac drives today tell us much more than the drives of yesterday. When the drive trips we can usually get a diagnosis from the unit itself. Many drives trip for the same reason, but the solution might mean looking at a certain terminal or adjusting a certain parameter for that particular drive. Therefore, always be sure to consult the specific drive manual. Common drive trips are as follows:

Motor runs at low speed only

Possible causes: Encoder signal has been lost; maximum and minimum speed settings might be incorrect; reference signal incorrect (noisy).

Solutions: Check and repair encoder if needed; if signal is lost, most vector drives default to a minimum speed. Check speed settings. Check speed reference signal for accuracy and integrity.

External fault condition not allowing vector drive to run

Possible causes: There is a problem with interlocks or permissives outside of the vector drive's internal circuitry. Vector drive main control board is bad.

Solutions: Locate external problem and fix. If all external signals and contacts are good, replace vector drive's main control board.

The motor will not run

Possible causes: No line supply power; output voltage from ac drive too low; stop command present; no run or enable command; some other permissive not allowing drive to run; drive could be faulty.

Solutions: Check circuit breakers and contactors on input power side. Check to see if stop command is absent when trying to run motor. Check for run, enable, or other permissives being present. Try another ac drive if the one in place does not respond.

Low line input, undervoltage, or power dip

Possible causes: Input fuse(s) blown; low input voltage; power outage.

Solutions: Check and replace fuse if blown. Check to see that input voltage is within drive controller's allowable high and low range. Check for possible problems with utility; sometimes lightning can cause this fault.

Overtemperature

Possible causes: Heatsink temperature sensor has caused trip; cooling fan has failed; drive ambient temperature has risen to extreme levels.

Solutions: Check thermistor with ohmmeter. Replace if bad. Check whether heatsinks are dusty and dirty. If so, clean. If drive is in an enclosure, check whether filters are clean. If not, clean or replace them. If there is a cooling fan, check whether it is operational. If not, repair it.

Overcurrent or sustained overload

Possible causes: Incorrect overload setting; motor won't turn or is overloaded.

Solutions: Check overload parameters and change to fit motor loading. Load might be too large for motor; a larger motor might be required or the gearing might have to be changed. If there's a jam in the drivetrain, clear it.

Motor stalls or transistor trip occurs

Possible causes: Acceleration time might be too short; high-inertia load; special motor.

Solutions: Lengthen acceleration time. Readjust volts-per-hertz pattern for application.

Short circuit or earth leakage fault

Possible causes: Motor might have a short circuit; power supply might have failed; excess moisture in inverter or motor.

Solutions: Check motor and cables for short; repair motor. Replace power supply.

Overvoltage

Possible causes: The dc bus voltage has reached too high a level.

Solutions: Deceleration time is too short for load inertia; supply voltage too high; motor might be overhauled by load.

Main fuse(s) have blown

Possible causes: Bad transistor(s) or diode(s); short or bad connection in power circuitry.

Solutions: Check devices and replace if found defective. Trace and check connections in power circuitry.

Speed at motor is not correct; speed is fluctuating

Possible causes: Speed reference is not correct; speed reference might be carrying interference; speed scaling is set up incorrectly.

Solutions: Ensure that speed reference signal is correct, isolated, and free of any noise; filter noise on line and/or eliminate noise; recalibrate speed scaling circuit.

Peripheral relays or control communication circuits tripping

Possible causes: Drive carrier frequency too high; other electrical noise present.

Solutions: Lower carrier frequency. Eliminate electrical noise.

Load commutated inverters (LCIs)

Load commutated inverters are used extensively with synchronous motors and are commonly found in high horsepower and lower speed packages. The synchronous motor's speed is directly proportional to the stator frequency. Whenever a load is introduced to the synchronous motor's shaft, a corresponding torque is developed immediately to maintain stability. Because the motor can be operated at a leading power factor, its inverter can function as a load commutator instead of using forced commutation (extra components required).

Synchronous motors have wound fields and incorporate a shaft-mounted brushless exciter for dc excitation (figure 6-29). Since the load commutated inverter does not have to have the elaborate commutation circuitry typically found in most ac drives, it becomes attractive as an efficient, simple, and reliable inverter. One slight disadvantage, though, is that the LCI cannot properly load commutate at speeds below 10 to 15% because there is not enough CEMF from the motor. However, by utilizing a shaft-mounted feedback device, a fourth commutation circuit can be introduced along with dc link current to provide commutation via thyristor firing.

The LCI is commonly used in many compressor and pump applications, 1,000 to 20,000 horsepower and at medium voltages of 2,300 V, 4,160 V, and 6,900 V. It can eliminate the need for a large gearbox, thereby saving space and initial installation costs. These types of drives, along with their synchronous motor, must be evaluated for the application; they are not well suited for frequent starts per hour. Additionally, a vibration analysis and special motor coupling might be a good idea for these synchronous motor installations.

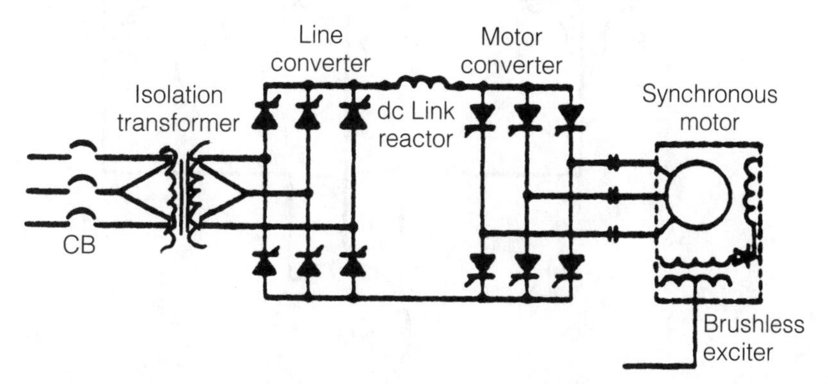

■ **6-29** *Load commutated inverter (LCI) circuit with synchronous motor.*

Cycloconverters

The cycloconverter has been around since the 1930s. The railroads used a version of the cycloconverter that had mercury arc rectifiers. Since then, because of the complexity of their circuitry and their cost, cycloconverter use has been sparse. Cycloconverters can be found in very large horsepower applications that require regeneration, frequent and severe load changes, and where lagging power factor is allowed. Many cycloconverter installations are also in place to step the line frequency down to ½ or ⅓ (e.g., running a plant's 30-Hz motors instead of replacing all the motors with 60-Hz units).

The cycloconverter is truly an interesting electrical device. It can control an induction motor, and the single-line diagram is shown in figure 6-30. Particularly interesting is that the cycloconverter "skips a step" in the traditional ac drive fundamentals. It takes ac line power directly to operating frequency without going through the converting-to-dc, inverting-back-to-variable-frequency-ac process. Figure 6-30 shows a three-phase, three-pulse cycloconverter with 18 SCRs.

Another interesting fact is that the cycloconverter circulates no current between positive and negative thyristors. The operation is

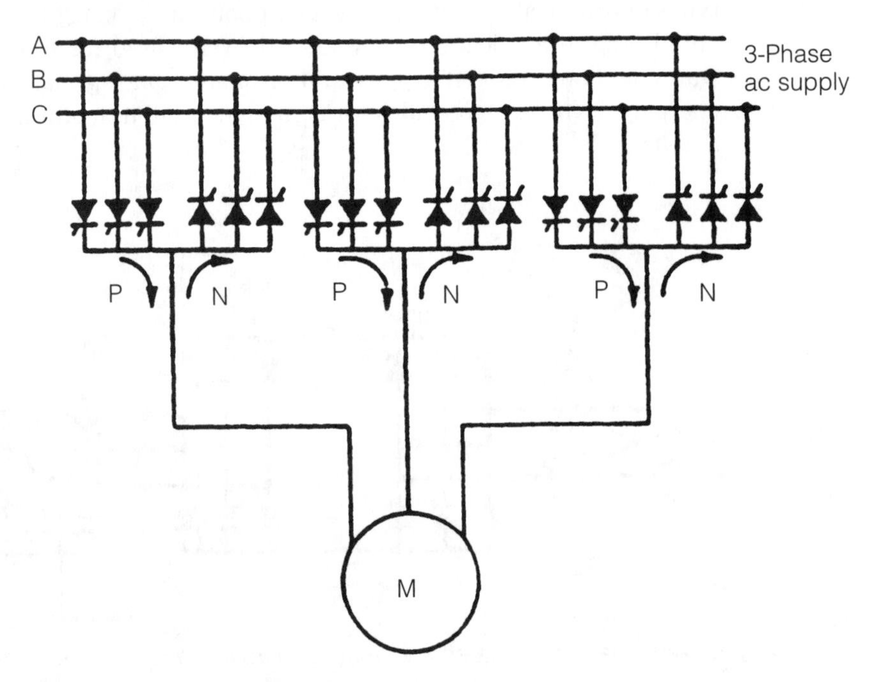

■ **6-30** *The cycloconverter.*

line commutated, phase controlled with firing angles on the power semiconductors modulated to simulate sine wave voltage. When this output is introduced to the motor, a nearly sinusoidal current with lagging power factor results.

Closing remarks on electronic ac drives

In the past five years, electronic ac drives have come down so much in cost that if there is an ac induction motor without a drive on it, that facility should seriously consider installing one. The ac drive can provide energy savings, and rebate programs are frequently offered. An ac drive will protect the motor from overloads and keep it from burning itself up. Another attractive feature is soft starting. This type of starting can add years of life to the ac motor and can also save wear and tear on the mechanical system to which the motor is attached. Also, if the application calls for a lower speed, the quickest and smartest way to vary motor speed is with an ac electronic drive. Another reason to consider an ac drive is that its unity power factor is in the high 90s. This alone could prevent the utility from penalizing your facility.

There are many very good reasons to incorporate an ac drive onto an ac motor these days. With surface-mount technology and several drive competitors vying for business, the user is presently in the driver's seat!

Another benefit of the ac motor over the dc is lower maintenance. Since there are no brushes or commutator, there is no need for cleaning or replacing them. The motor can go into dirty, hostile environments and run where the dc motor and its brush design would falter. The maintenance and downtime costs of the dc motor make considering the ac motor easy. Other advantages to going ac include size and environment. Because the ac motor is smaller, it can fit in confined spaces. It can also be used in environments where nonsparking motors are required. The ac motor can be overloaded, within reason. Multiple ac motors can be run simultaneously from one ac drive, sized to handle the total full load currents of all the motors (figure 6-31).

Another major feature of the ac drive is its ability to be bypassed in the event that there is a problem with the drive. Since an ac motor is being used that might have run full voltage and at line frequency (60 Hz) prior to the ac drive, a bypass system can be installed with the ac drive or even later. In an emergency, the ac motor can be run at full speed, as in the bypass scheme shown in figure 6-32. Contactors must be in place to isolate the drive from the supply line and the mo-

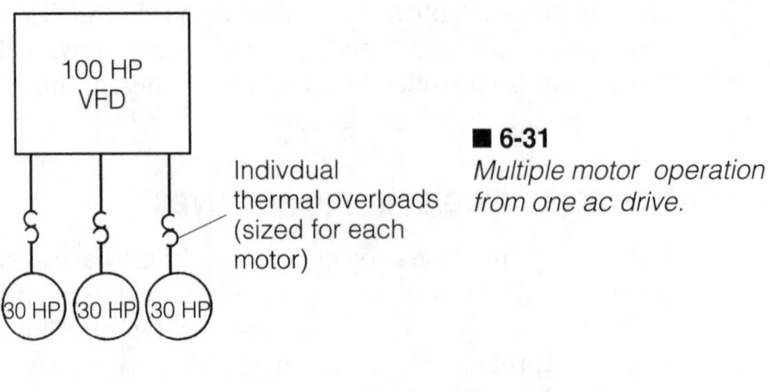

■ **6-31** *Multiple motor operation from one ac drive.*

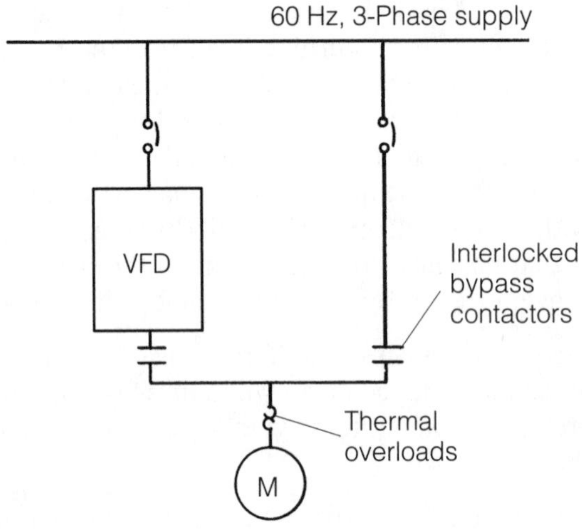

■ **6-32** *ac drive and bypass.*

tor whenever the motor is running in bypass and the converse is true whenever the motor is running off the ac drive. Also, motor overload protectors are recommended whenever the ac drive is not in control, electronically protecting the motor.

Throughout this chapter we have seen that electronic ac drives pay for themselves, save our ac motors and other drivetrain components from excessive wear, and give us process control that we never had before. Installing the ac drive is also not as troublesome as it once was. There is also an excellent historical base of ac drives in service, for many years, without incident. The reliability and integrity of the product is very good. And, since more people understand them better, misapplication has diminished. In the future, ac drives will be even more affordable and residential use will increase dramatically. Look for them in your local hardware store someday!

7

Servo, stepper, and specialty drives

SERVO, STEPPER, AND SPECIALTY DRIVES ENCOMPASS A class of drives used for high-performance, low-horsepower, and positioning-type applications. Servos and steppers are most often used in machine tool and robotic applications and thus have their own mystique. This chapter gives an overview of these intriguing machines.

Servo drive systems

Up to this point we have looked at various electronic ac and dc drives. Perhaps the electronic drive system that is the most versatile and can be used on the widest variety of applications is the servo. However, it is also potentially the most expensive and most complicated for comparable horsepowers. Also, servo systems typically do not allow for horsepowers above 50 HP.

A servo, or servomechanism, in its basic sense can be defined as an automatic device that can control large amounts of power by means of lesser amounts of power. Additionally, a servo system has the ability to automatically correct performance of a mechanism. A servo system can consist of a specific control component, a device to amplify signals, the device or actuator that receives the signal, and some type of feedback (figure 7-1). Servo systems are actually any programmable motion control systems; they can be found in many nonelectronic drive installations, such as hydraulic systems that use servo valves.

More appropriate to electronic drive controllers, servo systems are mainly comprised of a servomotor, amplifier, power supply, and controller (which many times closes the position loop). The permanent-magnet brushless motors are electronically commutated instead of using the conventional brushes and commutator. The amplifier is sometimes referred to as the drive, and it also can act

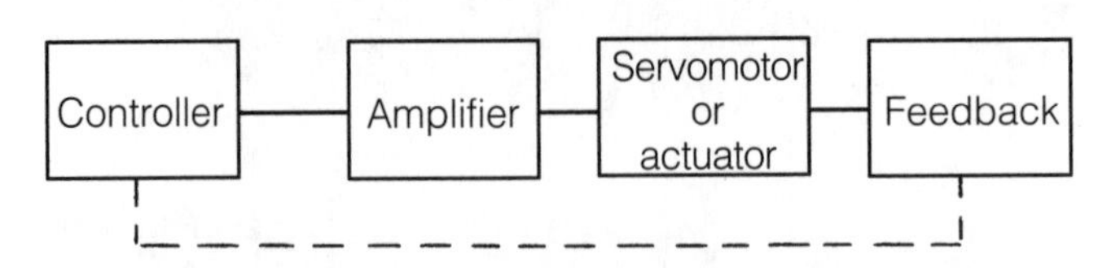

■ **7-1** *Servo system.*

as the positioner. Actually, the amplifier and controller together can equate to an electronic drive. The amplifier houses the power devices, and the controller houses the "smarts." The amplifier takes reference and feedback signals and outputs a specific, "amplified" speed and torque signal to the servomotor. The duration and timing of these output signals provide basic position control.

The last component is the power supply. Without this device, the amplifier cannot perform. We now have all the makings of a servo system. A further look at each of these components follows.

One of the first rules of speaking "servo-ese" is to discuss power requirements in terms of torque, not horsepower. This should be the rule for all applications, but often people seem to think of power, size, and capability of a motor system in terms of horsepower, which can lead to misapplications. Torque is the force that does the work as the motor shaft rotates. Servo people also talk in inch-pounds of torque. Sometimes the analysis can be made in foot-pounds; this is less common because servo systems usually are smaller, and foot-pounds of torque will have to be converted into inch-pounds or even ounce-inches later, anyway.

The upper limit in servomotor construction is around 800 inch-pounds. This is due in part to motor construction, the magnets, and the fact that, in high-torque applications, it is less expensive to provide a high-precision gearbox to get the output torque than to pay for a larger servomotor and amplifier. Also, as the power requirements get higher, high performance in positioning accuracy is not as critical. High-accuracy positioning is a big part of servo application.

Servo drive systems can be used to control speeds of motors down to a virtual crawl, with turn-down ratios of 10,000 to 1 and greater. Speed accuracy, or regulation, can be held to 0.01% or better. The response of the overall system, or bandwidth, is extremely fast. The servo system can also hold to very tight torque accuracy, meaning that as the load changes so too will the servo torque, or current regulator. It will correct quickly and can also provide more than adequate values of stall torque when its amplifier is sized accordingly.

A servo system is just that—a system. More precisely, it is a motion control system that provides high-performance, closed-loop speed or position control, and it requires several integral components in order to work. The servo drive can control a brush-type motor or a brushless motor. The closed-loop control of a brush-type motor is not done frequently and not done well. Therefore, for our purposes we will discuss strictly brushless systems. Yet, this still causes some confusion, especially in terminology. There are brushless ac and brushless dc systems. They are basically the same. The brushless motor is a permanent magnet-type motor with no brushes. A better description might be that the motor is a brushless synchronous motor with permanent magnets.

The servomotor is sometimes referred to as an "inside-out" motor because of the wound stator and permanent magnet rotor. Servomotor construction has been discussed in more detail in the previous chapter on electric motors. Some servomotors utilize permanent magnets, while others are of the switched reluctance type. Whatever the name, it is the function and interactivity with the servo controls and feedback devices that make the servomotor work.

Servo amplifiers are classified much like their ac and dc drive counterparts—by their particular output. The brushless dc amplifier is a linear amplifier, which generates a trapezoidal waveform to the servomotor for commutation. This waveform is compared to the feedback from the motor. Likewise, the brushless ac amplifier generates an ac waveform to the servomotor for commutation purposes, which is also compared to the feedback from the servomotor.

The servo system control aspects can be broken down into three distinct groups: speed control, torque control, and position control. In addition, we should look at coordinated control later as a separate entity of servo systems. First, looking at speed control, we find that in figure 7-2, the goal is to accelerate to a required speed in a certain amount of time, remain at that speed for a certain amount of time, and then decelerate back to zero speed. Every motion control application will have this type of profile; however, times and speeds will obviously change. The critical accuracies to be held in the servo system for this type of profile would relate to the quickness in changing speed states and holding at required states for the desired time, all with load.

As we discuss loading to a servo system, we have actually entered into a second type of control scheme. This is illustrated in figure 7-3 as the torque profile. There is an accelerating torque component to the curve, a running torque component, a deceleration component,

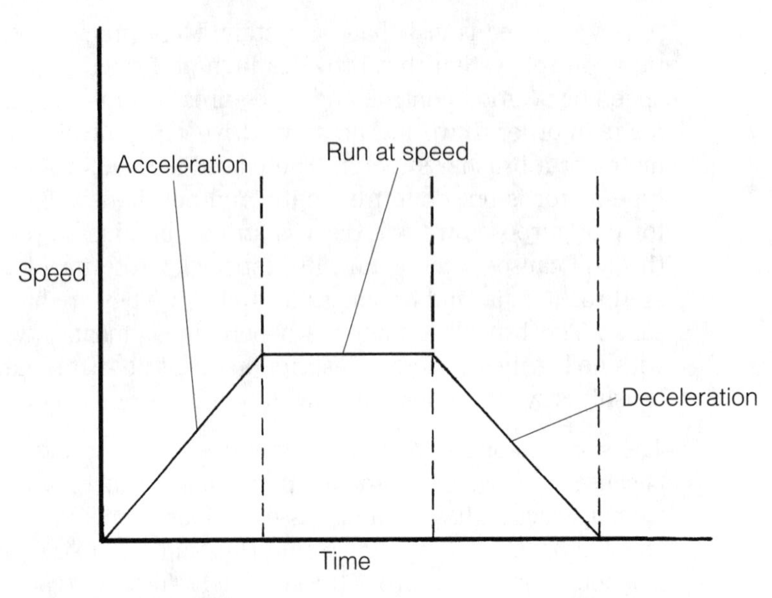

■ **7-2** *Velocity profile.*

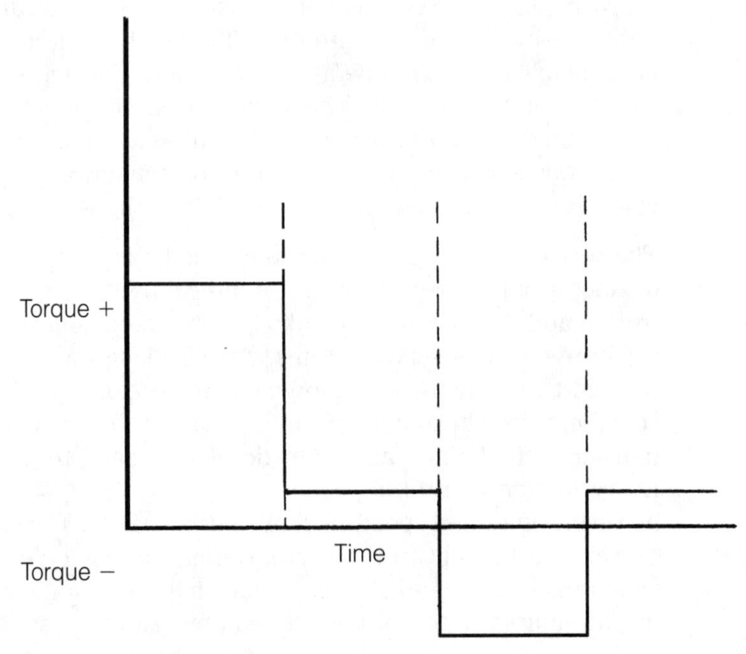

■ **7-3** *Torque profile.*

and a holding torque component (usually a zero speed). These, again, are based on time. Additionally, they typically relate directly to the velocity profile. This relationship is illustrated in figure 7-4. Therefore, we have a need for torque while we accelerate, run, and decelerate. A servo system's ability to maintain exact control of both velocity (the actual speed desired) and torque (driving the load) is its gauge as to accuracy and performance.

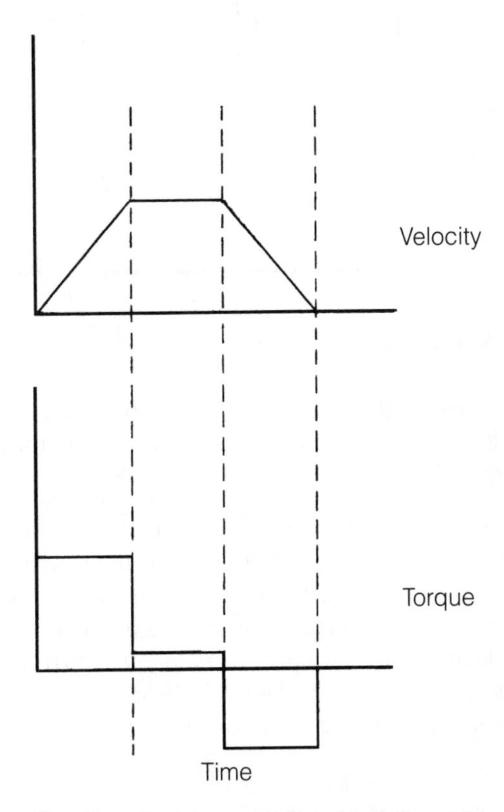

■ **7-4** *Velocity-versus-torque relationship.*

The last control loop is the position loop. While it is not always needed in a servo system with velocity and torque profiles, it is the control loop most often associated with servo control. A typical position profile indicating servomotor shaft position versus time is shown in figure 7-5. Depending on the control technique used for the particular application, the drive controller can incorporate velocity or torque inner loops if necessary. Most often these inner control loops are included, but it must be noted that for every control loop used, there is response time needed to perform that function. In servo systems, loop times are usually very critical to the

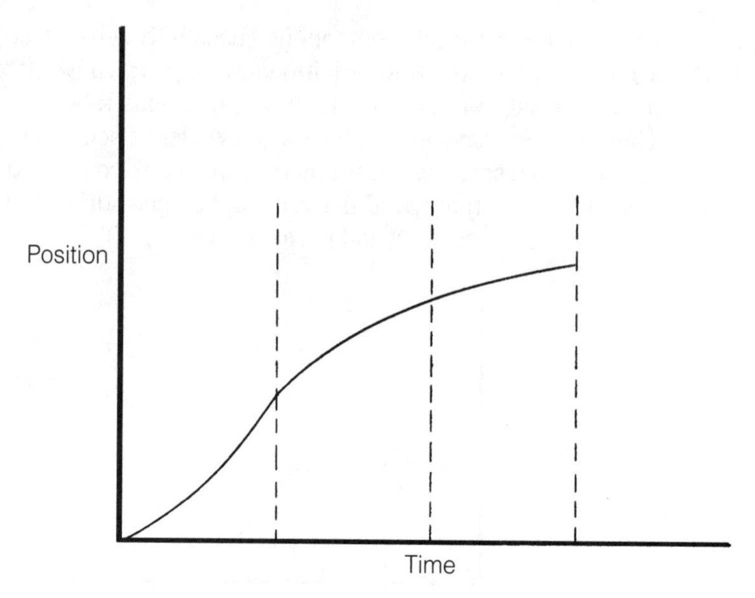

■ **7-5** *Position profile.*

success of the machine. The position loop is also the slowest of the three control loops. This is seen graphically in the block diagram in figure 7-6.

A servo system really gets dramatic whenever there are multiple servomotors to control. This is where the controller gets more complicated. It usually is comprised of a shared bus arrangement for attaching several servos, or axes to a common motherboard. In this manner, there can be a master controller telling the individual axes what to do, and each individual axis can solve for its own velocity, torque, or position loop (thus saving response time). A

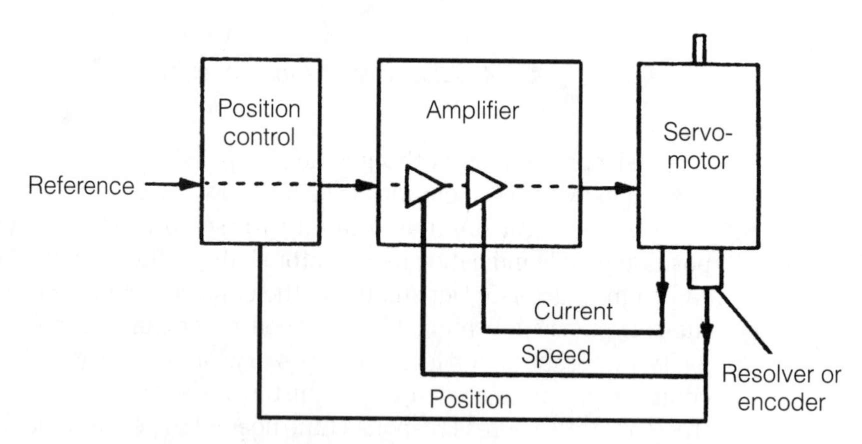

■ **7-6** *Relative speeds of various servo feedback loops.*

multi-axis, coordinated servo control scheme is shown in figure 7-7. These types of controllers are common in the machine tool industry where many servo drives and motors have to be controlled at the same time. In these types of systems the component count increases, but some devices, such as power supplies, can be sized for multiple loads and thus reduce the component count to a certain extent.

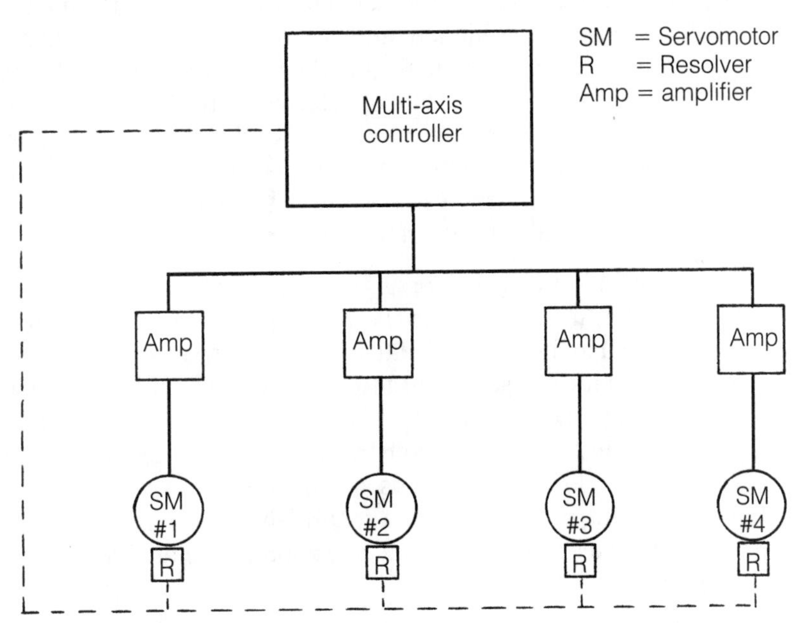

■ **7-7** *Multi-axis controller block diagram.*

In addition to the basic controller, power supply, amplifier, and servomotor, the peripheral components in a servo system include the following (some applications might not require all):

- ☐ Power cables from servo amplifiers to motors
- ☐ Feedback cables from motor to controller
- ☐ Isolation transformers (not always needed but recommended)
- ☐ Power supplies to provide appropriate bus voltage
- ☐ Braking resistors to handle regeneration and fast deceleration
- ☐ Heatsinks for the amplifiers and power supplies (some makes come built-in)
- ☐ Feedback devices, which might be the most critical component
- ☐ Resolvers, Hall-effect transducers, encoders, glass scales, tachometers, etc.

There are several other considerations when looking at the overall servo system. One is the system's bandwidth, the response factor of any drive system. Bandwidth is often described as the time a motor controller takes to correct from no-load to a loaded condition. It is usually expressed in radians per second, and while ac and dc speed control drives have to utilize bandwidth in their control algorithms, a servo system lives and dies by its bandwidth. It is a drive system that is installed by virtue of its performance. It usually costs more initially and is more complicated than a dc or ac drive system. Therefore, a higher bandwidth system is more immune to disturbances and performs better to produce better. One important factor associated with bandwidth is inertia, particularly load inertia and motor, or rotor, inertia. Other factors that affect system performance and bandwidth are backlash, belting, gearing, acceleration rates, overshoot, and torque ripple.

The fastest control loop is the current loop that is solving the new output, in amperage to the motor, based on actual motor current sensed in the servo amplifier by current transformer (referred to as CTs in most cases). It has to be the fastest because the load, a current dependent component of the system, has always got to be driven constantly and continuously. The next-fastest loop is the speed, or velocity loop, which needs speed feedback from the motor rotating element, or rotor. Either a separately mounted tachometer or similar device that can provide the appropriate pulses to translate into speed is needed.

The slowest loop is the position loop. However, good servo systems have position loop update times in the 5- to 10-ms range so as to hold to some very tight positioning accuracy. Of course, there are many variables that can affect each loop's performance such as temperature or sloppiness of the system mechanics; however, starting with a high response amplifier and controller is paramount.

The type of feedback devices used can vary. In a later chapter, feedback devices will be explored in greater detail. As for servo systems, the feedback device is an integral part. Some devices can provide feedback data that can be used to calculate new speed or new position requirements. An encoder is able to provide output in the form of pulses, which can be used to tell rotor position and also to count those same pulses for a given period of time for speed correction. Some encoders are absolute, which means they know their position upon power-up, while others are incremental.

Another device is the resolver, and its feedback must go through an R-D converter (resolver to digital) so as to be useful in the controller or amplifier. This R-D conversion also might be a limiting area in the servo system's resolution. Check the R-D's resolution as to how many bits that conversion solves to. It should be noted that encoder or resolver feedback can be accepted into a given amplifier or controller, but not all at once. Also, different manufacturers have different standards for their set algorithms and expect certain types of feedback into their products.

The stability of a servo system is dependent on how well the critical components are tuned and matched with each other. Many times this stability is found out during the initial startup of the drive equipment with the actual loads. This is when actual tuning of the gain circuits is best accomplished. As we have seen, any servo system's critical components are the servomotor, amplifier, power supply, and controller. These components accept some input signal and respond with some output signal. This relationship to signals can be referred to as the gain of the system. The gain is adjustable and will make the system's performance good or bad. Often called "tuning a system," adjusting the gains of the system determines such performance issues as smoothness, amount of overshoot, ringing, and oscillation.

Velocity gain is the system's speed error multiplied by this value to correct the loop (figure 7-8). The velocity loop gain provides a means of compensation whenever there is a wide range of load inertia found in the system. *Proportional gain* is used in positioning-type servo systems. This adjustment controls the overall response of the servo system and the magnitude of any positional following error. As seen in figure 7-8, the system positional error is multiplied by the proportional gain. *Integral gain* is more concerned with the positional error at zero speed. Thus, this gain is multiplied by the zero speed position error, and the net effect is that the motor stiffness is improved and the positioning accuracy of the moves is fully controlled.

The last gain, *derivative gain*, probably is the hardest to "tweak." This gain controls the dampening and ringing of the servomotor shaft during acceleration. The system positional error due to the position error rate of change is multiplied by the derivative gain.

Tuning the servo system is a straightforward task as long as all the system and machine components are proper, meaning that mechanical components are tight, backlash is minimal, and everything electrically is sound. Observing the machine performance with actual loading and making the necessary gain ad-

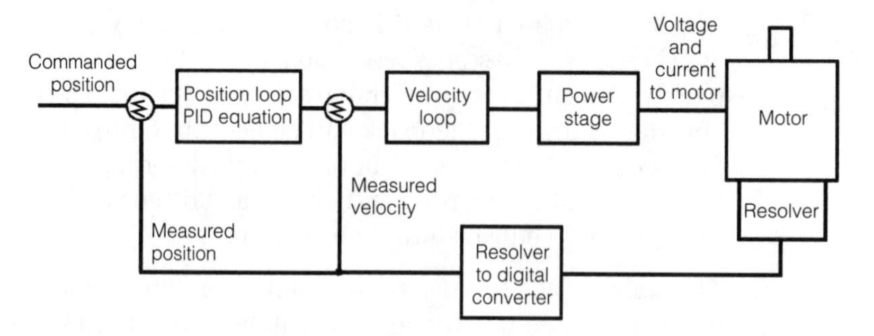

■ **7-8** *Typical gains in a servo system.*

justments in discrete increments is the method by which the most can be gained out of the system. However, advanced application sizing and attention to mechanical details will save time at startup.

A real-life example of a servo system with poorly tuned gains would be a robot whose individual axes are each servomotor-driven. The robot tries to move a delicate part from point "A" to point "B." Instability and unwanted mechanical vibrations can cause potential problems. The servomotors can overshoot, as shown in figure 7-9. This can lead into a condition called *ringing* where the servo system tries to correct itself but never quite does. A well-tuned servo system means that the critical components have been sized and matched with each other, have proper gain values programmed, and the user runs the system within the specified operating guidelines.

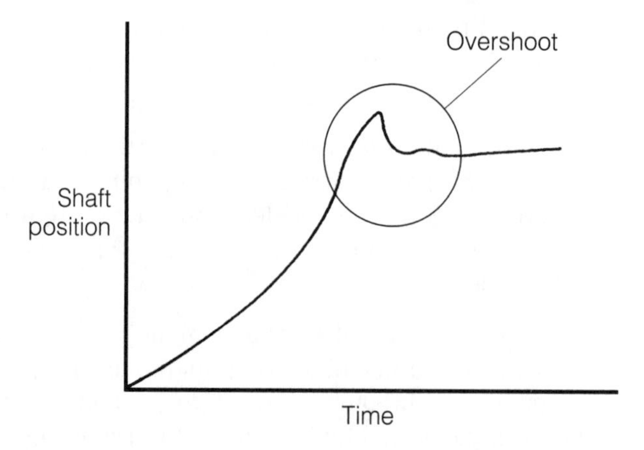

■ **7-9** *An example of overshoot.*

There are many other factors to consider when designing a servo system, as well as many other application requirements. Following is a partial list:

- ☐ What is the load, will it change frequently, and is it of high inertia?
- ☐ Is there a shock component associated with the load?
- ☐ How fast must the servomotor accelerate? Decelerate?
- ☐ What are the complete cycle times? (Heat dissipation and actual motor and amplifier size should be factored in.)
- ☐ If positioning, what is the positioning tolerance accuracy requested?
- ☐ What maximum and minimum speeds are needed from the servomotor?
- ☐ Is there known friction in the system?
- ☐ What is the load and machine inertia? How does this value compare to the rotor inertia of the servomotor? If acceleration is important, a good match is required. A one-to-one ratio is ideal for load inertia reflected back to the motor. A gear reduction scheme might be in order to better match that inertia.
- ☐ What inputs and outputs will be required of the servo controller? Discrete and/or analog?
- ☐ Communications now play an integral role. How will they apply in the application?

Matching the components in a servo system is always a challenge. There are many variables for a given application. Careful attention should be given to the inertia of the load being driven, the inertias of the mechanical drivetrain components, and that of the motor selected. This is called inertia matching and involves numerous calculations. The data is needed in order to match a servomotor to provide good accel and decel performance.

Figure 7-10 shows a gearbox and in figure 7-11, common gearing. Each figure contains formulas for calculating speeds, torques, and inertias. The symbols for these equations are defined in Table 7-1. Sometimes it is necessary to incorporate the gearbox or some gearing to greatly reduce the reflected inertia of the load back to the motor. As discussed earlier, inertia through any gear reduction can be reduced as a square function of the actual gear ratio.

Other common mechanisms found in servo applications include the rack-and-pinion arrangement (figure 7-12), the timing belt

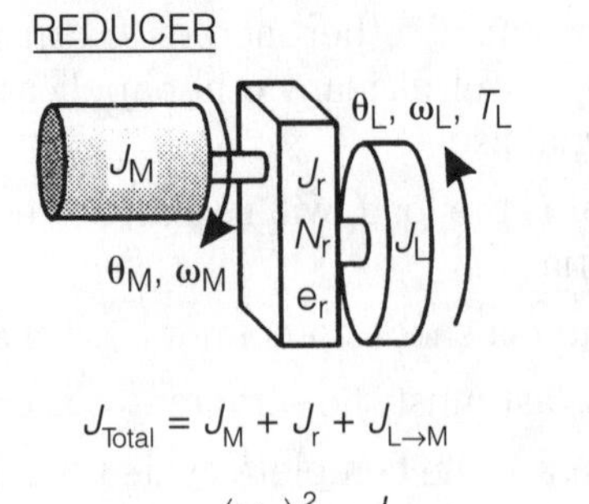

$$N_r = \frac{\theta_M}{\theta_L} = \frac{\omega_M}{\omega_L} \qquad J_{Total} = J_M + J_r + J_{L\to M}$$

$$\theta_M = N_r \times \theta_L \qquad J_{L\to M} = \left(\frac{1}{N_r}\right)^2 \times \frac{J_L}{e} \qquad T_{L\to M} = \frac{T_L}{N_r \times e}$$

$$\omega_M = N_r \times \omega_L \qquad J_r = \text{inertia of reducer reflected to input}$$

■ **7-10** *Typical gear reduction calculations.* Amechtron, Inc.

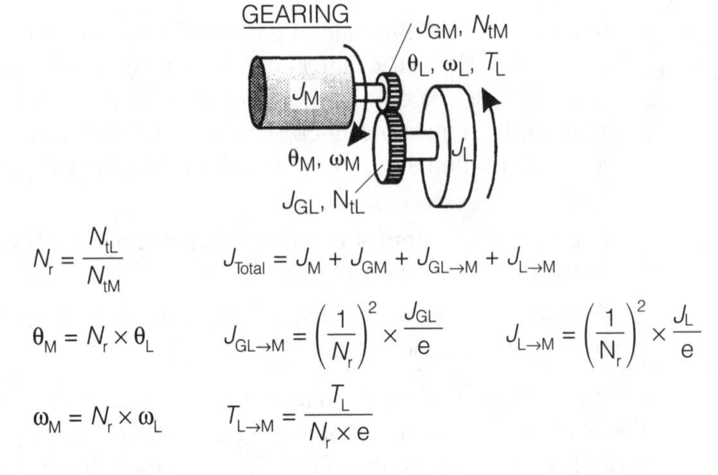

$$N_r = \frac{N_{tL}}{N_{tM}} \qquad J_{Total} = J_M + J_{GM} + J_{GL\to M} + J_{L\to M}$$

$$\theta_M = N_r \times \theta_L \qquad J_{GL\to M} = \left(\frac{1}{N_r}\right)^2 \times \frac{J_{GL}}{e} \qquad J_{L\to M} = \left(\frac{1}{N_r}\right)^2 \times \frac{J_L}{e}$$

$$\omega_M = N_r \times \omega_L \qquad T_{L\to M} = \frac{T_L}{N_r \times e}$$

■ **7-11** *Typical gearing calculations.* Amechtron, Inc.

(figure 7-13), and many conveyor systems (figure 7-14), and the most popular for servo applications, the lead screw (figure 7-15). The mechanics of these common systems have to be analyzed for every servo or stepper application. Likewise, when sizing and matching system components, the torque requirements must be considered. Peak torques and accelerating torques must be calculated along with the inertias. Once these values are known, a servo system can be pieced together. It might be better to leave component selection and the lengthy calculations to the manufacturers, since they routinely do this and can head off any glaring problems. Not to mention, this might make them responsible for the system's performance and integrity.

Servo systems and programmable motion control systems are most prominent in the machine tool and robotics industries. Grinders, lathes, milling, and boring machines are good applications for servo control in order to hold position for tight tolerances around a part. The same can be said of robotic applications. A servo system can be used for simple speed regulation, but much of its capability is wasted. Most likely, the cost of the complete system with all of its components dictates where it will be used.

Stepper drives

Generally speaking, a comparable stepper drive and motor system can displace an equivalent servo system for much lower cost. However, the performance has to be justified first. Often, a stepper motor system allows just that. A stepper drive system is basically similar to a servo system but often without the feedback. A stepper motor system, as shown in block form in figure 7-16, contains a step motor, an indexer, and the drive. The indexer can be substituted with a programmable motion controller, and the drive can be also called an amplifier. For higher accuracy requirements, a feedback device can be attached to the motor.

The stepper motor system works like this: The indexer gives pulse signals to the drive, or amplifier. These pulses represent the distance to travel. The frequency of these pulses indicates speed, or velocity. These pulse signals are then amplified and sent directly to the step motor. The motor, constructed specifically to convert electrical pulses into discrete moves, turns the load at the desired speed to the predetermined location. The indexer, or motion controller, does not accumulate error or drift, thus for every pulse output, the step motor should move the load to the desired location each time. Of course, closing the position loop with an encoder will assure moves to the exact location every time.

Stepper motors have two windings, or phases, and are located in the stator for maximum heat dissipation. Current is switched on and off in these two phases, which creates an electromagnetic field for rotation. The position of the motor shaft is dictated by the number of current switches and which phase has been turned off last. The frequency of the switching dictates the speed of the motor. There is some oscillation in a step motor with each step, and this is more noticeable at low speeds. Microstepping is one method of minimizing resonance due to oscillation. Because the increments are much smaller, the operation is smoother. Basically, there are more pieces to the microstep pulsetrain. A standard of 25,000 pulses per revolution of a motor is very common.

■ **Table 7-1 Symbols for motion control equations.**

Symbol	Definition	Units	
		SI	**English**
C_G	Circumference of gear	m (or cm)	in (or ft)
$C_{P:1, 2, 3}$	Circumference of pulleys 1, 2, or 3	"	"
D	Diameter of cylinder or ...	m (or cm)	in (or ft)
D_G	...(pitch dia.) of gear	"	"
D_{PL}	...(pitch dia.) of pulleys on load	"	"
D_{PM}	...(pitch dia.) of pulleys on motor	"	"
$D_{P:1, 2, 3}$	...(pitch dia.) of pulleys 1, 2, or 3	"	"
e	efficiency of mechanism or reducer	%	%
F	Forces due to...	N	lb
F_{fr}	...friction ($F_{fr} = \mu W_L \cos \gamma$)	"	"
F_g	...gravity ($F_g = W_L \sin \gamma$)	"	"
F_p	...Push or pull forces	"	"
a or d	linear accel or decel rate	$m\text{-}s^{-2}$	$in\text{-}s^{-2}$
α	angular acceleration rate	$rad\text{-}s^{-2}$	$rad\text{-}s^{-2}$
g	gravity accel constant	$9.80\ m\text{-}s^{-2}$	$386\ in\text{-}s^{-2}$
J	mass moment of inertia for...	$kg\text{-}m^2$	$lb\text{-}in^2$
$J_{B \to M}$	...Belt reflected to motor	or	or
J_C	...Coupling	$g\text{-}cm^2$	$oz\text{-}in^2$
J_G	...Gear	etc.	or
J_L	...Load	"	$in\text{-}lb\text{-}s^2$
$J_{L \to M}$	...Load reflected to motor	"	or
J_M	...Motor	"	$in\text{-}oz\text{-}s^2$
J_{PL}	...Pulley on the load	"	etc.
J_{PM}	...Pulley on the motor	"	"
$J_{PL \to M}$	...Pulley on load reflected to motor	"	"
$J_{P:1, 2, 3}$	...Pulley or sprocket 1, 2, or 3	"	"
J_r	...Reducer (or gearbox)	"	"
J_{Total}	...Total of all inertias	"	"
J_S	...leadscrew	"	"
N_r	Number ratio of reducer	none	none
N_t	Number of teeth on gear, pulley, etc.		
P_G	Pitch of gear, sprocket, or pulley	teeth /m	teeth/inch
P_S	Pitch of leadscrew	revs/m	revs/inch
t	time...	sec	sec
$t_{a, c, \text{or } d}$	...for accel, constant speed, or decel	"	"
t_m	...for move	"	"
T_{Total}	... for total cycle	"	"
t_h	...for hold time (dwell time)	"	"
T	Torque... (for "required" calculations)	Nm	in-lb
$T_{a, c, \text{or } d}$	...during accel, constant, or decel	"	or

Symbol	Definition	Units SI	Units English
T_{CL}	...Constant at load	"	in-oz
$T_{C \to M}$	...Constant reflected to motor	"	"
T_H	...Holding (while motor stopped)	"	"
T_L	...at load (not yet reflected to motor)	"	"
T_P	...due to preload on screw nut, etc.	"	"
T_{RMS}	RMS ("average") over entire cycle	"	"
T_{Total}	...Total from all forces	"	"
V_L	Linear velocity of load	$m\text{-}s^{-1}$	$in\text{-}s^{-1}$
ω_0	Initial angular/rotational velocity	$rad\text{-}s^{-1}$	rps or rpm
ω_M	Angular/rotational velocity of motor	"	"
ω_{max}	Maximum angular/rotational velocity	"	"
W_L	Weight of load	N (or kg)	lb
W_B	Weight of belt (or chain or cable)	"	"
W_T	Weight of table (or rack & moving parts)	"	"
X_L	Distance X traveled by load	m (or cm)	in (or ft)
θ	Rotation...	radians	revs
$\theta_{a, c, \text{ or } d}$	...Rotation during accel, decel, etc.	"	"
θ_L	...Rotation of load	"	"
θ_M	...Rotation of motor	"	"
θ_{Total}	Total rotation of motor during move	"	"
π	"PI" = 3.141592654	none	none
2π	Rotational unit conversion (rads/rev)	rad/rev	rad/rev
μ	Coefficient of friction	none	none
γ	Load angle from horizontal	degrees	degrees
	The following definitions apply to the torque vs speed curves		
	...Typical torque terms used with servos...	Nm	in-lb
T_{PS}	Peak torque at stall (zero speed)	"	or
T_{PR}	Peak torque at rated speed	"	in-oz
T_{CS}	Torque available continuously at stall	"	"
T_{CR}	Continuous torque rating (@ rated speed)	"	"
	...Typical torque terms used with steppers...	"	"
T_H	Holding torque (at zero speed)		
ω_R	Rated speed (servos)	$rad\text{-}s^{-1}$	rps or rpm
ω_M	Maximum speed (servos & steppers)	"	"
ω_1	Speed at peak torque (not commonly used)	"	"
ω_H	"High" speed...real maximum (not common)	"	"

Source: Amechtron, Inc.

RACK AND PINION

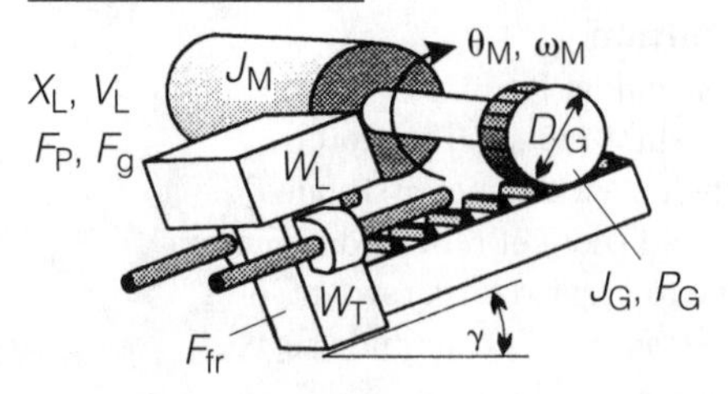

$$C_G = \pi \times D_G = \frac{N_t}{P_G} \qquad J_{Total} = J_M + J_G + J_{L\to M}$$

$$\theta_M = \frac{X_L}{C_G} \qquad J_{L\to M} = \frac{(W_L + W_T)}{g \times e} \times \left(\frac{D_G}{2}\right)^2$$

$$F_g = (W_L + W_T) \times \sin\gamma \qquad F_{fr} = \mu \times (W_L + W_T) \times \cos\gamma$$

$$\omega_M = \frac{V_L}{C_G} \qquad T_{L\to M} = \left(\frac{F_P + F_g + F_{fr}}{e}\right) \times \left(\frac{D_G}{2}\right)$$

■ **7-12** *Rack-and-pinion calculations.* Amechtron, Inc.

TIMING BELT

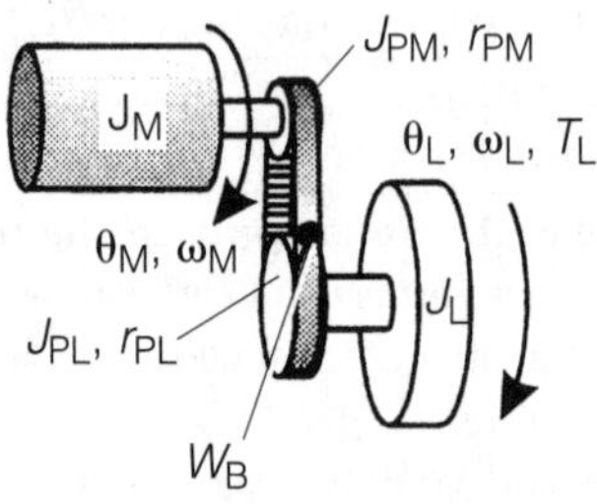

$$N_r = \frac{N_{tL}}{N_{tM}} = \frac{D_{PL}}{D_{PM}} \qquad J_{Total} = J_M + J_{PM} + J_{PL\to M} + J_{B\to M} + J_{L\to M}$$

$$\theta_M = N_r \times \theta_L \qquad J_{PL\to M} = \left(\frac{1}{N_r}\right)^2 \times \frac{J_{PL}}{e} \qquad J_{B\to M} = \frac{W_B}{g \times e} \times \left(\frac{D_{PM}}{2}\right)^2$$

$$\omega_M = N_r \times \omega_L \qquad J_{L\to M} = \left(\frac{1}{N_r}\right)^2 \times \frac{J_L}{e} \qquad T_{L\to M} = \frac{T_L}{N_r \times e}$$

■ **7-13** *Timing belt calculations.* Amechtron, Inc.

CONVEYOR

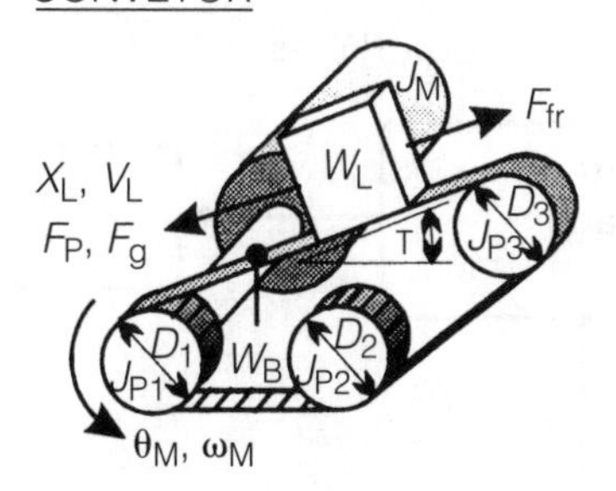

$$C_{P1} = \pi \times D_{P1} = \frac{N_t}{P_G} \qquad J_{Total} = J_M + J_{P1} + \left(\frac{D_{P1}}{D_{P2}}\right)^2 \times \frac{J_{P2}}{e} + \left(\frac{D_{P1}}{D_{P3}}\right)^2 \times \frac{J_{P3}}{e} + J_{L\to M}$$

$$\theta_M = \frac{X_L}{C_{P1}} \qquad J_{L\to M} = \frac{(W_L + W_B)}{g \times e} \times \left(\frac{D_{P1}}{2}\right)^2$$

$$F_g = (W_L + W_B) \times \sin\gamma \qquad F_{fr} = \mu \times (W_L + W_B) \times \cos\gamma$$

$$\omega_M = \frac{V_L}{C_{P1}} \qquad T_{L\to M} = \left(\frac{F_P + F_g + F_{fr}}{e}\right) \times \left(\frac{D_{P1}}{2}\right)$$

■ **7-14** *Conveyor calculations.* Amechtron, Inc.

LEADSCREW

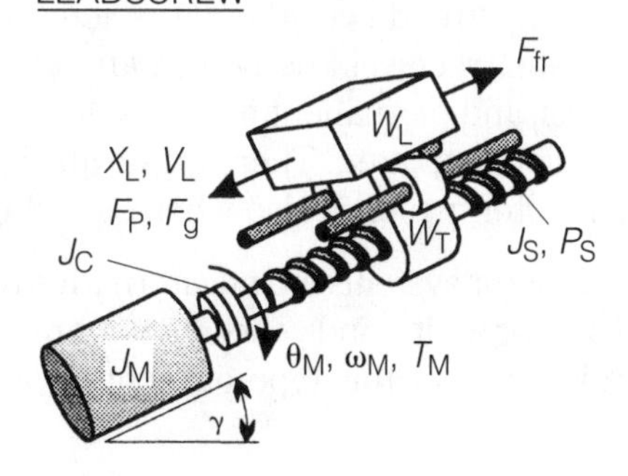

$$J_{Total} = J_M + J_C + J_S + J_{L\to M}$$

$$\theta_M = P_S \times X_L \qquad J_{L\to M} = \frac{(W_L + W_T)}{g \times e} \times \left(\frac{1}{2\pi \times P_S}\right)^2$$

$$F_g = (W_L + W_T) \times \sin\gamma \qquad F_{fr} = \mu \times (W_L + W_T) \times \cos\gamma$$

$$\omega_M = P_S \times V_L \qquad T_{L\to M} = \left(\frac{F_P + F_g + F_{fr}}{2\pi \times P_S \times e}\right) + T_P$$

■ **7-15** *Lead screw calculations.* Amechtron, Inc.

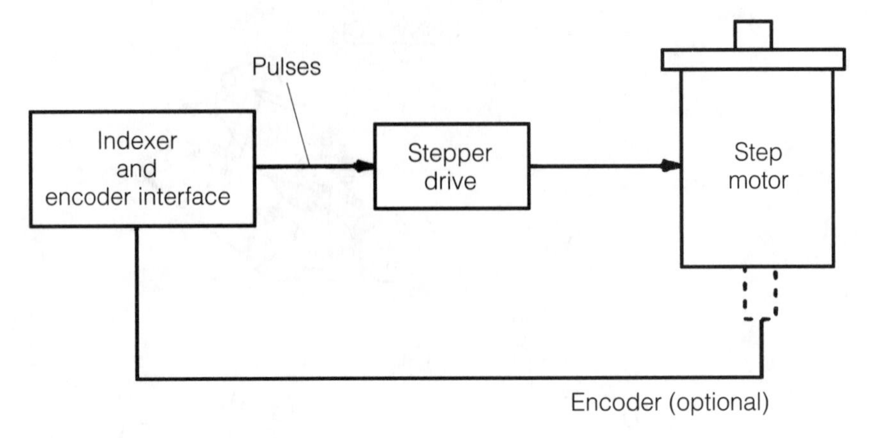

■ **7-16** *Step motor system.*

Linear motor systems

Linear motor systems operate on the same principles as those found in rotational equipment. Electromagnetism and permanent magnet properties are found in both linear induction and linear stepper motor and drive systems. Figure 7-17 shows a linear step motor in simplified form. There are two main components, the platen and the thrust mechanism. The platen can be likened to the stator of an electric motor, rolled out into a plate configuration. The teeth extend over the entire length of the platen. The thrust component receives the power, and contains the electromagnets, bearings, and permanent magnets. It moves in both directions over the stationary platen. There exists an air gap, common to the rotational motor, which is also where the magnetic flux is maintained.

Linear motor systems are an alternative to those applications where lead screws, belts, and gears are not practical. But perhaps the most prevalent use of this type of technology can be found on a larger

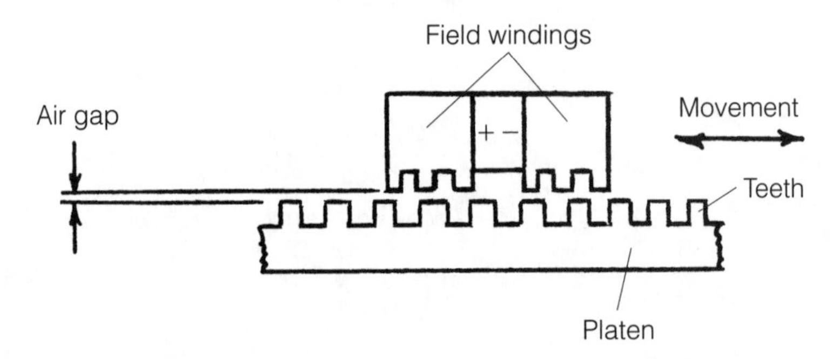

■ **7-17** *Linear step motor.*

scale in magnetic levitation train systems. The same principle is being applied in a much larger, modified scale. The bearing system is a critical factor in both the large-scale systems and the common low-mass, high-speed applications. Air bearings (which require an air source) and mechanical bearings are commonly used.

Spindle drives

A spindle is often referred to as the shaft in a lathe or in a capstan. This spindle, if it is to turn, must be motor-driven. The motor must be controlled by a specialized device called a spindle drive. This drive is commonly seen in the textile and machine tool industries. Typically, a spindle is required to turn a part at high speeds so as to machine it. Normally, when turning the part at high speeds, often to 7,000 rpm, power requirements are constant. Usually, the part spins at high rpm's to remove metal from it. In the process it is usually necessary to run at lower speeds and provide constant torque. The duty cycle of the spindle is contingent on the associated motors and drives coordinated with it. As the spindle spins at high speed, the other axes of motion are either moving in or out and traversing the part. Depending on the machining, the spindle can slow or remain at a constant speed when not doing any work.

In the machine tool environment, smaller, compact designs are needed, since space is usually at a premium on the plant floor and on the machine itself. Spindle motors are usually of a longer stack length, more square, and are completely enclosed with a separate fan for cooling throughout its speed range. Over the years, the industry has stressed low maintenance, and therefore no brushes or commutators are incorporated into the design. The most common sizes of typical spindle motors will go to roughly 30 HP and can be handled by ac drive technology. Larger spindle drives have traditionally been left to the dc drive, but this scenario is rapidly changing as ac drive current and switching capabilities increase.

Closing remarks

Servo drive and stepper drive systems have niches in industry. They can perform in many applications where ac and dc drives dare not tread. Cost plays an important role in when and where to use ac or dc drives over servos. An ac drive is not going to position the gripper arm of a robot as a servo drive system will. And a servo will be overkill in an energy-saving pump application, usually reserved for the ac drive. Likewise, cost plays a role in choosing stepper drives over servo drives for the same application. The choice is ultimately up to the user. The ease of use, programming, interconnectivity, and reliability factor heavily into the user's decision.

Servo and stepper applications remain plentiful in the machine tool, robotics, feed or cut-to-length applications, and most any other positioning motion requirement. Building these devices for operation in harsh environments and making them less sensitive to outside electrical noise will increase their use in industrial applications. The construction of servo and stepper motors was covered in more detail in Chapter 4. Feedback devices are discussed in Chapter 9.

8

Drive peripheral equipment, control, and communications

AN ELECTRONIC DRIVE BY ITSELF IS MERELY A POWER conversion unit between the main power supply and an electric motor. And without proper control wiring, power cabling, and so on, the drive really is just a "black box." The black box, or drive, often must be housed in some type of enclosure. There are also different power disconnect and power isolation schemes, along with voltage matching requirements. The drive has to be connected to and function with many other devices. These other devices compose an extensive list of drive peripheral equipment without which the drive would be virtually useless. Likewise, these peripheral devices allow for additional control and communication schemes utilized by the drives. However, additional electrical equipment also makes for more potential problem areas. This chapter looks at that equipment, how to interface it with electronic drives, and how to avoid problems by using sound wiring and interconnecting techniques.

Often these external devices or the wiring to and from becomes a problem area for one reason or another. Wiring and cabling are perhaps the most important issues when discussing any electronic application. Whenever two or more electrical devices are required to work together, the traditional method of getting commands, power, or other information back and forth is via a conductor, either wire or cable. And since we are dealing with power conversion products in the drives, power delivery and distribution is just as vital to the application's success as is the control wiring integrity. We will look at power wiring issues, how control and signal wiring practices interrelate, power supply issues, communication schemes utilized by drives, and common drive control methods.

Power wiring and supply power considerations

Electronic drives need power—power to convert, power to invert, and power to control electric motors. Their internal components need power, as do their external pieces of electrical equipment. Without the hundreds, and even thousands, of feet of wire and cable to deliver electrical energy to and from devices for every installation, we end up with an awful lot of hardware not moving, blinking, or making any sound. Thus, the power wiring is a sometimes overlooked component in the application. And if not given the proper attention, it can cause more problems later.

Wire sizing and proper wiring methods start with the National Electrical Code (NEC). Plant practices and local building codes also serve as reference points when it comes to wiring electrical equipment. However, drive technology has brought with it the need to pay attention to other details. We have to look at cable-size selection for minimum voltage drops and various methods with the conduit to eliminate or reduce magnetic radiation or electrical noise. This in addition to the actual current carrying capacity of the cable based on drive size and loading. Table 8-1 lists common diameters, resistance values, and weights for various sizes of copper wire. Equivalent data is also shown for aluminum wire in Table 8-2.

In variable-speed applications using electric motors, motor voltage is proportional to speed, and current is almost proportional to torque. When the speed is near full or base speed, the voltage is near maximum. Thus, at full speed from a 460-V supplied system, there should be approximately 460 V applied to the motor. There will be some voltage loss due to the voltage drop in the cable to the motor. However, at full speed, the voltage drop in the cable to the motor is less critical than when the speeds are lower. Therefore, at low operating speeds, the voltage to the motor is low, and a few volts' drop in the cables to the motor will prevent the motor from providing full capacity. The severity of that voltage drop and the need to run at these lower speeds might affect performance.

Those applications requiring full torque at low speed (constant torque) will be sensitive to voltage drops in the cable caused by rated motor currents. A normal 5% voltage reduction at full speed could result in a 50% reduction in voltage at 10% speed. This is where a voltage boost function in an electronic drive is needed.

However, that voltage boost function must be used carefully because it can overcompensate at higher speeds. Also, the voltage boost function is not always sufficient to overcome too severe of voltage drops. For high starting torque and constant torque applications, a good practice is to size the cable so that no more than ½ V is dropped in a single cable when carrying the full-load amps of the motor. The maximum resistance of the cable is defined, and the cable size can be selected based on the total length of cable between the drive and the motor.

■ Table 8-1 Diameter, resistance, and weight of copper wire.

Wire size/ AWG.	Diameter/ nom. inch	Resistance in ohms per thousand feet	Weight	
			lb. per m feet	Feet per lb
1	0.2893	0.1239	253.3	3.947
2	0.2576	0.1563	200.9	4.978
3	0.2294	0.1971	159.3	6.278
4	0.2043	0.2485	126.3	7.915
5	0.1819	0.3134	100.2	9.984
6	0.1620	0.3952	79.44	12.59
7	0.1443	0.4981	63.03	15.87
8	0.1285	0.6281	49.98	20.01
9	0.1144	0.7925	39.62	25.24
10	0.1019	0.9988	31.43	31.82
11	0.0907	1.26	24.9	40.2
12	0.0808	1.59	19.8	50.6
13	0.0720	2.00	15.7	63.7
14	0.0641	2.52	12.4	80.4
15	0.0571	3.18	9.87	101
16	0.0508	4.02	7.81	128
17	0.0453	5.05	6.21	161
18	0.0403	6.39	4.92	203
19	0.0359	8.05	3.90	256
20	0.0320	10.1	3.10	323
21	0.0285	12.8	2.46	407
22	0.0253	16.2	1.94	516
23	0.0226	20.3	1.55	647
24	0.0201	25.7	1.22	818
25	0.0179	32.4	0.970	1030
1/0	0.3249	0.09825	319.5	3.130
2/0	0.3648	0.07793	402.8	2.482
3/0	0.4096	0.06182	507.8	1.969
4/0	0.4600	0.04901	640.5	1.561

■ Table 8-2 Diameter, resistance, and weight of aluminum wire.

Wire size/ AWG	Diameter/ nom. inch	Resistance in ohms per thousand feet	Weight	
			lb per M feet	Feet per lb
1	0.2893	0.2005	77.02	12.98
2	0.2576	0.2529	61.08	16.37
3	0.2294	0.3189	48.44	20.65
4	0.2043	0.4021	38.40	26.04
5	0.1819	0.5072	30.47	32.82
6	0.1620	0.6395	24.15	41.41
7	0.1443	0.8060	19.16	52.19
8	0.1285	1.016	15.20	65.79
9	0.1144	1.282	12.05	82.99
10	0.1019	1.616	9.56	105
11	0.0907	2.04	7.57	132
12	0.0808	2.57	6.02	166
13	0.0720	3.24	4.77	210
14	0.0641	4.08	3.77	265
15	0.0571	5.15	3.00	333
16	0.0508	6.50	2.37	422
17	0.0453	8.18	1.89	529
18	0.0403	10.3	1.50	666
19	0.0359	13.0	1.19	840
20	0.0320	16.4	0.943	1060
21	0.0285	20.7	0.748	1340
22	0.0253	26.2	0.590	1690
23	0.0226	32.9	0.471	2120
24	0.0201	41.5	0.371	2700
25	0.0179	52.4	0.295	3390
1/0	0.3249	0.1590	97.14	10.29
2/0	0.3648	0.1261	122.5	8.163
3/0	0.4096	0.1000	154.4	6.478
4/0	0.4600	0.07930	194.7	5.135

Electronic ac drives control the electric motor by maintaining a constant volts-per-hertz ratio. This ratio maintains the air gap flux in the motor at the desired value. When the voltage at the terminals of the motor varies, the air gap flux in the motor will also vary. Controlling the air gap flux in the ac motor controls the speed and torque performance of the motor. In some applications, any variation in the air gap flux will result in rotational variations or varying torque capability as the shaft of the motor rotates.

An electronic ac drive transfers the line power to the electric motor. In most cases, that output voltage will vary if the input line voltage varies. Any variation in the output voltage will result in a variation in speed. At lower operating speeds, a significant speed variation can occur due to any changes in the output voltage. Since percent slip is constant, when a constant air gap flux is maintained in the ac motor, the actual rpm change will be greater as the speed of the motor is reduced.

It is very important that the terminal voltage be maintained within ±1% to obtain the best motor performance. Some ac drives, such as the pulse width modulating (PWM) type, regulate the output voltage by correcting the modulation to compensate for input voltage variations. By correcting for these incoming power variations, an electric motor can maintain its desired performance despite variations. With this input voltage compensation by the electronic drive, variations in speed, current, and motor temperature can be minimized.

Conventional cable sizing is usually based on current-carrying capacity. The electrical codes (local and NEC) provide guidelines for the selection of wire and cable. However, these guidelines are based on fixed voltage supplies, of which electronic drives are not. Whenever a variable-frequency drive is used, the output voltage changes with the output frequency. Since ac motor performance is affected by any voltage variations, the selection of the proper cable size is important to the application's ultimate success. At low speeds (low frequencies), the corresponding reduced voltage will compound the effect of any voltage loss in the cables between the drive and the motor.

To minimize the effect of the voltage loss, the resistance value of the cable should be kept as low as possible. The maximum resistance in ohms will be defined by the length and American Wire Gauge (AWG) size of the cable (see Tables 8-1 and 8-2). The value of the maximum resistance is equal to ½ V divided by the nameplate amps of the motor. The longer the cable length, the larger the cross-sectional area of the cable. The low voltage drop in the cable will ensure that the motor receives as much of the voltage present on the output terminals of the ac drive.

In many facilities, commercial and industrial, it has become common practice to use aluminum wire instead of copper. This is due mainly to the cost differential per lineal foot of each material.

However, wherever a termination is made between a copper component and an aluminum component, the possibility of future problems exists. The dissimilar materials can, over time, cause bad connections, excessive, local heating at the connection, and even voltage transients. Whenever this condition exists, routine checks of the connections should be made and, if possible, a complete change-out of one the material types should be considered.

Since today's electronic drive equipment is designed as solid-state, the voltage supplied to the drive must be consistent—hopefully, without line spikes. This is the ideal situation for a power distribution system to a drive, but unfortunately, the reality is that there will be instantaneous voltage transients, and the voltage might not be consistent. Voltage transients or "spikes" are basically any deviation from a true voltage sine wave, as illustrated in figure 8-1.

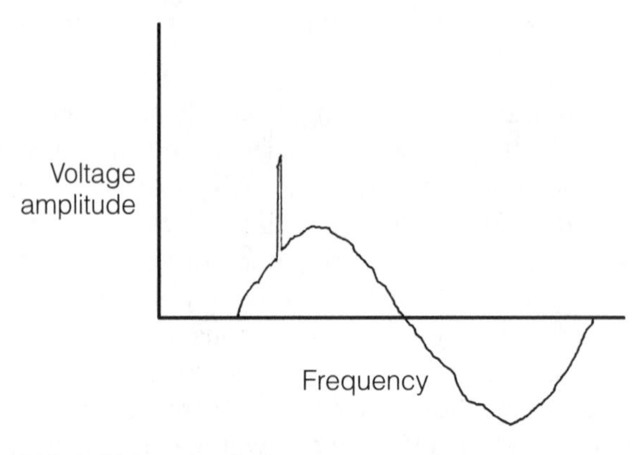

■ **8-1** *Voltage spike.*

Drives today can tolerate quite a bit. A ±10% swing in input voltage is commonly requested from the electronic drive. Many drives can even tolerate a complete outage for a cycle or two. But we must separate the different aspects of the supply power issue. Control power within an electronic drive has different needs from the supply than does the rectifier circuit. The control portion may or may not be able to tolerate undervoltage, overvoltage, or complete loss of incoming power. This higher voltage power is most often stepped-down through a control transformer into a usable lower voltage. This transformer aids in smoothing the transients somewhat. However, the rectifier circuit is only concerned with the instantaneous voltage value. Current will flow as long as the input voltage instantaneous value is greater than the voltage in the dc bus.

Obviously, there are limits. Once attaining a maximum voltage, the dc bus voltage will cause the drive to fault in order to protect the valuable power semiconductors and other power-conversion devices within. In addition to the nuisance drive trip, another method of limiting current adds impedance to the circuit. This impedance delays how much input current is allowed while the instantaneous voltage transient exists. Typically this impedance is introduced as an inductor to the filtering circuit. This choke or inductor prevents current from increasing rapidly, delaying an increase in the capacitor voltage. The inductor does not prevent the voltage from changing, but simply lengthens the amount of time for the change to take place.

As discussed earlier, with today's solid-state electronic drive equipment, the voltage transients found on the typical ac line will not be so severe that the drive trips. Many drives come equipped with or can be fitted with line reactors to handle the voltage transients. These impedances can be in increments of 2.5% but should not exceed a total of 10% as a rule of thumb. Another more expensive solution is to add an isolation transformer at the input. This transformer should also be shielded. Both options are shown in figure 8-2. More discussion regarding transformers follows later in this chapter.

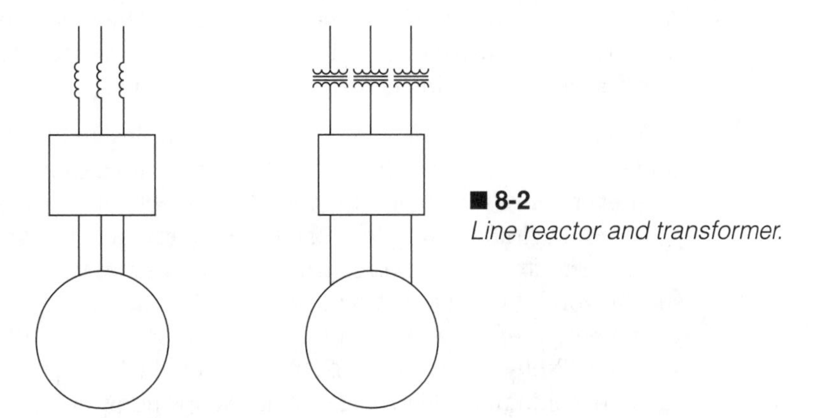

■ **8-2**
Line reactor and transformer.

Another method of suppressing voltage spikes is that of line filters. These can clamp the transients to be almost equal to the incoming voltage, but they can be a more expensive alternative.

Today's drive equipment is being merged with many old power distribution systems. The good news is that the drives being installed are very adaptable and very flexible when it comes to existing power systems. However, there will always be some electrical issue that surfaces and the drive will often be the focal point. The drive will be blamed for creating the transient, or else it will be blamed for being too sensitive. Either way, there are methods of dealing with just about any problem in the distribution system. The most important consideration is that whenever ac or dc drives are being considered for a facility, you should look closely at the existing system and predict the problems. They are easier to address before hardware is installed!

Control and signal wiring

The "little brother" to the electronic drive's power wiring is the control and signal wiring. However, it should not be taken lightly. As a matter of fact, problems associated with control and signal wire are harder to diagnose, most of the time fatal to the process, and sometimes difficult to prevent. Since many drives are microprocessor-based, the information going to and from the control board (which normally houses the microprocessor) either gets to it destination or it doesn't. If it doesn't, then the drive hangs up and stops. Likewise, if feedback from a motor or auxiliary device doesn't reach its destination, other nondesirable actions can take place (drive stops, motor runs at wrong or dangerous speeds, etc.). Neither of these conditions is tolerable. Therefore, a little common sense along with some practical guidance can go a long way to making a drive installation a success. This section discusses many issues concerning control and signal wiring.

The place to start is with the overall system design and how the control and signal wire will be run. Steel enclosures should be used whenever there's a choice for mounting electronic drives. Steel provides shielding from electrostatic, electromagnetic, and magnetic noise. It can also better withstand industrial-type accidents such as forklifts running into it. As for withstanding the actual environment, there are different classifications for enclosures, as shown in Table 8-3 on page 232. Try to avoid installing windows in cabinets containing electronic circuits. If a window is unavoidable, particularly in electrically noisy locations, a grounded copper mesh behind the window will provide some protection. Drive enclosures and ratings are discussed at greater length later in this chapter.

When routing low-voltage wire, use steel cable troughs and metal conduit to route wire between control cabinets. As with cabinets, steel cableways provide shielding from electrostatic, electromagnetic, and magnetic noise. Using dissimilar metals for cabinets and raceways can result in eventual corrosion at joints, with eventual degradation of the electrical connection. Steel also provides better physical protection than softer materials. Avoid plastic pipe whenever possible.

Always route power, control, and feedback signal wiring separately. Route different types of wiring through separate conduit or cable raceways. In a typical electronic drive control system, there are three major types of wiring: Power wiring (discussed previously), which is high voltage (120 V, 240 V, 460 V, 575 V, and higher) and high current rating. Such wiring is insensitive to noise but can generate large amounts of it. *Digital* or *discrete signal wiring*, which is low voltage (24 V, 12 V, etc.) and low current, operates very fast. It is very sensitive to noise from power conductors but also generates high-frequency noise that can affect analog signal wiring. *Analog signal wiring* is very low voltage (0–5 Vdc or 0–10 Vdc) and very low current; it is the most sensitive in the overall system and the most likely to have problems. Many times this signal can be in the form of a 4–20-mA current, which is less susceptible to noise.

Another good practice is to avoid tightly bundled cables, even if they are of the same type. Never route power (higher voltage) wire with control/signal wire. But also try to avoid bundling low-voltage control and signal wires whenever possible. Tight bundling helps couple noise between cables. This is very important for digital and analog signal wiring. Normally, power wiring can be bundled together.

Also, many times it is convenient to use the conduit or metal wireway as a ground point. Do not use the wireway as a ground, although metal cableways should be grounded. Never use a metal cableway as a cabinet or chassis earth ground connection. Try to use a separate ground connection. Within an enclosure, since it is not practical to route the wire in metal conduit and raceway, try to keep analog signal, digital, and power cables as far apart as possible. Avoid parallel runs and make them cross at right angles to each other.

The next subject is grounding. Grounding is done mainly for two reasons: To protect operators in case of an electrical malfunction and to minimize and protect equipment from electrical noise. However, improper grounding can lead to other problems and, in particular, ground loops. All equipment in a control system has to be properly grounded. All drive chassis and cabinet grounds should connect to a central ground. Connect all electrical chassis and cabinets to the facility's central earth ground. (As mentioned previously, *earth ground* refers to the central ground point for all ac power and electrical equipment within a factory.) Every facility with electrical equipment has an existing grounding electrode system. Local building codes will address grounding requirements, and for further definition the *National Electrical Code* can be consulted.

Figure 8-3 shows a typical and effective ground drive enclosure system. Figure 8-4 shows a ground system with multiple drive controllers. Shown are the proper ground method for these and the improper method that will cause a ground loop. The minimum conductor size for ground wire is #6 AWG copper. These conductors should be properly sized to provide full protection in case of an ac wiring fault. Know the current and voltage ratings of the drive equipment installed. Also, learn how each drive handles a ground fault. Again, consult your local building and wiring codes. Also, the drive manual will show how and where ground terminations are to be made.

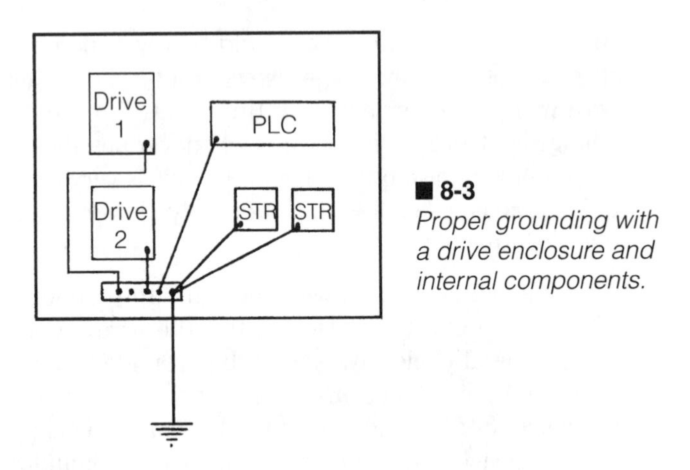

■ 8-3
Proper grounding with a drive enclosure and internal components.

For practical reasons, only copper conductors should be used for ground wiring. Aluminum is subject to corrosion and has a high electrical resistance compared to copper. Any ground electrode must be connected to any main ground bus bar. Install a grounding conductor between the grounding electrode and an earth ground bus bar in the ac power distribution panel. The earth ground bus bar is typically a copper bar with threaded holes for connections. Securely attach the earth ground bus bar to the distribution panel cabinet. Do this either by bolting the bus bar to the cabinet or using a very short piece of copper wire (#6 AWG or higher). Bond this securely to clean, bare metal, not painted surface.

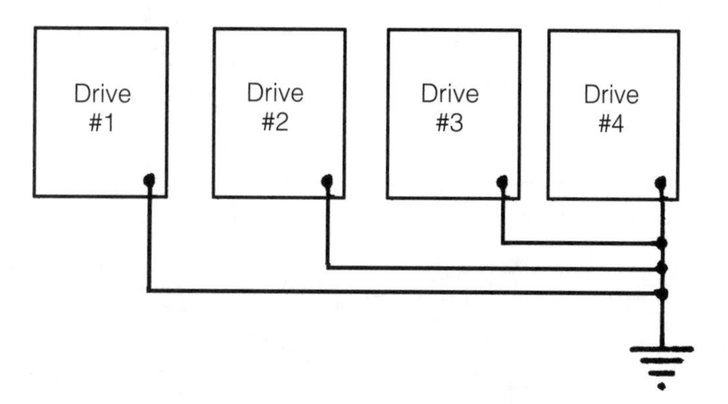

■ **8-4** *Grounding multiple drives.*

When laying out or designing the wiring scheme for the electronic drives, give the grounding circuit special attention. Doing so might save time and headaches later. As a rule, ground connections must not be broken by any switches. The main disconnect switch must never break any ground connection. Each drive enclosure should have its own ground. Every equipment cabinet must have a ground bus that connects directly to the earth ground bus bar in the power panel. It can, however, be lighter. Bond it directly and securely to the cabinet. Once again, be sure that all paint is scraped from the bond area. Connect the cabinet ground to the ground bus bar in the power panel using #6 AWG or larger cable and ring connectors. This cable should have green (or, in some areas, green with yellow stripe) insulation, and can be routed through the same cable trough as ac power lines. Keep it as short a run as possible.

Ground loops have been mentioned and illustrated in figure 8-4. Never connect grounds in series. Except where specifically allowed, never connect drive or cabinet ground leads in series (daisy-chained). This approach makes noise signals cumulative, turning several low-level noise signals into one large noise signal. This can result in noise problems and improper controller operation. Instead, use what is called a "star" system, which utilizes individual ground cables going to the system ground. See figure 8-5.

A common mistake is to equate the ground circuit with the common or return in the wiring system. This is not so. The ground is not the common, nor is it the return. The term *ground* should not be used interchangeably with common or (power) return. Although commons or returns may connect to chassis or earth ground at some point, they should not be used as drive chassis or drive enclosure ground connection points.

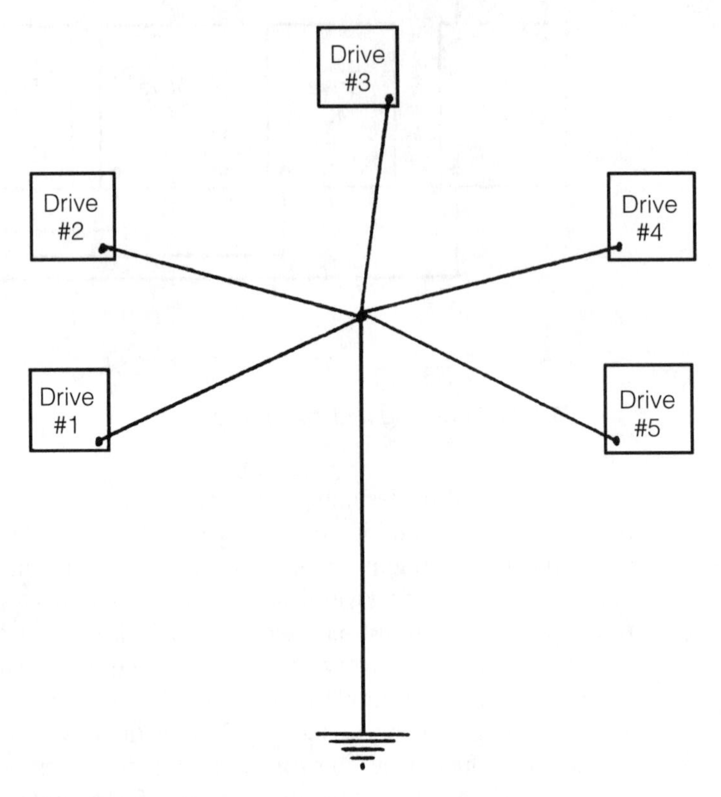

■ **8-5** *Star grounding system.*

Using a separate ground cable for each drive in a multiple drive enclosure is strongly recommended. Even if the steel enclosure itself is grounded, it is good practice to do this because copper is much more conductive than steel. Electricity, in taking the path of least resistance, will flow the route we design, drives before enclosure or enclosure before drives. Also, it is easier to properly attach to a visible, accessible drive component than to connect and maintain a blind connection between drive and enclosure. Lastly, the enclosure might not be a continuous piece. It might not be welded at every seam. This could cause the electrical ground to be unstable or not be proper at all. Remember, electricity flows the path of least resistance, and it will not flow at all if connection is not made.

Grounding and shielding are often confused with one another. Shielding is just that—a shield or protection to sensitive wiring. Use shielded cable for low-level analog signal cables. Shielded cable is particularly effective in reducing electrostatic (capacitive) coupling between parallel cables running together in a wiring trough. Shielded cable also provides protection from electromagnetic noise in the RF (radio frequency) range. Shielded cable is usually not effective against magnetic noise or low-frequency electromagnetic noise. All analog signal wiring should be shielded, including motor feedback wiring (encoders, resolvers, etc.) and serial communication wiring if longer than 10 feet. Occasionally, digital signal wiring must also be shielded (shielded ribbon cable is available). Figure 8-6 shows a typical wiring method for shielded cable. Shielded cables should only be connected at one end.

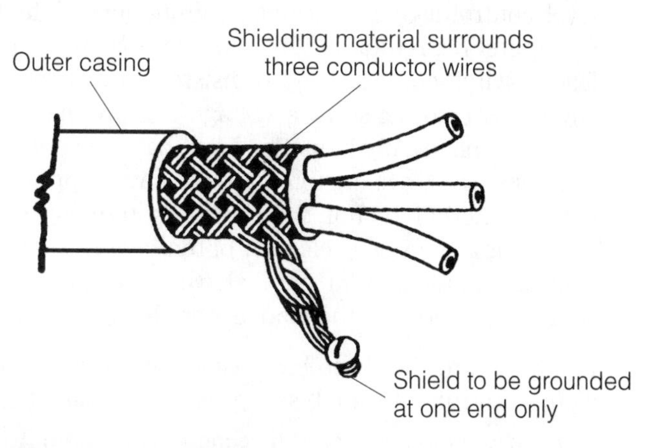

■ **8-6** *Cable and wire shielding.*

In addition to shielding signal and control wire, try to keep any exposed signal wire runs as short as possible. Although this is common sense, it needs to be stated. Keep both exposed signal cables and the bare shield drain cable as short as possible. If there is a possibility of shorting, insulate the bare drain cable with plastic sleeving or tape. Also, avoid excessive terminal strips. Terminal strips have connections and these connections can loosen or corrode. Any terminations can be a trouble spot. Use full lengths of wire whenever possible. If terminal strips are used, try to maintain shields through the terminal strip.

As stated earlier, connect shields to ground at *one end only*. Connect shield drain wire to ground at exactly one end. The best end to connect is the one at the drive so as to minimize any open spots. At the other end, insulate the shield with shrink tubing or tape to prevent it from shorting to ground or other wiring. Some controls do not provide terminals for shield connections. In such cases use heavy, solid copper cable or light copper strap to make a shield bus. Remember, shields must not be used as commons. They are not to be used as returns, either. Use them strictly as shields!

Control and signal wiring is also susceptible to noise generated from inductive loads around the drive equipment. Contactor coils, solenoids, and relays are examples of inductive loads commonly found in or around a drive enclosure. Typically, noise suppression techniques are incorporated into these inductive load networks in order to protect the control and signal wiring.

Inductors typically generate large electrical noise transients when contacts controlling them open. These transients can affect low-level control signals or cause malfunctions in high-speed digital (computer) circuits. High-voltage transients can also damage loads, switch contacts, and transistor outputs. A suppressor usually limits peak noise to a voltage slightly higher than the peak voltage applied. Any inductive load that can generate a noise transient near the drive equipment must have suppressors. It does not matter whether or not the device is actually controlled by the drive. If it is physically close, it potentially can cause problems. Installing suppressors in parallel with the inductive load and installing it as close to the load as possible is one solution.

Following are the three basic types of noise suppression devices along with rule-of-thumb sizing for suppressing the noise.

1. *Diodes*. One amp rated for small loads and 5 A for larger loads. The peak inverse voltage (PIV) rating should be at least twice the voltage. A heatsink is generally not required for this component.
2. *Resistor/capacitor (RC) networks*. Capacitors should be sized to be approximately three times the RMS ac voltage. Resistors should be ½ W for small loads and 1 W for larger ones.
3. *Metal oxide varistors (MOVs)*. Should be used in conjunction with a resistor/capacitor network, 120-V and 240-V systems.

These noise suppression devices can be used individually or in combination, depending on the voltage and the inductive load. Remember, we are trying to minimize or eliminate noise to control and signal wiring. Higher voltages and ratings associated with power cabling should not be addressed in a similar manner. Use these noise suppression techniques for low voltage wiring schemes only. The basic types of applications for noise suppressors include small and large single-phase ac inductive loads, three-phase ac inductive loads (usually larger than single-phase loads), and small and large dc loads.

Unfortunately, even when all the shielding, grounding, and noise suppression methods are followed, our electronic drive systems can still be in a problematic application. By following the recommended and proper wiring rules, we certainly minimize most of our noise problems. Yet, they still exist out there. Radio frequency interference (RFI) and harmonic distortion are two other common occurrences continually being blamed as culprits in electronic drive installations. More discussion on RFI and harmonics follows in a later chapter.

Some electrical noise problems can be easily traced, while others are a mystery—especially in those cases where intermittent malfunctions occur at the drive controller. These are very difficult and expensive to trace in a finished system. Therefore, it is extremely important to design a system with sound practices and integrity to keep problem areas from emerging in the system. Typical symptoms of noise-related electrical malfunctions are

- ☐ *Drive CPU faults*. No further explanation given (some drives can tell us when the fault happened and what some voltage and current conditions were at the time of the fault).
- ☐ *External drive faults*. Something happened in the circuitry outside of the drive.
- ☐ *Complete drive lock-up (freeze)*. No fault displayed, no data recorded, nothing at all.
- ☐ *Scrambled display data on drive local display or remote monitor*.
- ☐ *Drive system spontaneously resets*. Check to see if manual or automatic reset is set in the drive. Check the run or enable contact circuit.
- ☐ *Motor runs unstably*. Noise in the drive reference signal will cause this and can be often hard to diagnose. Most troubleshooters will simply blame the drive.
- ☐ *Loss of data in battery-backed memory*. This should not happen when the battery is charged; however, data stored and retrieved from storage can be mixed with "garbage."

Servo drives can experience positioning errors and unstable servomotors. They are much more susceptible to noise than are other ac and dc drives.

The basic types of electrical noise include the following:

- ☐ *Harmonic distortion*. Caused by other power conversion equipment in the facility. Any time ac power is converted to dc, there is the possibility of line disturbances. This is especially evident with higher current capacity devices.

Harmonic distortion is so important an issue that IEEE has dedicated a standard to defining it and has provided guidelines on dealing with it.

- *Magnetic*. Caused by a changing magnetic field inducing voltages into adjacent wiring or components. A transformer is an example of a device that deliberately uses magnetic coupling.
- *Electrostatic*. Caused by capacitance coupling from a changing electric field.
- *Electromagnetic*. RF (radio frequencies or "radio waves") transmitted by a device and received (even if unintentionally) by other electrical components.

Sources of industrial electrical noise

There are many sources of electrical noise in an industrial environment. These include:

- Relay panels
- Welding equipment
- The dc and ac motors, especially high horsepower
- Lighting systems, especially fluorescent
- Generators
- Motor starters
- Contactors

When determining the source of noise, be sure to check the following:

- Are all ground cables large, solid, and routed as directly as possible to earth ground?
- Do all inductive loads have suppressors?
- Are all shields grounded at exactly one end?

In some cases, you might have to experiment with disconnecting one end of common or return cables. Multiple parallel connections of commons or returns (especially if the equipment connects these to chassis ground) can create large, noisy circulating electrical currents called *ground loops*. Ground loops can create hard-to-trace noise problems. If noise is a problem, try disconnecting duplicate returns or commons one at a time to see if noise lessens. A multipurpose volt-ohmmeter is useful for tracking down multiple grounds and commons.

With newer techniques in use today to convert ac power to dc and with computerized equipment being so sensitive and prevalent, noise in a facility is more than a nuisance. The absolute best scenario is to install equipment so that the potential for noise problems is all but eliminated. But this is easier said than done, particularly because not everything is known about the subject. This is why consultants and specialists are called in to help diagnose problems. Also, not every installation can be brand-new from the ground up. Most installations of electrical equipment are retrofits, modifications, and plant upgrades. Thus, electrical noise will probably become an issue at some point.

Whenever noise becomes a problem in an existing installation, some initial areas that can be suspect are

- ☐ High-frequency switching devices
- ☐ Cable shield shorted to ground or another shield
- ☐ Broken shield connection
- ☐ Faulty noise suppressor
- ☐ Corroded ground connection
- ☐ Loose connection
- ☐ New wiring nearby, especially higher voltage wiring
- ☐ New (or portable) electrical equipment nearby
- ☐ Outside of the facility; monitor incoming power from utility since problems can actually start here.

Electrical noise is fast becoming the major factory problem of the 1990s. Almost all equipment has a microprocessor content to it, making the equipment sensitive. Manufacturers of this equipment can add filtering and protection—but this is always added at a price. Therefore, solid research into the application beforehand will probably save downtime and extra costs later. For further discussion on drive harmonics, radio-frequency-interference, electromagnetic interference, grounding methods, and other electrical noise-related issues, refer to Chapter 12.

Transformers and line reactors

Electronic drives are nothing more than robust electrical circuits with a microprocessor in charge. This is truly an oversimplification, but when we look at all the individual pieces of hardware within the drive we find that many resistors, capacitors, and inductors are in use, along with the power semiconductors. The inductors are of specific interest in this section. The inductors appear before, within, and after the actual power conversion sections in the electronic drive. These inductors are provided in the form of transformers, line reactors, and chokes. Depending on the type of drive, these devices can appear at different physical locations, and in many instances, they are used for similar purposes.

When discussing transformers and electronic drives, some clarifications and distinctions must be made. Let's start with the incoming power line to the drive. Here, we can see transformers and/or line reactors. Looking at figure 8-7, we find that this is the probable location for some type of inductor. The original purpose of the transformer was to buffer the ac line from the effects of the conversion equipment and to help keep transients from entering the drive. This type of transformer is referred to as the *isolation transformer*. Its use today is debated, since diode converters don't disturb the supply like older drive converters. However, when in doubt and if cost isn't an issue, adding the isolation transformer makes sense, especially with servo drive equipment.

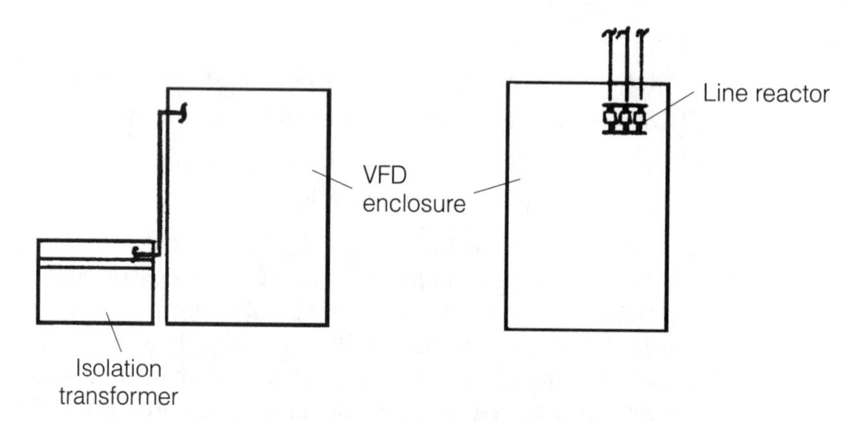

■ **8-7** *Transformer and line reactor physical locations with drives.*

In some instances a transformer has to be incorporated on the supply merely to match line voltage with the drive voltage. This has to be done in order for the electronic drive to function properly; this transformer is commonly called the *step-down transformer* (usually taking a higher voltage at the primary and stepping it down at its secondary, e.g., 460 V to 230 V). In this case the step-down transformer acts as an isolation transformer, thus the line buffer is an inclusive feature. With some higher horsepower-rated and higher voltage-rated drives, there is sometimes the need to utilize a transformer with more than one secondary. This allows for "phase shifting" on the three-phase input to the drive's converter section. This advancing and retarding of the incoming power to the drive makes for fewer disturbances and minimizes possible supply-line problems. The use of phase-shifting transformers is common in drives the size where thyristors are used. However, this is a special transformer and will cost more.

As shown in figure 8-8, Wye or Delta systems, which could be grounded or ungrounded, are typically used. With ac drives that rectify the ac line and store power in a dc bus, the ac line current waveform is pulse-shaped and consists of the fundamental current (50/60 Hz) and many harmonic currents. To minimize the harmonic currents, a Wye configuration is used to eliminate any harmonic current whose frequency is divisible by three. By using a fourth cable for neutral or ground in a Wye system, all current paths will be defined, minimizing voltage unbalances that occur when currents are conducted through an earth ground "conductor."

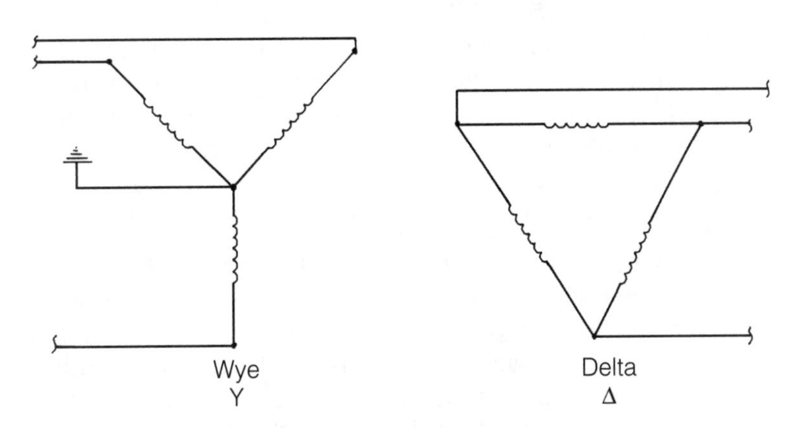

■ **8-8** *Wye and Delta transformer systems.*

The harmonics that are caused in a distribution system are more often due to any unbalance between the phase voltages than by the power equipment taking power from that distribution system. To minimize harmonics, the ac line must have equal voltage waveforms in the positive and negative cycles and must have the same form or shape. Any deviation will create harmonic currents when power is drawn from the distribution system. If a Delta configuration is used and one phase is grounded, the equivalent Wye circuit is no longer balanced. The resulting line currents will not be equal. This can cause harmonic heating, premature line fuse failure, and failure in the input rectifiers used in drive equipment.

When cost is a factor, voltage matching is not an issue, and supply line disturbances must be minimal, a line reactor can be used in place of the isolation transformer. This inductor is dedicated to adding line impedance and is usually not as substantial as an equivalent transformer. It weighs less and is less expensive. The line reactor is also used on the output of an electronic drive to protect the output transistors in times of short circuit on the output line and to reduce the standing wave condition whenever long cable lengths are seen on the output to the motor. The reactor minimizes voltage transients.

Internally, the electronic drive has to have lower voltages to operate its control circuitry. There is commonly a control transformer, which steps incoming 230- or 460-V power down to usable voltage levels throughout the drive's control scheme. Since maintaining control power at an electronic drive is so critical, a constant-voltage transformer is sometimes used as the control transformer. It will cost more initially but might allow the drive to ride through voltage deficiencies. The other alternative is to bring separate control power directly into the drive's control circuit.

Another inductor sometimes seen within the drive is the dc bus choke. This inductor is often used with PWM-type drives and must be used in current source-type drives. It provides a buffer between the converter and inverter sections of the drive and smooths the voltage going into the inverter. It protects against ground faults and provides necessary inductance in the bus circuitry.

As mentioned earlier, between the drive's output and the motor, there is sometimes the need to incorporate an output reactor for protection purposes in certain instances. Another situation where an inductor might be required on the output is to match the motor voltage. While it is much better to have the drive's output voltage match the motor's, it is still possible to use a step-up, voltage-matching transformer between the drive and motor. This really can only be done in applications where the current requirements are low at low speeds, such as in variable-torque applications. When the loading is constant torque, the output transformer could saturate with current and an unwanted condition then exists. Consult with the drive manufacturer when trying to accomplish this.

Most transformers used today have to be designed for both linear and nonlinear loads. Incandescent lighting and line-started motors are examples of linear loads. Nonlinear loads are typical of any electrical equipment that has a power-switching component associated with it. In addition, nonlinear loads include lighting ballasts (fluorescent lighting), metal-oxide-varistors (MOVs), arc equipment, and so on. Thus, nonlinear loads have put an additional demand on transformers. This increased demand can be addressed in new installations, but adding rectifying and phase controlled converters to existing transformers is a topic in need of investigation. Nonlinear loads demand current from the utility that creates higher frequencies. The waveform or shape of the load current is no longer defined by a single frequency. It is a complex shape that contains many frequencies.

When compared to the fundamental, or waveform, we can expect these higher frequency currents to behave differently at 60-cycle current. High-frequency currents will attempt to flow through the surface area of a conductor. When the cross-sectional area of that conductor becomes too restrictive, the conductor eventually becomes hot. When the conductor is packaged inside layers of wire, as is the case within a transformer, the temperature of the transformer begins to rise, and hot spots are created, which will ultimately lessen the life of the transformer.

Transformer manufacturers, knowing when nonlinear loads are going to be present, take this into account when sizing the transformer. The rule of thumb is to increase the size of the transformer, sometimes by 10%, which in effect derates a transformer with a higher rating. Using a larger transformer does not always guarantee that it will run at a lower temperature. A larger transformer will use wire with a larger cross-sectional area. Increasing cross-sectional area does not provide a proportional increase in surface area. Harmonic currents can cause hot spots in oversized transformers. Harmonics associated with drives also increase the eddy current losses in a transformer, and this loss is difficult to predict. Therefore, there are two major issues when dealing with nonlinear loads such as those generated by electronic drives.

One procedure to deal with the hot spots within the transformer that is going to run nonlinear loads is to shape the wire so that there is as much surface area as possible to offer the least resistance for the higher frequency currents. The other is to attempt to specify the amount of harmonics by use of the K-factor. Today's transformers are now classified with a K-factor. The K-factor defines the transformer's ability to handle harmonic currents while operating within the thermal capability of the transformer. Linear load transformers are classified with a K-factor of 1. Transformers with a K-factor of 4 are suitable for moderate levels of harmonic currents. A K-factor of 13 is suitable for greater levels of harmonic currents, as might be seen with current source drives. Likewise, the typical pulse width modulated (PWM) drives used today will carry a K-factor rating of 6. The K-factor is merely an attempt by industry to define and address the issue so that equipment lasts longer and catastrophic failures are kept to a minimum.

There are other issues to consider whenever nonlinear-rated transformers are used. The associated distribution equipment and switchgear such as circuit breakers must be sized to the transformer rating. However, when oversized linear-rated transformers are used, electrical codes dictate that larger, more costly circuit breakers be used. The same can be said for contactors and other electrical devices. Therefore, the issue is important as well as costly. Likewise, when considering line filters to reduce the level of harmonic currents transferred back to the distribution system, other voltage problems can be created with nonlinear equipment. Additional losses can be introduced into the system.

Enclosing electronic drives

Electronic drive equipment needs a safe, sturdy mounting. It is not acceptable to simply start mounting these electrical devices, relays, starters, and other switches on machines and on the walls in the factory or office building. For one thing, this procedure is not safe. For another, it is unsightly and will probably lessen the life of the electrical product. Many industrial users, especially those with not a lot of electrical equipment in their plant, will take the easy route and mount an electronic drive in the most convenient place in the building, which is usually right to the wall. This will drive building code inspectors crazy. However, there exist several alternatives.

Many facilities have dedicated space for electrical equipment. Control rooms that can be given restricted entry are common in facilities that have a substantial amount of electrical product in use. Electrical rooms, mechanical rooms and vaults, along with outdoor buildings all can serve as ideal locations for drives and peripheral equipment. Depending on the size of the drive and the quantity, there might be a need to further enclose the drive or drives. In facilities where specific rooms or vaults are not available and wall space might be at a premium, a drive enclosure might be the only alternative. Some type of a box, floor- or wall-mountable and made of a metal material to withstand the local environment will provide a degree of protection to the workers and the product within. As mentioned, steel is preferable because it provides electrical disturbance protection. This has become the accepted solution for housing most electrical products in today's factory.

Electronic drive equipment does not necessarily have to be located right next to the motor it is controlling or right next to the supply transformer (although keeping the distances short is recommended). Sometimes it is not possible to physically locate the drive right next to a piece of equipment. This is what wire and cable is used for—to connect the two pieces electrically. Therefore, enclosures can be located away from the traffic and actual production areas in the plant, within reason. It still costs extra to run cable longer distances, but there usually exist logical locations to place enclosures. Once a suitable location is chosen, select the appropriate enclosure type for the application and for the actual environment. To aid in this process, standard ratings have been established by NEMA for degrees of protection, both for indoor and outdoor enclosures. These are shown in Table 8-3.

Several factors to consider when selecting an enclosure for electronic drives:

☐ *Total heat content*. Drives produce heat. If multiple drives are to be placed on a panel and into an enclosure, then some calculations must be made in order to determine if the cabinet will be ventilated or cooled. Some drives can be configured such that their heatsinks can protrude out the back of an enclosure. This will virtually eliminate most cooling and ventilation requirements. Consult the drive manufacturer.

☐ *Ventilation*. As is common to most drive enclosures, there will probably be slots and openings for heat discharge. Some cabinets will also incorporate forced-air ventilation with fans. Explore whether or not this is practical for the ambient conditions surrounding the enclosure. Dirt and dust being brought into the enclosure will collect on components and cause overheating. Likewise, if filters are used over any slots or openings, they will have to be cleaned routinely or similar overheating problems will occur.

☐ *Size*. Floor-standing or wall-mount units are available, as well as different heights and depths. Doors should be able to be opened without obstruction.

☐ *Cooling*. When air cannot be brought into the cabinet, then a closed-loop air-conditioning scheme must be considered. This is expensive and also requires some maintenance.

☐ *Layout of drives*. Drives within the enclosure should be located on the panel such that a maintenance person can gain quick and convenient access.

■ Table 8-3 Environmental protection classifications.

Type	Enclosure
1	Intended for use primarily to provide a degree of protection against contact with the enclosed equipment.
2	Intended for indoor use primarily to provide a degree of protection against limited amounts of falling water and dirt.
3	Intended for outdoor use primarily to provide a degree of protection against windblown dust, rain, sleet, and external ice formation.
3R	Intended for outdoor use primarily to provide a degree of protection against falling rain, sleet, and external ice formation.
3S	Intended for outdoor use primarily to provide a degree of protection against windblown dust, rain, sleet, and provide for operation of external mechanisms when covered with ice.
4	Intended for indoor or outdoor use primarily to provide a degree of protection against windblown dust and rain, splashing water and hose-directed water.
4X	Intended for indoor or outdoor use primarily to provide a degree of protection against corrosion, windblown dust and rain, splashing water, and hose-directed water.
5	Intended for indoor use primarily to provide a degree of protection against dust and falling dirt.
6	Intended for indoor or outdoor use primarily to provide a degree of protection against the entry of water during occasional temporary submersion at a limited depth.
6P	Intended for indoor or outdoor use primarily to provide a degree of protection against the entry of water during prolonged submersion at a limited depth.
7	Class 1, Group A, B, C or D hazardous locations, air-break - indoor.
8	Class 1, Group A, B, C or D hazardous locations, oil-immersed - indoor.
9	Class 11, Group E, F, or G hazardous locations, air-break - indoor.
10	Bureau of Mines.
11	Intended for indoor use primarily to provide a degree of protection against dust, falling dirt, and dripping noncorrosive liquids.
12	Intended for indoor use primarily to provide a degree of protection against dust, falling dirt, and dripping noncorrosive liquids other than at knockouts.
13	Intended for indoor use primarily to provide a degree of protection against lint, dust, seepage, external condensation and spraying of water, oil and noncorrosive liquids.

The NEMA ratings will state the degree of protection required but do not always imply the recommended means of achieving that protection. For instance, an enclosure must be completely sealed and placed indoors. There might be heat-producing devices within the enclosure that will not work well (or at all) when the temperature increase within gets too severe. Thus, an air conditioner may be added to the enclosure for local cooling of the enclosure contents. Thus, the integrity of the sealed, indoor enclosure is not diminished. Likewise, the same rating enclosure could also tolerate simple muffin fans pulling air in, over the devices, and back out. This in lieu of an air conditioner. Costs, interpretation of standards, and common sense should prevail.

Switchgear

Power in and out of an electronic drive must have a suitable means of disconnect. Any device that functions in a switch capacity for a given circuit is commonly referred to as *switchgear*. Switchgear will include circuit breakers, high interrupting circuit breakers, fused disconnects, contactors, starters, and so on.The ac and dc drives all utilize some type of switchgear in every application. Every drive needs a way of killing incoming power in order to perform maintenance on the drive. Likewise, many drive installations require a contactor, sized for the current capacity of the drive, to isolate the drive from the motor (ac) or merely to allow proper operation between drive and motor (dc). Many local codes mandate that certain disconnect methods be in place and how and where to implement them. The *National Electric Code* also goes into great detail on the subject.

Drives will utilize a standard circuit breaker or fused disconnect switch to both open the circuit coming into the drive and to protect the rectifier devices. This unit should be sized properly and should be interlocked with any drive enclosure door. Also, the handle should be accessible through the drive enclosure door. Most circuit breakers are defeatable with a standard screwdriver, however, and many drive enclosure doors are opened with high voltage power still on. **This is not a safe condition at all and is definitely not recommended!** After all, the circuit breaker is installed for the protection of the drive and the *worker*! A high interrupting circuit breaker will carry a much higher value for short circuit current and will cost more. However, it might be a necessary device for an installation.

Many drive installations will incorporate a contactor of some kind. These contactors are mainly used to isolate power from a drive or a drive from a motor. They should be rated according to the amperage and voltage rating of the drive. These are common in drive systems that utilize a bypass scheme with a starter. NEMA sizes for some common starters and contactors are shown in Table 8-4. Contactors are typically interlocked with the drive logic and their coils are usually powered by a low voltage, supplied by the drive.

The contactor can be physically heard pulling in most of the time whenever it is energized.

■ Table 8-4 NEMA starter sizes.

NEMA size	230 V	460/575 V
00	1½	2
0	3	5
1	7½	10
2	15	25
3	30	50
4	50	100
5	100	200
6	200	400
7	300	600
8	450	900
9	800	1,600

Typically, the switchgear is mounted within a drive enclosure with the drives. Often it can be mistaken as a part of the drive, when in reality it is just another piece of peripheral electrical equipment. More importantly, if the switchgear is not sized properly (as must be a consideration whenever harmonics is concerned), it can lead to nuisance tripping and failures. Consult the proper codes, switchgear manufacturers, and the drive manufacturers whenever switchgear is involved with a drive installation. Often it might be recommended that the drive supplier provide the switchgear as part of their offering. This makes them responsible.

The drive control

Today, with most electronic drives having one or more microprocessors built-in, control methods have greatly advanced. In fact, the electronic drive now is asked to perform other control functions that traditionally were reserved for programmable logic controllers and other computers. The drive is not just a motor speed controller anymore. But before we delve into PID setpoint control and advanced communication interfacing with the drive, it is important to completely understand what the control circuitry of the drive is and how it functions.

All drives are made up of a power section and a control section. The control section performs several functions. It controls the actual firing of the power section's semiconductors. It tells them when to turn on and turn off. In addition, the control circuit acts as the drive unit's traffic cop. All signals coming in to and going out of the drive are monitored by the control circuit. These signals might be from another drive, computer, programmable logic controller, or from an operator directly. Figure 8-9 shows in block form the many signals transmitted and received by the electronic drive controller. Use of digital data processing has enhanced the drive's performance greatly over the older-style analog techniques of earlier drives.

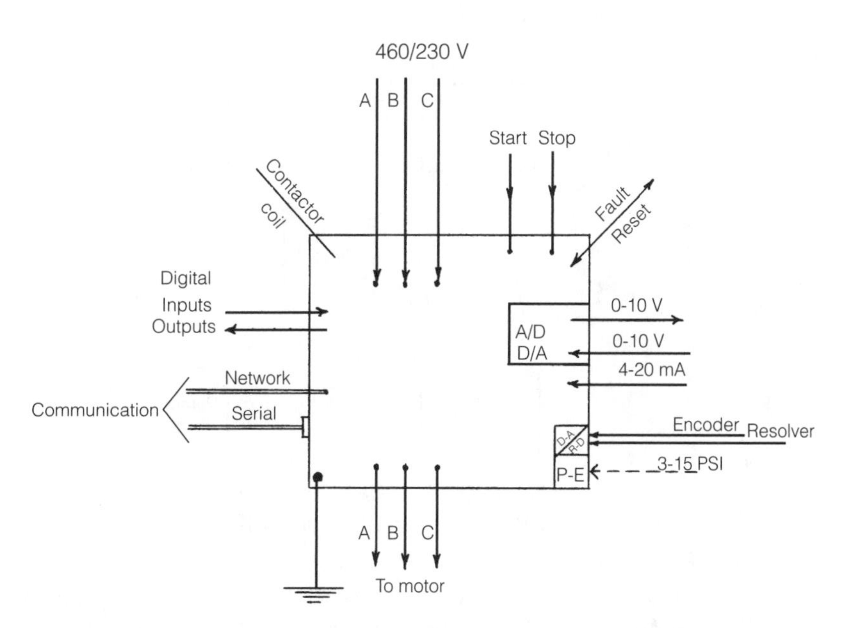

■ **8-9** *Typical variety of signals into and out of an electronic drive.*

First as a motor controller, the optimum pulse and output pattern is calculated by the control circuit. This is based on the motor feedback and intermediate circuit voltage. When compared to a command, or reference signal, the control circuit responds in milliseconds to update the output to the motor. This is the microprocessor's primary role within the drive, and other functions are treated as secondary. However, with today's high-speed, high-throughput microprocessors, many more secondary functions can take place with little processing time taken from the gating of the power devices. Beyond the controlling of the power circuit, the drive's on-board computer system is made up of other elements.

The simplified control scheme for a typical electronic drive is shown in figure 8-10. The control circuit is basically made up of the CPU, or microprocessor, data storage (memory), and an input/output module. These individual components are tied together by parallel conductors which is likened to a bus—specifically, a data bus. The microprocessor is the heart and soul of the electronic drive. It calculates, processes data, and runs the main drive program. Without it and its high-speed performance, the typical assembly line or machine, driven by an electric motor controlled by the electronic drive, shuts down. Thus, our microprocessor is more valuable to the factory than we think.

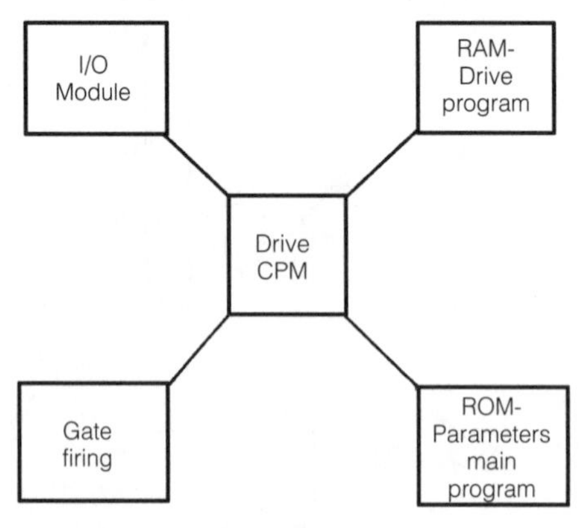

■ **8-10** *The CPU, memory, and I/O modules of an electronic drive.*

The microprocessor gets its instructions from a program stored in memory. Usually a drive will carry two types of memory: EPROM (erasable programmable read-only memory) and RAM (random access memory). The main drive program is usually stored on EPROM. The program is compiled and "burned" into the EPROM, thus keeping it from being lost during power outages to the drive. Other data is placed in RAM for later use by the drive's microprocessor. This other information is typically the specific drive operating parameters and any current pieces of data received from the I/O module. These parameters give the particular drive its characteristics. This can be data stored relative to the horsepower size, base speed, voltage, and full-load current rating of the motor.

Additionally, the drive has to be given information relating to its overload setting, minimum and maximum speeds, and several other pertinent factors important to proper and safe operation of the motor. This RAM is normally battery backed so that this data, typically entered at startup, is not lost when voltage is not present. While running and controlling a motor, the microprocessor is taking data from RAM and interjecting it into the main program as necessary.

The last component in the control circuit is the input and output module. This allows the electronic drive to communicate with peripheral electrical equipment, sensors and feedback devices, other drives, and other computers. Depending on the application and the amount of control given to the drive, there can be many inputs and outputs. It should be noted that the drive should be treated as a computerized motor controller and not a PLC. The input and output capability makes the drive extremely flexible in many applications, but the microprocessor has limits and can "time-out" waiting for an input to satisfy the program. This can lead to motor and drive shutdowns and leave the technicians wondering what happened.

The inputs and outputs at the drive are classified as analog and digital (sometimes called logic or discrete I/O). These can be seen graphically in figures 8-11 and 8-12. The analog signals are commonly in the form of a voltage or current value for a given range. The most common analog signals are 4–20 mA, 0–10 V, and 0–5 V. These signals can be appropriately scaled in the drive and typically are used to set the drive's speed reference. A negative signal can also trigger a reverse speed as with a –10 V. The digital signals have only two conditions: ON or OFF. Any external device's (PLC) output must match the drive's input voltage to function, and vice versa. Typical voltages of digital inputs and outputs are 24 Vdc.

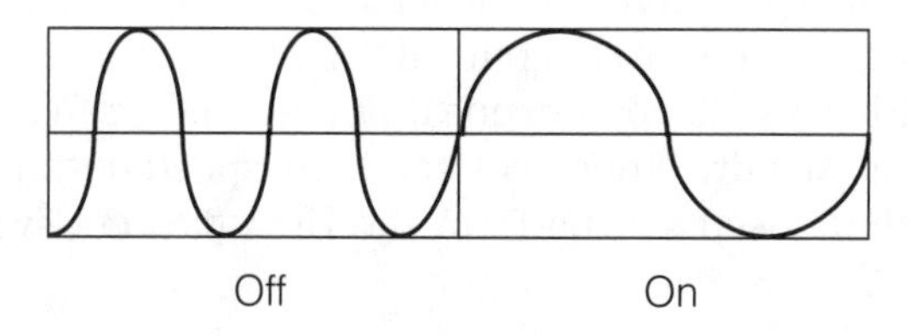

■ **8-11** *An analog signal.*

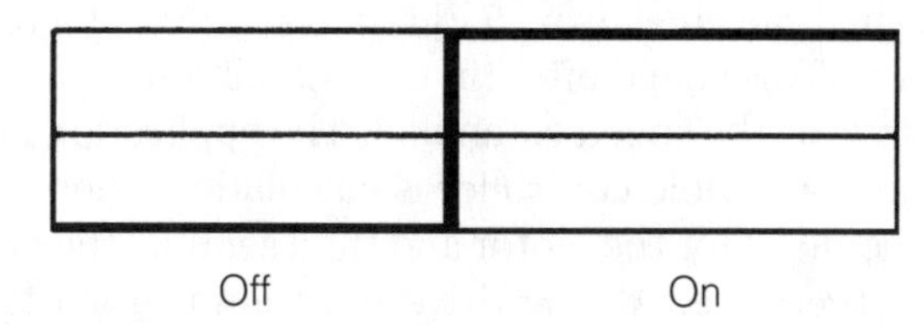

■ **8-12** *A digital signal.*

Drive control schemes

With the control circuitry and control hardware in place, the electronic drive becomes more powerful. As it controls the speed and torque of a motor, it can be monitoring position, temperature, tension, pressure, time, and many other conditions. By utilizing the I/O module the drive can transmit and receive data useful in overall application and process control. But there are traditional drive control schemes worth looking at and many schemes common to many drive applications. As with most factors concerning drive applications, define exactly how the drive and motor are to operate, how they will be controlled, and what peripheral equipment will be connected.

The first concern is how will speed of the motor be controlled, locally at the drive or outside of the drive's control scheme. Most drives have the capability to set the speed directly from the drive enclosure, either by a single-turn or 10-turn potentiometer or by entering a speed via a keypad. This is called *local control*. Local control means that there will be no external loops to be concerned with in determining motor speed. An individual will set the speed at the drive; this is the speed that the motor will run until the potentiometer or keypad setting is changed. Setting the speed at the drive always has to be an option because whenever a service engineer is working on the drive, it might be necessary to set the speed at the drive in order to dynamically see what happens. There is usually a three-position selector switch at the drive, which is called an H-O-A (Hand-Off-Automatic) or M-O-A (Manual-Off-Automatic) switch. Whenever the switch is in the HAND or MANUAL position, the drive controller looks for the speed command from either the keypad or local potentiometer. When in the OFF position, the drive does not send any speed output to the motor. The other speed setting from the HOA or MOA switch is A, or AUTOMATIC.

The automatic mode for speed control tells the drive controller to look at certain terminals for a speed signal. This signal is typically an analog 4–20 mA or 0–10-V signal proportional to a speed. This signal can come from a PLC, a computer, or a remote operator station. This signal oftentimes is conditioned prior to arriving at the drive as the speed command. The application might be such that a programmable controller is calculating where the optimum speed should be for the motor and then sending the corresponding signal to the drive. All the drive does is maintain the speed based on where this analog signal is.

Prior to allowing any motion at the motor the drive must receive a permissive, OK-to-run signal. This might be in the form of a dry contact closure or a digital input from a PLC. With this RUN or ENABLE signal present, the drive automatically goes to whatever speed the analog speed signal is showing at the analog signal terminals. The drive controller also will look continuously at the fault circuit for any external faults, E-stop, or other alarm that indicates it cannot run or should stop running. These fault inputs are also digital. Likewise, the drive might also send various status outputs to other controllers. Whether or not the drive is faulted is a digital signal, while sending an analog signal (4–20 mA or 0–10 V) to another device can be representative of running speed, output current, or power.

In many applications, there is no other controller and the drive is left to do some calculations on its own to determine the optimum speed that the motor should run at for the given conditions. Many fan and pump applications run faster or slower based on temperature or pressure. Thus, the drive becomes a setpoint, or PID (proportional-integral-derivative) controller. This might involve the addition of an add-on module to the drive control circuit so that the proper calculations can be done.

In these types of applications, there might be an air signal (which usually is a 3–15-psig signal needing conversion into an electrical voltage) or a straight electrical signal directly from a transducer or sensor. The drive must be programmed with a setpoint that it can make a comparison with, in order to increase or decrease the speed of the motor. Sometimes the drive will receive just an analog signal for motor speed but be asked to invert this signal in order to treat the higher voltage or current from that reference not as high speed but as low speed (figure 8-13). This applies in air-handling applications where a high-pressure signal means that the drive is requested to actually slow down. Most microprocessor-based drive controls can accommodate this request.

Another control scheme commonly used involves setting speeds for a process ahead of time and logging them into the drive controller's memory. In doing this, an external device or operator control station can select the speed by a push button. This looks like a digital signal to the drive and instructs the drive to go directly to one of these *preset speeds*. With a programmable logic controller sending this digital information, several combinations of just a few digital inputs will correspond to many different predefined speeds (figure 8-14). This can eliminate the need for an analog signal be-

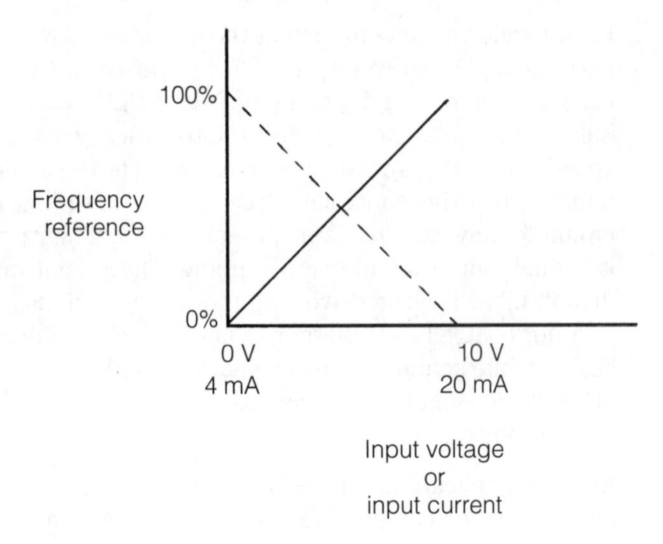

■ **8-13** *An inverted reference signal.*

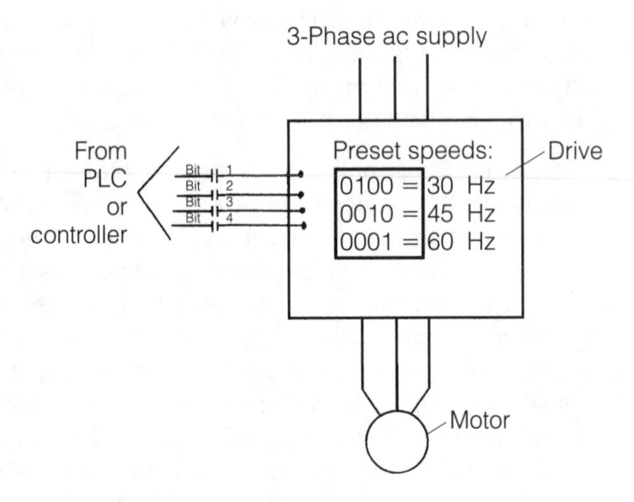

■ **8-14** *Various preset speed configurations.*

ing used for setting the speed, thus ensuring that the speed requested will be the speed the motor runs. Any time a signal is used, it should be isolated; it can possibly be affected by noise and might not always be scaled exactly at the drive.

There are a multitude of other control schemes that utilize special control algorithms and custom modules for electronic drives. Some drives, such as the flux vector, require pulse feedback in order to perform. Thus, modifications must be made to the drive to accept special signals or do special routines. Most drives can accomplish the functions described earlier in this section without special accommodations. In addition, selecting a direction for motor rotation (forward or reverse), a jog or creep speed, or a special volts-per-hertz pattern are common control requests asked of the drive.

Another function that the drive is programmed to do is *ramping*. Ramping or acceleration is a settable feature within a drive. It is usually set as a time-based function and starts a motor with enough current to move the load and then furnish enough voltage and current to continue bringing the motor up to the desired speed. Ramping can be set as fast as possible, or it can be set to take minutes (as in starting high inertia loads). Some drives set the deceleration rate the same as the acceleration rate, while others have separate settings for both. This feature gives the applier of the electronic drive good control of the starting and stopping aspects of the motor. Additionally, some drives can allow for more than one setting for acceleration. This is a nice feature whenever two-stage ramping is required.

Drive communications

In addition to the basic digital and analog means of sending and receiving data between drives and their peripheral equipment, there are other ways of communicating. Whenever two or more devices must communicate with one another, there are basically three data types transmitted as signals: basic control and speed data (run/stop, speed reference, forward/reverse, jog, etc.), status information (drive OK–no faults, drive running, a certain speed has been reached, frequency output signal, etc.), or drive fault or alarm (motor stopped running, motor overtemperature, internal drive faults, E-stop, etc.). Thus, a drive needs to communicate to another device especially in a process or multiple-drive installations. The amount of wires, the distance the wiring runs, and the quantity of drives determine the method of communication used.

The traditional communication method is to run several discrete wires from one device's terminals to another's. Analog and digital control can be achieved in this manner, and this approach has worked well for decades. However, if proper wiring, isolation, and grounding methods are not followed, the potential for noise and related problems exists. With newer high-speed, microprocessor-based electronic drives in service, communications has gotten more elaborate. And even with fiberoptic technology, discrete point-to-point wiring is much better.

With ac, dc, and other types of electronic drives today, serial communication in some form is the most commonly used. There are different versions of serial communications. The RS-232 standard is a common point-to-point scheme where one transmitter and one receiver is utilized. The limitation with this version is the maximum distance of 50 feet. Typically, the voltage level of the signal is between 5 and 15 V. Another point-to-point serial communication scheme is RS-422. This standard can handle 10 receivers with one transmitter. It can also tolerate cable runs up to 4,000 feet long. The voltage level for the RS-422 signal is between 3.5 and 6 V.

Perhaps the most capable serial communication scheme is that of the RS-485 standard. It can handle up to 32 transmitters and 32 receivers. This allows for several drives to be attached to the same bus. Additionally, cable runs of 4,000 feet are feasible. The signal level is also much lower than the others at approximately 1.5 V. Serial communication means that the transmitter and the receiver each must have a driver in place to set up the serial port to the correct protocol. Once this is accomplished and the proper serial standard selected, the two-wire communication scheme works very well.

There is a trend, however, to go a step further from serial communications and utilize the network. The local area network (LAN) is fast becoming a viable communication standard. It utilizes a coaxial cable arrangement able to withstand electrical noise over long distances. The cable runs are apt to run several thousand feet, and transmission speeds are beyond one million bits per second. The drives, when connected to the LAN, can place data onto the bus and receive data virtually at will. The LAN makes for good, real-time updating of drives for master/slave arrangements, peer-to-peer communications, and controller area network (CAN) technology. In a CAN topology each node on the network attempts to transmit whenever the bus is free.

In addition to drives, other devices that appear frequently on a network are digital soft starters, sensors, PLCs, computers, smart monitors, and bar code scanners. With all these devices able to tie together, the facility can be automated rather easily.

Once the drive communication scheme is selected and the hardware is in place, it now must be decided when and what data is passed from drive to drive, drive to plc, sensor to drive, or any other combination. Every application must be analyzed completely to determine the most efficient method. In a master/slave drive arrangement, the master drive can send a speed value to each of the slave drives to follow. In this case, the speed signal is a numeric value rather than a voltage value.

Another benefit that a network affords the drive user is the ability to get much more information whenever a drive faults. Normally when a drive faults, that is all the information another device receives—that the drive fault contact has indicated the drive is faulted. We don't know the type of fault and might want to for resetting purposes. The network can send all this data quickly to make the determination. Thus, drive diagnostics becomes a major benefactor of the network. In monitoring a drive system with a host on a network, trends can be detected and potential failures can be diagnosed before a catastrophe occurs. For instance, motor currents can be gathered and stored over a period of time. If a particular motor's current is seen progressing upwards, faster than the others, this might be an indication that a bearing failure might occur.

Additional electronic drive diagnostics include the logging of all faults with times, how they were reset, and what some current voltage and amperage conditions were at the time of the fault. The future of drive diagnostics will see the drive telling the host that an internal component is going bad or if there is a failure, where to start troubleshooting within the drive.

Setting up multiple drives also becomes easier with a network. "Recipes" of running speeds and acceleration rates can be downloaded from a host computer to be received at each drive. In this way, all the drives on the network can receive the same information instead of having an operator manually going to each drive and entering all the information. This is the preferred method of setting up many similar drives; however, not all facilities will have the luxury of elaborate communication "centers." Another common approach of downloading utilizes a laptop computer, some software, and a serial cable and performing the download function at each drive. This still beats manually inserting hundreds of values.

Closing remarks on peripheral drive equipment, drive control schemes, and drive communications

This chapter has covered many facets of drive technology. Without these peripheral components, switchgear, transformers, all the wire and cable, the drive is just the "black box" many make it out to be. And if these peripheral devices are applied or installed haphazardly, we have the potential for failures and problems. Putting all these physical pieces together requires many disciplines (electrical, mechanical, computer, etc.), many manufacturers (motors, drives, transformers, etc.), and others working together.

It is very easy to blame another electrical device for a problem. Sometimes proving whose equipment is to blame in an electronic drive application is almost impossible. This is why there is a strong movement to place drive system orders with manufacturers who will take full system responsibility or to system integrators who will bring different pieces of equipment from many manufacturers together into a working drive system for which they are completely responsible.

9

Drive closed-loop control and feedback devices

SOMETIMES APPLICATIONS REQUIRE VERY EXACTING SPEED and torque control. To give the electronic drive controller the information necessary to meet these requirements, feedback devices must be implemented into the system. Additionally, any positioning drive system must utilize feedback in order to close the position loop. Ironically, these feedback devices are not going to be the most expensive piece of the system, but they might loom as the most important. This chapter looks at those feedback devices, along with closed-loop drive control schemes in general.

Visualize this scene: A plant manager walks out to meet the plant engineer at their million-dollar "widget" machine. It is not running—not making any widgets. Other workers are in the area also. The plant electrician, mechanic, and machine operator are all looking for the cause of the shutdown. Fortunately, the widget machine has electronic drives on it, and the workers are able to get some diagnostic information off of one of the displays. One of the drives indicates that there was a loss of tachometer (a drive feedback device) signal. The drive, instructed by its program, faults to protect from a motor runaway condition, and the drive also tells the widget machine's master controller that it has faulted. With this drive faulted and not running, the master controller has to stop the entire line.

Obviously, stopping any production line will get many managers excited. Thus, everyone gathers at the "crash-site" to expedite the machine's restarting. Finally, the cause is located: a coupling between a tachometer and motor shaft has broken. Luckily, there is another coupling in stock and it is installed on the motor. The drive is reset and the entire line starts back up and, again, widgets are being produced. Afterward, the plant manager is heard to say, "How could such an inexpensive component in our factory cause a full machine to completely shut down?" No one has the answer.

These sensing and feedback devices are very important to the uptime and performance of machines. It is a catch-22 issue; the plant needs high-performance speed regulation to make great widgets and the way to get the regulation is by utilizing speed feedback into the drives. At present, this is the best method. These components are vital to electronic drive operation. They must be mounted properly to the motor (or anywhere else in the drivetrain), and their signals must get to the drive without interruption. These components provide the valuable information that allows any drive system to correct itself and better regulate a motor.

Open- and closed-loop control

To better understand higher performance motor speed and torque control, we must first review open-loop control versus closed-loop control. Open-loop control, in its most basic form, does not lend itself well to fully automating anything. In figure 9-1, we find that an input signal, or command, is given and a direct resultant output is achieved. We don't affect the signals, and we don't know if what we requested actually happened. It might be safe to assume that if an open-loop control scheme is at work, there are no sensors or feedback devices present in that system.

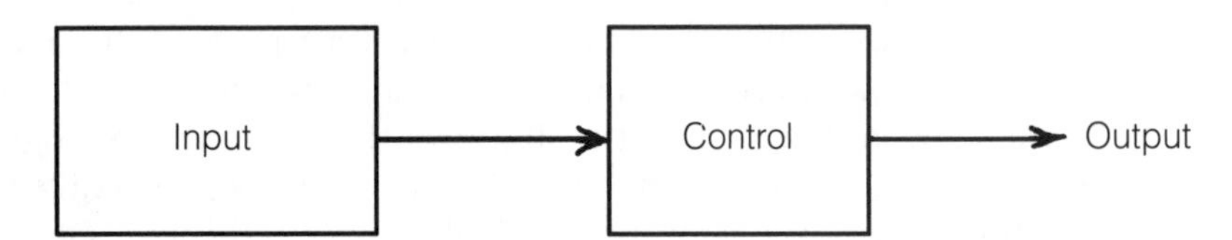

■ **9-1** *An example of an open-loop control system.*

A good example of feedback is the human operator. How often have we asked, "How is the machine being speed controlled, now?" The answer comes back, "Our operator goes over to the machine and turns a speed potentiometer a half-turn or so, and the machine is running where we like it!" The operator is the feedback device. He is continually correcting the system. If the operator forgets to adjust the potentiometer, the machine runs as "open loop." The speed remains wherever it is commanded to be, unaware that there might be a speed-related problem. This is a prime candidate for the implementation of some automatic sensor or feedback device to close the loop and automate the speed regulation loop.

The closed-loop system is shown in figure 9-2. Note that there is now another component in the diagram that is providing informa-

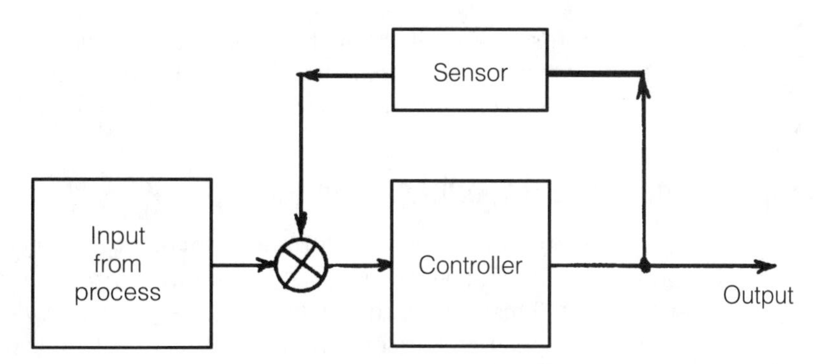

■ **9-2** *An example of a closed-loop control system.*

tion about the process. This information returns to a summing, or comparison, point for analysis, and then new output is provided based on the comparison with the process input signal, or the "what-should-we-be-doing" signal. This is a cause-and-effect relationship between devices. The feedback device tells of any actual error in holding to the prescribed condition, and the controller makes the attempt to correct for that error. A closed-loop control scheme is constantly correcting itself.

A good example of a closed-loop system is an automobile's cruise control function. A speed-sensing device located about the car's wheels lets us know how fast the car is traveling, and that information is fed back to a control point to compare to the desired mph's we have set. The more often the error signal is sampled (sent, received, and corrected to), the better the overall system's accuracy.

With electronic drives, many types do require feedback in order to function at all. Others, such as the common ac variable-frequency drive, do not. The ac variable-frequency drive is given a speed command, and it sends a corresponding frequency and voltage out to the motor. There is a pseudo-feedback loop in that the counter EMF from the motor is being monitored against the speed command and there is some correction for speed error. However, this type of drive is not considered a closed-loop system. Once a pulse generator is attached to an ac motor and accommodations are made at the ac drive to accept that feedback, then we have a closed-loop ac drive (some form of ac vector drive).

Likewise with the dc motor; there is usually a tachometer to provide armature speed feedback or even an encoder added to the dc motor to provide more exact feedback to control motor speed better. As for servo systems, feedback is the essence of these applications. The term *servo* implies closing a loop in some way. Either controlling the speed extremely well or positioning a motor shaft, the servo drive (and even stepper drives) depend on feedback.

Speed and current feedback methods

Since motion usually involves a rotating element, or electric motor, the feedback devices are most often directly connected to a motor shaft. This way, as the shaft spins, the sensing device gathers speed or position data or both. Prior to feedback devices in the electric motor control arena, there were attempts to control the motors with the electrical data available. The ac motors can provide a form of feedback in the electrical circuit called counter-electromotive force (CEMF). This is a form of voltage that the electronic drive monitors. The drive takes this voltage through onboard devices called *potential transformers (PTs)*, and it is converted into a usable form for the controller's logic section.

Likewise, in current sourcing drives there are *current transformers (CTs)*, shown in figure 9-3, that provide a current reading from the wire running through it. This current value is also used in the drive's control logic section. This electrical data can be used by a motor controller to determine how fast a motor might be running, with some inaccuracies, and thus provide correction to the output signal to the motor. Depending on how well the speed regulation has to be for the given motor and application, this means of feedback can be adequate.

In addition to current transformers, other methods utilized by electronic drives to sense current are Hall-effect sensors, ac shunts, and dc shunts. The dc resistive shunts are very similar to ac resis-

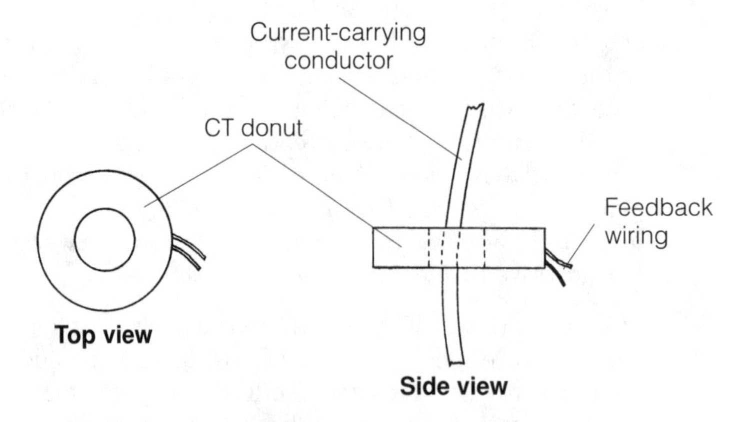

■ **9-3** *Current transformers.*

tive shunts but are mainly used in dc motor and drive systems. Both are low-cost methods of measuring current, and easy to understand because they merely apply Ohm's law. They require no external power to operate, are very reliable, and there is no zero offset (at zero current there is zero output). Although they are used from time to time in drive products, they are not electrically isolated and therefore are susceptible to electrical noise. The output commonly has to be amplified to be used in the control circuit. For these reasons they are not widely used.

A Hall-effect sensor is shown in figure 9-4. A magnetic core with a coil surrounds the wire. The Hall generator measures the value of current and then this signal is amplified. A separate power supply is required, unlike the current transformer, which does not need one. Similar in appearance, Hall-effect devices and CTs are often confused with one another. They are different in that the CT only can measure ac current whereas the Hall-effect device can measure both ac and dc current. The Hall effect is similar to the current transformer (CT) in that it is reliable and electrically isolated. Many times the Hall-effect device is used to measure current within the electronic drive.

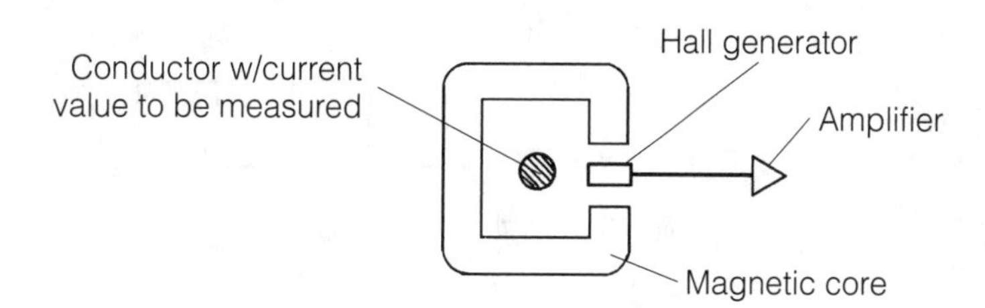

■ **9-4** *The Hall-effect current sensor.*

The dc motor, when controlled by a rectifier circuit, can provide armature voltage feedback to the dc drive for speed regulation. As the controller sets the output to the motor, the electrical circuit in it is monitored to try and maintain the desired speed. As speed regulation has to be held tighter, better feedback means must be incorporated into our system. Rather than relying on armature feedback, we might install a simple tachometer, an analog speed feedback device. This device will attach to the motor and produce a voltage output to the controller in the form of so many volts per 1,000 revolutions per minute of the motor. With this information we now have a more accurate way of comparing the actual speed with the commanded speed and can correct accordingly.

Other devices, discussed below, are available that can provide even more precise information for motor speed, and their use will only better the speed regulation in a particular process.

Other feedback devices often used in conjunction with electronic drives include Hall-effect sensors, and ac and dc shunts.

Optical encoders

A device to enhance speed and often shaft position is the *encoder*, or *digital pulse tachometer*. These are shaft-mounted devices that work on the principle of electrical light pulses through a disk with coding, or slots, over one revolution of the disk; they then utilize a light pickup sensing device to count the light pulses. In this way, data relative to where the disk is and how fast it is turning can be sent to the electronic drive for corrective action.

The encoder basically translates mechanical motion into electronic signals. A depiction of this electronic signal is the pulsetrain, shown in figure 9-5. This feedback device is called variously an encoder, optical encoder, digital tach (short for tachometer), pulse tach, or pulse generator. Understanding what it can do for an electronic drive application and how it works is more important.

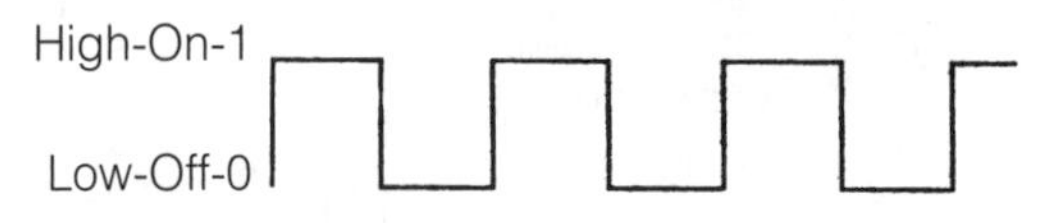

■ **9-5** *A digital pulsetrain.*

There are basically two types of encoders: the incremental encoder and the absolute encoder. There are two versions of each, rotary and linear. When electronic drives and rotational equipment such as electric motors are involved, we are mainly concerned with the rotary type.

The most commonly used and least expensive is the incremental encoder. The *incremental encoder* is made up of the light source, disk, light receptor, grid, and amplifier. As shown in figure 9-6, a circular disk has several slots along its perimeter. This disk is typically made from a glass material with marks imprinted as the slots, or it can be made of metal with precision, machined slots on the outer edge. The light is emitted through the slot and received by the receptor through the grid assembly. The signal is then amplified and sent to a host for further use. This entire sequence can be seen in figure 9-7.

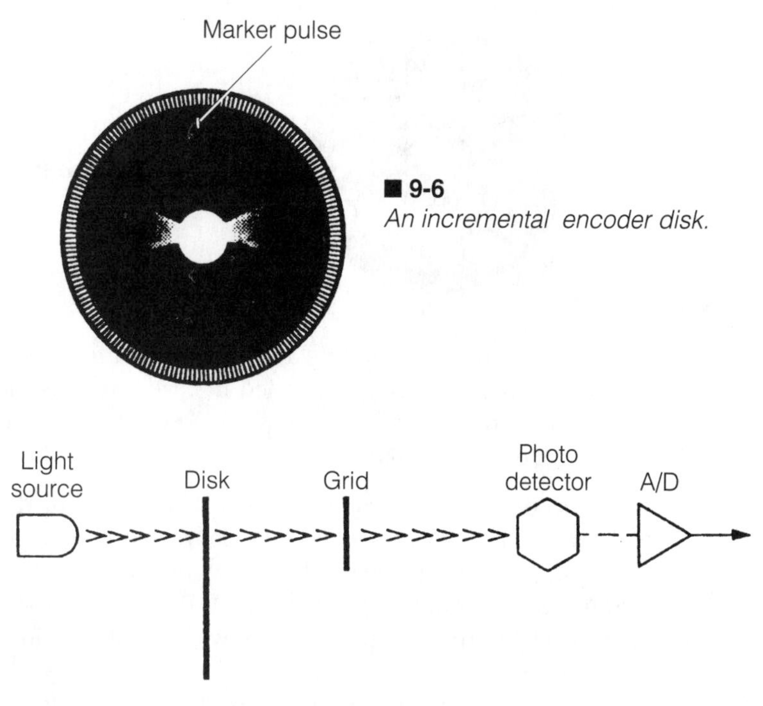

■ **9-6** *An incremental encoder disk.*

■ **9-7** *The optical encoder system.*

The actual number of slots on an encoder disk is critical to the drive application. Encoders are specified as having 240, 600, 1,024 or 2,048 pulses per revolution, most often in powers of two, which keeps us consistent with binary math for the drive's onboard microprocessor and which will aid in the absolute encoder disk design later. The more pulses per revolution, the more precise the system's accuracy. In a high-precision drive application, we had better request more pulses per revolution on the encoder and be able to properly decode that signal within the electronic drive. Also, keep in mind that more slots on the disk probably will mean more expense and a physically larger-diameter disk. Additionally, there must be a marker pulse somewhere on the disk to use as a reference.

An *absolute encoder* disk is shown in figure 9-8. Here, the disk has a unique slot design, and there is a definite reason for the pattern. These slots are actually concentric, getting larger as they get closer to the center of the disk, and again, there is a binary relationship to the pattern. By passing light through all the slots at a given instant, the encoder can provide an exact position. This method of absolute decoding solves for the binary patterns on the

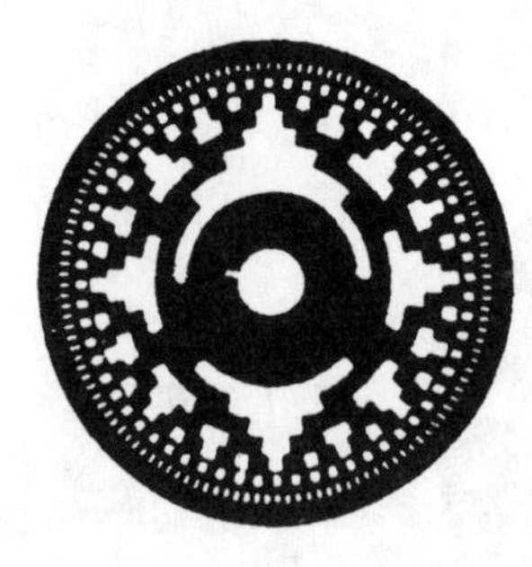

■ **9-8** *An absolute encoder disk.*

disk. These types of encoders are mainly used for positioning applications where position is needed in case power is interrupted. An absolute encoder does not require a "homing" sequence to find a starting reference point.

Most encoders employ a quadrature form of decoding. Here, a second channel employs a second light source and receptor. The second channel is physically positioned within the encoder a half-slot distance away from the other channel. By doing this there will be more available data points to work with, as shown in figure 9-9. This can provide better resolution (i.e., more pulses per revolution). This quadrature approach also allows for the detection of rotational direction—forward or reverse.

These units can be large or small and can be mounted in different ways. Attention should be given to applying these devices. Glass disks can sometimes collect a film on their surface thus impeding

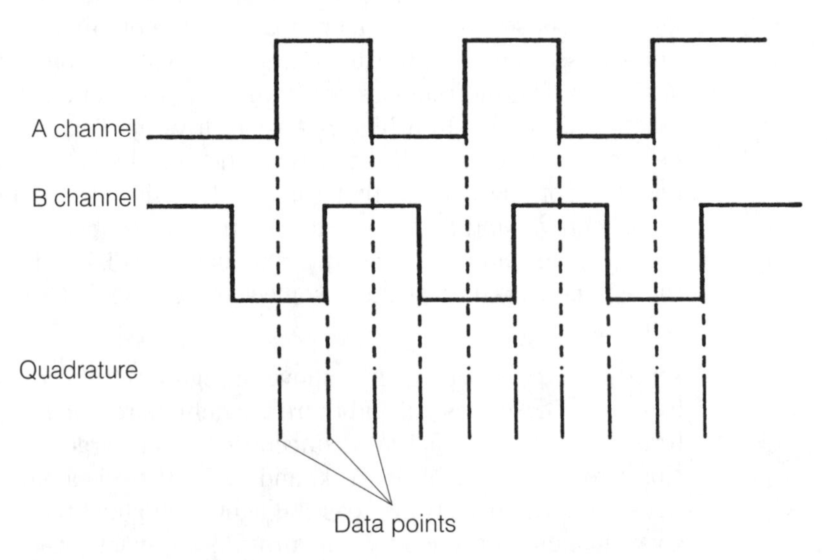

■ **9-9** *Quadrature signal decoding.*

good light transmission. They also can break. Metal disks, although more expensive, might be a better choice. Various housing methods are used to protect the electronics and disks within.

The mounting and housing of the encoder is very critical to its operation and extended life. If mounted loosely or misaligned, premature bearing failure on the rotating element of the encoder can occur. Likewise, the coupling from the encoder shaft has to be flexible and aligned as well as possible. The encoder shaft is to be as free-wheeling as possible for optimum feedback performance. Any drag indicates a future problem could arise. Additionally, rigid couplings should be avoided because they can break easier. Earlier we provided an example of a plant manager who was angered by a simple feedback device shutting down his entire plant. These situations can be avoided if sound connection methods are utilized.

With devices such as encoders providing more accurate speed information and providing that same information faster to the controller, motor shaft speeds can be maintained to even 1 or 2 rpm's in a steady state load condition. Of course, loads tend to change, which makes our encoder/control correction system work harder.

Pulse generators, or encoders, carry with them many electrical and mechanical characteristics. Electrically, they are usually supplied with a low voltage of 12 to 24 Vdc. This voltage value must be held within 5% plus or minus, or else the output can suffer. The output is a low-voltage pulse in the form of a square wave. This square wave is made up of two channels, 90 electrical degrees out of phase from each other (quadrature). The signal is differentially driven with a marker pulse (home or null). The pulse-to-pulse accuracy is equal to plus or minus 2 arc minutes in 1 revolution (RMS, or root-mean-square). It has a workable frequency range of 0 to 100 kHz, which refers to the clock, or crystal, speed for stable velocity determination. Typical operating temperature ranges are 0 to 70°C, or 32 to 178°F.

Mechanically, the encoder has to have suitable housing. Depending on the motor surroundings, a particular housing might be more suitable than another. For harsh, dirty environments a typical housing can be of machined anodized aluminum, with shafts made from type-416 stainless steel. Both metals are corrosion resistant. The shaft rotation should contain no backlash. Zero backlash is the absence of mechanical "play" if the shaft were turned. The shaft can turn in both the forward and reverse directions with no effect to the motor to which it is attached. The whole mechanical and physical purpose of the pulse tachometer is to basically ride "free-wheeling" with the motor shaft, adding as little inertia and drag as connected. The encoder will have low torque and inertia values.

Bearings are very important with any feedback device that is attached and expected to run on a motor. If the bearings are a weak link, the entire process can go down without notice. Typical bearing life expectancy of an encoder should expect 15 to 20 million rotations. The lifetime in hours depends on the speeds that the motor and feedback device will run. The bearings are usually sealed with grease, thereby making them basically maintenance free. The cover should be also made from extruded aluminum and the connectors are MS, or military style, which screw down and lock for extra protection. Encoders must have good shock and vibration characteristics because they ride along with the motor.

Unfortunately, encoders, so vital to a machine or process uptime, are often abused. Sometimes individuals will actually step on an encoder housing to reach something above the motor. Therefore, the mounting of the encoder and housing has to factor into the equation the human factor. Making the connection and housing mount substantial is the best solution!

Resolvers

The resolver, or rotary transformer, is another type of feedback device used primarily in motion control applications. More drive applications see it used with servo and stepper systems. The resolver is different from the encoder and pulse generator devices in that there are no electronics at the resolver; it is separately excited from a source external to the physical resolver itself. The resolver is a rotary transformer, with a rotor component and a stator component. It is in essence a small ac motor, excited by a voltage to itself. Figure 9-10 shows a graphical interpretation of a resolver, mounted to a motor.

Like the encoder and its mounting, resolver alignment with the motor shaft is very critical. A flexible coupling can provide some cushion, but this feedback device needs to be mounted as concentric to the motor shaft as possible to minimize premature failures or bad readings. The wave shape to the resolver is actually what is being looked at in the circuit. Because the resolver is excited by a sine wave, low-voltage signal, this signal can be decoded within the controller (with power source). This decoding is accomplished through a resolver-to-digital (R/D) conversion. The converter is

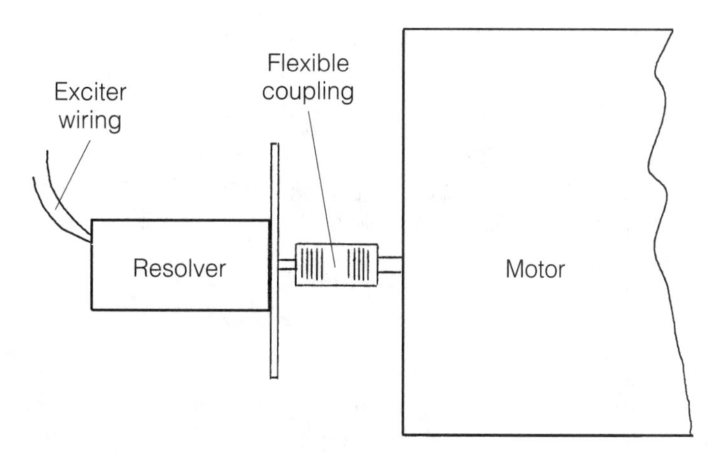

■ **9-10** *The resolver, or rotary transformer, feedback device.*

actually looking at displacement in the returning sine wave to determine shaft location. In this way, the loop is closed to the motor, or rotating device. Accuracies for resolver systems are presented in arc-minutes and usually are in the 6 to 7 arc-minute range for most resolvers.

Because there are no electronics within the resolver, it is more suitable for nasty, dirty environments. Also, the resolver is an absolute measuring device. It can retain the exact position of a motor shaft even through a power outage. The feedback from the resolver can also be driven great distances, over 1,000 feet if necessary, with good resilience to noise. However, it is always recommended that signal wiring be routed separate from any power wiring wherever possible. A single resolver can provide resolution accuracies in some cases up to a 14-bit number, or 16,384 counts. Dual resolver packages can also be utilized and can greatly increase the measurement accuracy in a particular feedback system. In these cases, a fine resolver is geared to a coarse resolver at some ratio. Common ratios for dual resolver packages are 99:1, 64:1 and 32:1.

Magnetic pickups

The magnetic pickup is a feedback device package made up of a magnetic sensing component and a toothed wheel. It can act much like a resolver. The gear, or toothed-wheel, is installed in conjunction with a magnetic pickup sensing device, as shown in figure 9-11. The objective is to get rotating shaft position data accurately and consistently as the magnetic pickup sensor, working in unison with

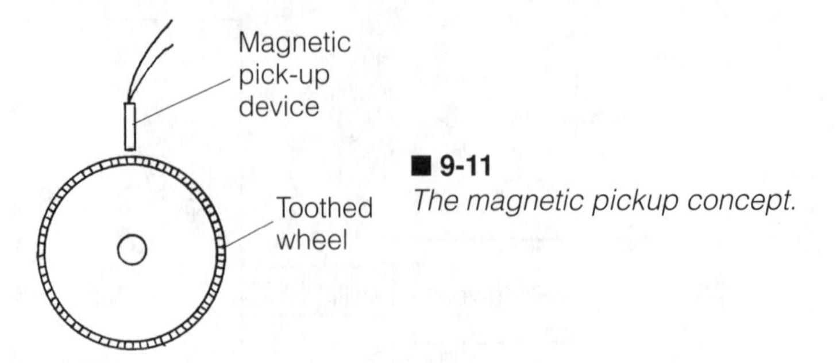

■ **9-11** *The magnetic pickup concept.*

the toothed wheel, counts the teeth on the gear as they pass. Once set up and properly aligned, this magnetic sensor is fairly reliable. Depending on the environment, the initial cost expectations, and motor/controller application, the magnetic pick-up device selected is a critical component to the process.

Peripheral feedback and control devices

In addition to drive and motion control feedback devices, the process or machine might have other feedback devices directly interfacing to the electronic drive. These devices might not have anything to do directly with the speed, current, or torque of the motor but rather are needed to start and stop the drive, accelerate the motor at a certain rate, or select a preset speed at which the process now wants the motor to run.

Different types of feedback and control devices have dedicated functions, including temperature, pressure, flow, and ultrasonic sensors. Primarily to the drive, these devices will trigger whenever a drive event is to occur. Their feedback signal can run directly into the drive. That signal can be as simple as a digital on/off state or as complex as having the drive invert the analog signal to decrease speed as pressure increases in a fan system. These other sensors will have very similar needs to the motion sensors previously discussed. They will need a good, steady-state power supply. They will also have to be noise immune and reliable.

Peripheral feedback devices most often are used in a PID (proportional-integral-derivative) control scheme, where the electronic drive is a major component. Without these transducers or feedback devices, there can be no proportional, integral, or derivative loops to control. There is no closing of any loop whatsoever. Once again, open-loop control is the absence of feedback devices. There are two types of closed-loop feedback, negative and positive. Negative feedback will reduce the variable in the controller's routine, which will make a change to the process or machine. The direction of the signal—for instance, negative or down in voltage—will be interpreted by the controller as reason to change the output accordingly. Conversely, for positive feedback, the feedback signal's direction requests an increase to the variable that the feedback represents.

Transducers

The *transducer* is a common feedback device used in gauging, measuring, and detecting displacement. A transducer usually is defined as a device that converts mechanical energy into electrical output. A better, more appropriate definition is that variations of this device can convert any input energy, including electrical, into output energy. The output energy will, in turn, be different from the known input energy, thus supplying useful feedback information from which closed-loop control and correction can be achieved. There are literally hundreds of different kinds of transducers and just as many ways to provide the input and output to the transducers. The most common include pneumatic, hydraulic, electrical, and ultrasonic types.

Transducers sending an electrical output signal sometimes send that signal as a voltage, 0 to 10 Vdc, and sometimes as current, commonly 4 to 20 mA with 4 mA being an off-condition, 20 mA being full-on, and any value in between being the range of the transducer. This range is generated from a stimulus to the transducer and corresponds to some direct action from which the transducer's output signal is going. For instance, a process involving the unwinding and rewinding of a material, as shown in figure 9-12, utilizes a linear displacement transducer, sometimes called a dancer. Changes in the dancer position correspond to a voltage drop across a resistor. This voltage output value from the transducer indicates how loose or tight the web material is and sets the span accordingly. From here, a correction can be made by the motor drive controllers to increase or decrease speeds and/or torques.

A piezoelectric device is another type of feedback device that produces motion from an electrical stimulus or conversely, produces an electric signal from a motion, usually a strain, or load. This type includes strain gauges and accelerometers. Another device is the RVDT, or rotary variable differential transformer. This device is a unit employing a rotor and stator relationship to produce voltage. No slip rings are used because the electrical output is via the elec-

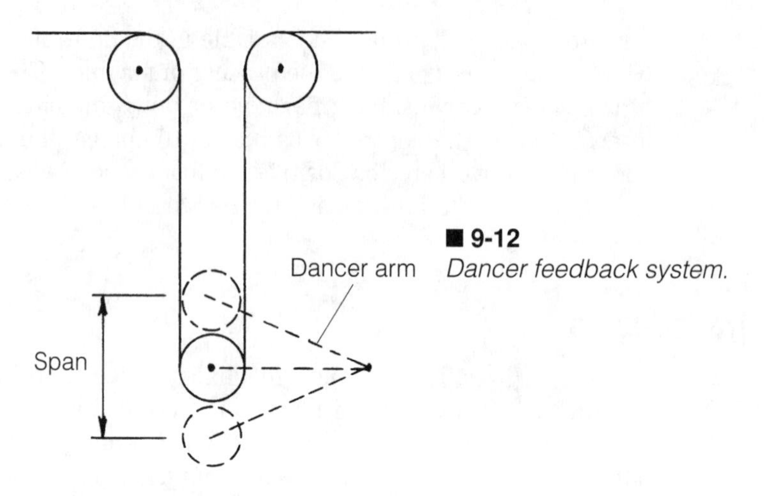

■ **9-12** *Dancer feedback system.*

tromagnetic relationship of the stator windings and the rotor. It produces a voltage, usually 0 to 10 Vdc, whose range varies linearly with the angular position of the shaft. Figure 9-13 shows a typical RVDT device's performance curve for output voltage versus input shaft position.

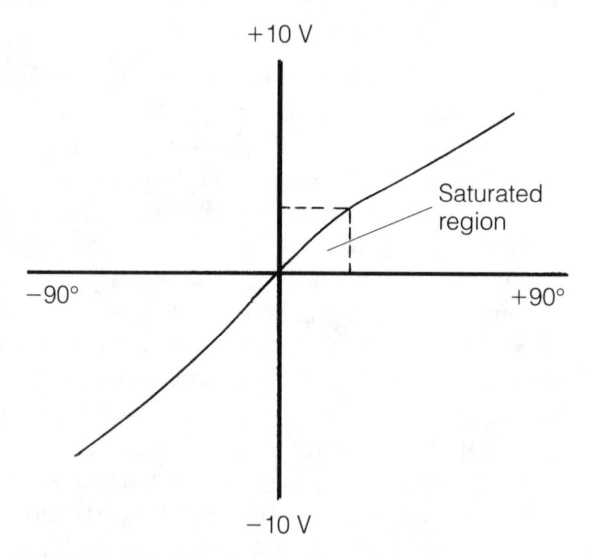

■ **9-13** *Rotary variable differential transformer (RVDT). Output voltage versus input shaft position.*

Temperature control and pressure sensors

Many times with process control there is the need to control temperature. Sometimes the temperature can attain extremely high levels. Thus, an industrial grade of thermometers has emerged. The class of temperature-sensing devices used that generate electrical output are called *thermoelectric devices*. One common device is the thermocouple, which utilizes two metallic components and will conduct an electric signal when their junctions are at different temperatures. It is sometimes referred to as a thermal junction device.

Other temperature-sensing devices use a direct output linear to the measured temperature whenever feasible. A thermistor is an electrical-resistive device that relies on its resistance to vary with a change in temperature. A thermostat is yet another temperature-sensing device which often utilizes a bimetallic scheme to sense when a predetermined temperature has been met. At this point a single output signal is provided, thus starting or stopping another piece of equipment. A good example of this process is a furnace's operation.

Temperature-sensing equipment in the factory of today has to be able to interface with many higher-level systems. This means that the simple thermometer doesn't suffice, and digital displays, along with microprocessor-based units, come into play for communicating with other devices and for charting and recording temperature data in a process.

The same is true for the present-day pressure-sensing devices. They must not only be rugged for industrial use but also must be fast-acting and high-precision in order to be useful. These kinds of devices, much like their temperature-sensing counterparts, get installed virtually in-line with the fluids that they are measuring or sensing. Thus, the integrity of the unit is challenged as it physically sits in the wet and hostile environment. Common pressure ranges go up to 500 mb (millibars, the unit for measuring pressure), and the accuracy of the pressure sensor is rated by its linearity, or error from steady-state input/output, its hysteresis (the difference in response to an increase or decrease in the input signal), and its repeatability (the deviation over many readings that should be the same).

Many times the pressure sensor converts the pressure signal into an electrical signal, and other times this conversion is done in an external controller. This is sometimes an overlooked facet of the pressure-control loop. This conversion module has to physically reside somewhere, and the application and environment will typically decide where that location is. Once located, the conversion is simple, usually taking a 3- to 15- psig or 0- to 100- psig pressure signal (scaling is important) and changing it to a 0- to 10-Vdc or 4- to 20-mA electrical signal. From here, the controller scales the electrical signal further for its use, and a resulting output signal is generated.

10 Drive integration and machine control

THROUGHOUT THIS BOOK WE HAVE ANALYZED AC AND DC drives, electric motors, drive peripheral devices, feedback devices, and ways to size and select the drives. Putting all the ac and dc drive components together into a working system is called *drive integration*. This might mean tying multiple drives together or merely matching one drive with other electrical pieces of an installation. Many times it is hard to predict how each electronic component will interact with another and with the mechanical counterparts. Experience helps. So does in-depth familiarity with a particular manufacturer's product. But most successful drive system installations take a combination of knowledge, experience (the school of hard knocks), perseverance, and a little luck. It takes a team of dedicated players to integrate drives.

With electronic drive integration have come new disciplines such as the automation engineer, the motion specialist, and the system integrator. These individuals have emerged to provide the specialized expertise necessary to combine, program, and coordinate all the pieces. Because the technology is so diverse, these specialists are asked to make systems out of many dissimilar components. Often individual components are specified and even purchased for a particular machine or purpose. Then the integrator is called in. The integrator must make everything work. The better scenario is to go back to the initial steps and involve the integrator from the beginning. Better yet, many projects are awarded to a systems integrator based on the integrator's ability to provide the solution, for a fair price.

This makes for the turnkey system, which means the system is built or installed ready to operate; the customer just needs to "turn the key" and he or she is in business. Many installations don't go that smoothly. Most often, we are dealing with precious design and capital procurement dollars. Thus, customers demand what they paid for, a working system.

Often the system integrator's job is difficult when machine components are reused to save money (sometimes the job costs more anyway because of lost time when new parts have to be purchased after all). This is the classic situation called the "machine retrofit." And whenever electronic drives are concerned, the philosophy is that the motors, drivetrain, and all interconnecting components can be reused. If a formal analysis has been made of this existing hardware, and it has been deemed functional and proper, the drive installation probably will go well. The system integrator is employed to ensure that it does.

Drive integration

As we have seen throughout this book, electronic drives by themselves are virtually useless. Their main purpose is to control an electric motor and its subsequent load. But often the drive must connect to several other computerized pieces of equipment in the process, air-conditioning system, or factory. This connectivity means that power and signal wiring, communication protocols, and actual control techniques have to be considered. Couple all of these electronic devices with all the mechanical components for one drive and motor system, and we have an extensive package. True drive integration encompasses all of this and then some. The next step is to integrate several drives, several mechanical sections of a machine, and even many machines.

Two scenarios for integrating drives are shown in figures 10-1 and 10-2. The first depicts multiple drives connected within one machine. The second shows two machines, a pick-and-place robot and automated roll grinder, working in unison. In figure 10-2, servo drives drive the robot, dc drives, and ac drives within the roll grinder. In these drive applications, there has to be some type of master controller to coordinate the machine activities. The individual drives can efficiently perform their independent functions of controlling motor speeds, torques, and positions when necessary.

Installing a single ac or dc drive can be classified as drive integration. To the end user placing a single drive into service, this can be a major task, as shown in figure 10-3. That single drive might be the heart and soul of their factory, and therefore it has to connect to the other electrical switchgear (circuit breakers, fusing, etc.), transformers, plant computer system, a program-

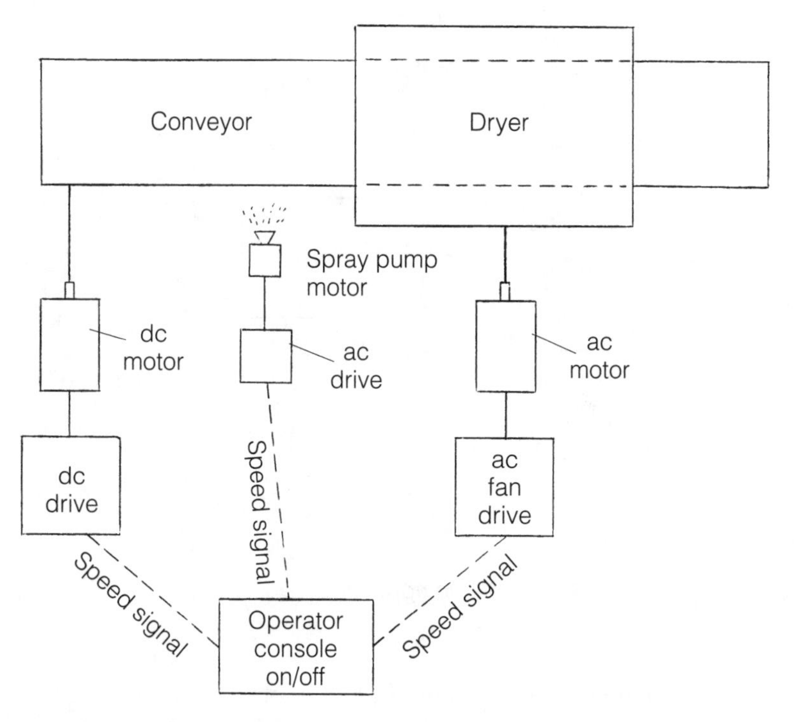

■ **10-1** *Multiple drives in one machine.*

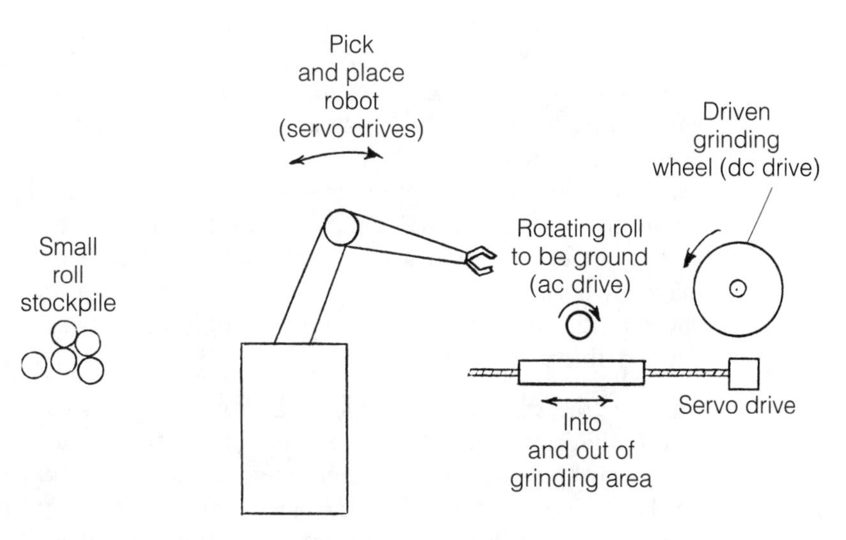

■ **10-2** *Two machines, multiple drives.*

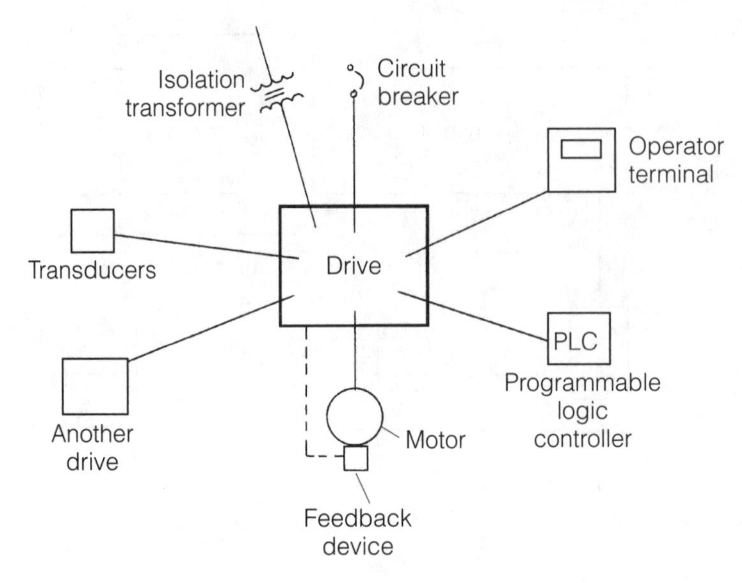

■ **10-3** *A single, integrated drive.*

mable logic controller (PLC), an electric motor, a drivetrain, and a machine. This is a lot of equipment centering around the electronic drive. If it is misapplied or sized incorrectly, we have problems. Whether involving single or multiple drives, computers or PLCs, LANs, or WANs, the system integration comes with a great deal of responsibility.

Who are these system integrators and where do they come from? They are engineers and technicians, frequently former employees of a large corporation's engineering group who got caught in a corporate downsizing or simply got frustrated. They are in business for themselves to do whatever it takes to make a drive system installation successful. The system integrator might also be a resident employee, perhaps as part of a design and engineering group within that company. However, with so many corporations downsizing their engineering staffs and outsourcing these services, the system integrator is usually a contracted entity.

The integrator is often fully specialized on certain applications or electronic products. Often the system integrator is a former employee of the company for which the contracted engineering is to be done. The integrator knows that company's rules and procedures and even knows quite a bit about the machines and their controls already in the plant.

System integration is fast becoming a big industry. A manufacturing facility gets new budgets every year. Projects are planned to be done within a given year's budget. However, the facility no longer has the expertise in employees to implement the system required. Either a system integrator or a supplier of electronic product with system engineering capabilities is called. Each will have expertise in certain product, software, and machine application, which comes into play more so for retrofits of old or existing machines. And this can be a nightmare. If the wrong integrator is selected or if the electronic equipment is misapplied, then the project can take forever to finish and might not work the way everyone had hoped.

A well-known case in point is the fancy, high-tech baggage-handling system that was to be implemented into a new airport. The baggage-handling system could not be made to run properly, and this held up the opening of the airport. In this case millions of dollars were lost. This kind of thing happens very often on a lesser scale at many plants and factories around the country. That is why retrofits, turnkey systems, and integration must be fully qualified beforehand.

A system integrator checklist, or spreadsheet, similar to the one previously shown in figure 1-1, might be in order when embarking on a project. So many times, system integrators imply that they are well-versed in a particular software language, when in reality they have a manual, some source code, and have scrolled through some programs. This does not mean that they can quickly and effectively get a machine software program written, debugged, and running in short order. This is a tough item to qualify with an integrator unless he or she is known to the end user or comes highly recommended. Software and programming hours are hard to track, as is whether or not any real progress is being made. That is why interrogating integrators regarding their expertise and application experience is always recommended.

Retrofits are always interesting because so many things can be overlooked. They also carry with them the surprise factor. Since most are older machines, many components that are to be reused don't satisfy electronic controls. If a positioning system is dependent on feedback from a mechanized device and that device has backlash between components, there is no possible way for the feedback to be consistent. The controller only corrects to the signals it receives. It's easy for someone to say that the controller simply doesn't work; however, further investigation by someone more knowledgeable will provide the proper solution.

The systems integrator will typically be a person or group of persons capable of acting as a contracted third party to implement a system. These individuals will be engineers or programmers who have certain specialized skills for the application. Since so many disciplines interact in machines and systems work these days, integrators can and do have many specialized talents. For instance, motion control applications involve knowledge and expertise in many areas. The American Institute of Motion Engineers offers a certification for individuals in this industry. What is interesting is that the certification process does not take into account education and experience with, say, only electric motors. Computer controllers, PLCs, drives, sensors, and many other technologies are part of the scope. Remember the term *automation engineer*. This is the new breed of specialist needed in industry today.

Systems integrators provide solutions. They do not usually promote certain types of hardware unless the end user wants it or the integrator feels that it is the only acceptable hardware. The integrator promotes the hardware that not only can do an effective job, but also hardware they are comfortable with and can confidently implement. They can offer a turnkey solution or work together with plant electricians and technicians to install electronic equipment. A turnkey system is often preferred by the end user but frequently harder to implement by the integrator. This, too, is often a subject of misunderstanding. The end user expects to go out to the machine and "turn the key," and the machine starts making parts. The integrator might have a different interpretation. Therefore, a complete written specification of what is required should be agreed to by both parties.

Machine systems and applications

Behind so many computer-driven machines are electronic drives. Without an effective means of controlling the electric motor, there would probably be no robots, no high-speed coordinated process lines, and automation would not exist in the magnitude we know today. Again, if it's in motion in the factory or plant and that motion is being controlled, then somewhere in that factory is an electronic drive or controller. Electronic drives and the machines they drive and control allow better product due to drive consistency and performance, more product due to less waste and higher motor speeds, and more efficient use of the energy to produce all these products.

Many electronic machines are capable of higher speeds and producing more product than before. This involves integrating electronic drives, running them at and beyond their design capacity (yes, many users of drive products push those drives to their limits and beyond, and get away with it in the form of more throughput). However, drive life can be shortened when stressing the power components. The trade-off is more product over a shorter period of time.

Some facilities would rather purchase new hardware in order to keep producing at extreme levels. All this because today's needs come from a world economy and marketplace, and there is more demand. One day a machine will produce a certain product, and the next day a different product is produced on that same machine. This is production planning and is based on new orders and demand. Even though electronics has brought with it the ability to minimize retooling times often by just changing the software, there can be other factors that don't allow instant changeover from product to product. Thus, the plant production planner will run as much of one certain product for as long as possible when the machine and drives are running good and cranking good product out.

Tension control

One particular use of electronic drives involves tension control. More specifically, these applications call for regenerative drives capable of holdback torque. Industries such as paper, film, foil, packaging, and web converting all demand this requirement from the drives. This makes these applications predominantly dc, although ac drives are making some inroads especially with flux vector types.

A partial web converting line is shown in figure 10-4 and has to coordinate the activities of many motors and other machine functions. The line, or machine, is converting a web of material as it unwinds a roll of raw material. The web can be slit into narrower strips, coated and dried, laminated to another web, painted or printed on, and then wound into a final roll. There are zones between certain sections where tension must be controlled. These are the tension zones.

The main control theme in lines such as this is tension control. The tension between sections of the machine has to be not too tight and not too loose. This can be accomplished by precisely speeding up a downstream motor or by maintaining better control of torque at each motor. In these systems torque equates to tension. A line

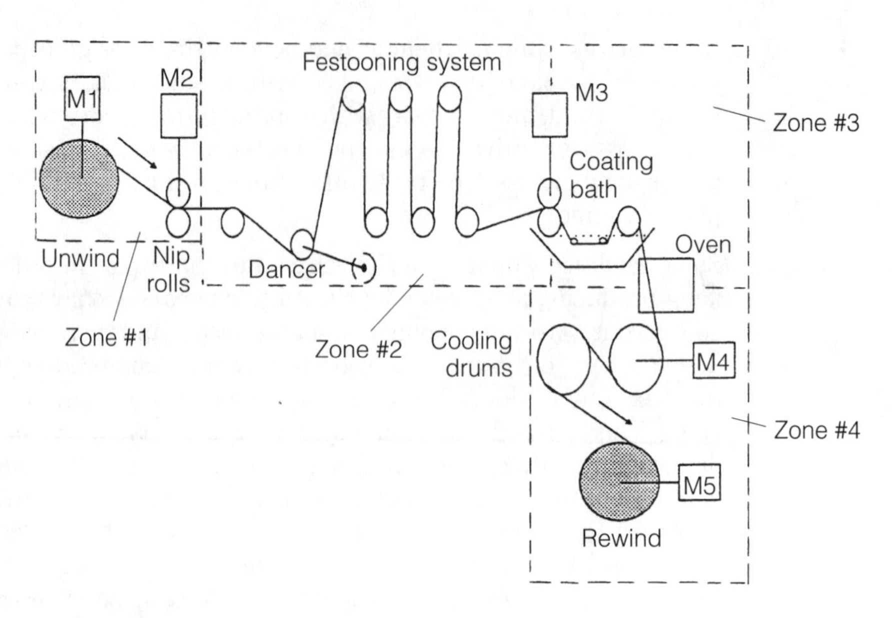

■ **10-4** *Web converting line tension zones.*

such as this needs to have high-speed, high-response control equipment running the motors, electronic drives, and other machine controls. If not, webs break or unroll all over the place, making a mess. The goal is to run the process at the fastest speeds possible. Therefore, controllers must coordinate between devices sometimes down to the millisecond range for electronic and processing response.

Many sections in the web converting line will derive transducer feedback from the line to use in controlling tension. This can be in the form of voltage from a dancer, load cell, strain gauge, LVDT, or ultrasonic device. This signal is fed into an electronic drive or other master controller for use in high-speed processing. Once processed, the tension feedback is now in the form of a new torque output to the motor. Additionally, line speed and overall line tension are set by a local operator via a remote operator station or by a master controller.

These types of lines depend quite a bit on remote control and regenerative motor control, both of which are discussed later in this chapter.

Coordinated drive machine control

Machine control applications involving coordinated electronic drives are those mainly of the machine tool, NC, and CNC industry.

By the mere virtue of their nomenclature (CNC, or computerized numerical control), these machines have computerized front ends and have to exhibit coordinated control between drives. This coordinated control can mean coordinating speeds, torques, or positions between drives.

Sometimes it is more complicated than that whenever multiple functions are concerned. A good example is a lathe. Here, a part must be presented to a tool that will do work to the rotating part. The motion is the rotation of the part, the tool's in-and-out movement, and a traversing motion that the tool must make up and down the length of the part. Obviously, the three axes of the machine must work coordinately, especially to minimize the number of passes, back and forth, that have to be made.

Sometimes the three-axis machine tool lathe is part of a complete machining center. This can be found in a typical modern manufacturing facility. Parts move from different machining centers down the assembly line. There will have to be some master, central, or cell control for this work area of multiple machines, each with multiple drives. A master controller is in charge; think of it as the traffic cop. In this way, there should not be a bottleneck to the assembly line. Each electronic drive performs its own routines and motor control, thus allowing the master controller the freedom to coordinate each machine, rather than being overwhelmed with controlling each drive.

Adding machines and retooling for different product runs can be accomplished quickly when needed. Changing speeds, accel rates, and individual drive programs beats changing out mechanical gears, adding belts, linkage, and other machine components in order to run a new product. This is a perfect example of exploiting electronic drive controls in the factory.

Remote control

Often electronic drives are located away from the motor they are controlling. The drive, or drives, might be located in an enclosure in a control room or on a mezzanine. Therefore, it is common to locate an operator console at the motor or machine so that as adjustments are made, the operator can physically see the change at the motor. This operator console can be as simple as a start-and-stop push button with a speed potentiometer or as complex as a full-color monitor with touch-screen capability. The latter is obviously much more elaborate and expensive. It will require special communications schemes, local microprocessor ability at the console, and possibly additional cooling at the console. These systems are typical in multistory office buildings that might have an energy management system to give a central command center, via a main console, the ability to view any fan or pump motor drive. By polling the drive we can find out drive status, speeds, currents, and so on.

A typical operator's console is shown in figure 10-5. It houses some basic electronic drive control devices for a two-drive system. Those devices include push buttons for starting and stopping either drive, a maintained jog push button, and a drive reset push button. Many times it is convenient to have these push buttons lighted so they can be seen from a distance. Also on the console

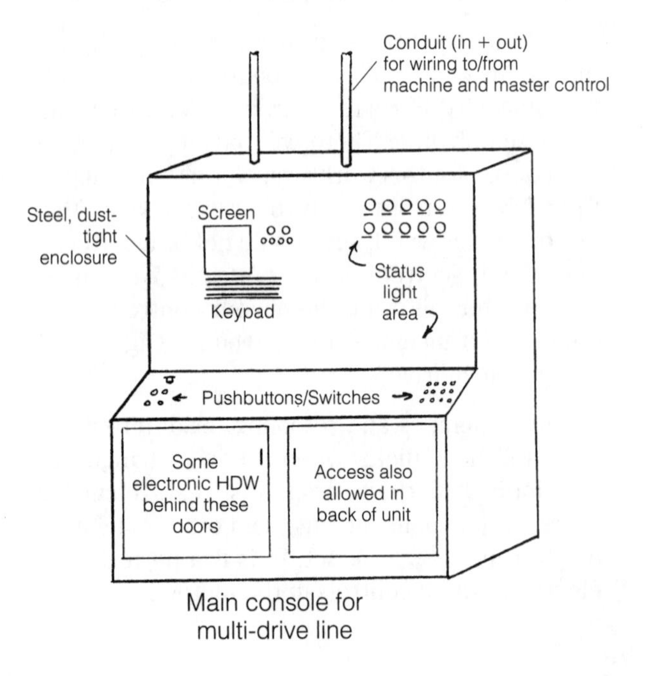

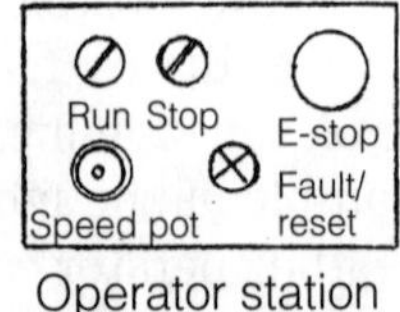

■ **10-5** *Typical operator consoles: one for a large, multidrive line and another for a single drive.*

will be a device that can dial a speed into the drive. The speed potentiometer is very common and provides an analog signal proportional to requested motor speed. Every application's control needs are different and so, too, will be its operator's station. All stations must have an emergency stop (E-stop) push button with a large, red, mushroom-shaped head for safety. Other applications can utilize reverse switches, tension potentiometers, various lights as alarms, and so on.

Many times this operator console information goes to a host computer before it is, in turn, sent to the drive. If other drives or other machines are affected by the operator command, the main computer might need to alert the other devices accordingly. Sometimes the host controller monitors other conditions about the overall machine. It might calculate what speeds and torques individual drives should run at based on various inputs into it. Likewise, the main controller can download "recipe" information for various runs of product for that particular day.

Sometimes the electronic drive is the slave in the system and other times it is the master. In either case, the electronic drive is not the "dumb" device. Its microprocessor capabilities are being utilized somehow, to what extent is up to the user.

Braking and regeneration

One of the often-confused topics in motor control and in electronic drive applications is braking and regeneration. How does it work whenever an electronic drive is used? What type of electronic drive should be used? If the drive has regenerative capabilities, is dynamic braking required, also? Which method is the fastest at stopping the motor? Which is the safest? Does the application even require braking? Should a mechanical brake be used? Many are good questions, and some have more than one answer.

An electric motor, whether dc or ac powered, moves its desired load and demands that amount of power to do its job. However, when the load decides to drive the motor, the motor now is a generator, a generator of power. Where does this energy go? This energy coming from the motor is called CEMF, also known as back electromotive force. The magnitude of this electromotive force is the main concern in braking or regenerative situations. If this CEMF has a channel to get fully back to the source of supply power, then it is called *regenerative*. If this channel is used to slow or stop a motor and its associated load then it is called *regenerative braking*. A high-inertia, low-friction load, an overhauling load, a crane or hoist—all are applications in need of stopping. The choices are controlled stop, dynamic braking of the motor, or regenerative braking of the motor.

Regenerative braking assumes that there is a means of getting the load-generated energy back to the supply mains. It also assumes that this means, or channel, is operating. If a current source ac drive or a regenerative dc drive has faulted and is no longer in control of the motor because it has lost its control power, regeneration through its power bridges via the firing of SCRs is not going to happen. Thus, it is usually customary to include dynamic braking as a safeguard in those applications needing ensured stops. This involves adding a circuit (contactor-activated), which upon power loss will dissipate the motor-generated energy to a resistor bank. This will stop the motor quickly.

Dynamic braking takes mechanical energy that is being backfed through the system as electrical energy and dissipates it as heat at a resistor bank. Voltage source drives, diode rectified drives, and PWM inverters utilize this approach to braking an electric motor. This method of dissipating regenerative energy is also used to slow a motor so as to provide back tension, or holdback torque, as in winding and unwinding applications. This is used, of course, when regeneration back to the mains is not possible. It should be noted that when a drive that is not fully regenerative tries to control a motor in a generating state, the dc bus voltage rises quickly and an overvoltage fault will occur. If the regeneration rate is too dramatic, power devices can actually explode!

Obviously, dynamic braking wastes energy as heat. Therefore, if the application brakes often and quickly and the loading is heavy, then alternate methods should be explored. The regenerative drive is one. It allows this electrical energy to go back to the mains. Regenerative drives must have the appropriate number and type of power semiconductors in order to accomplish this. There is a premium for this extra hardware, and it must be decided if the energy losses in heat outweigh the up-front costs of the hardware in a regen system.

Another method used as an alternative to dynamic braking is *common bussing*. In this scheme, regenerative power can be used to power another motor that is in a motoring state. This power is resident on the bus for the appropriate use. It can only be used in those instances where the machine has a motoring component and a generating component. This can typically be found on a line that has an unwind motor and a rewind motor.

To our electric motor, it doesn't matter if braking is accomplished via resistors or through regeneration. The motor has become a generator and that energy is going somewhere. The route in the motor and drive circuit for this energy is shown in figure 10-6. The motor controller must have the necessary logic and means to handle this energy. Today's dynamic braking utilizes solid-state componentry and drive-integrated logic to ensure proper sequencing. The conventional contactor to remove the armature power supply in a dc system can now be replaced with an SCR.

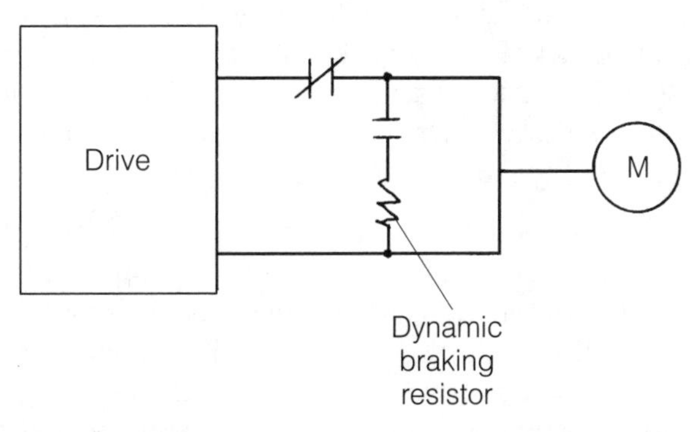

■ **10-6** *Dynamic braking (DB) theory of operation.*

We are trying to take this motor-generated current and reuse it as braking torque. Deceleration is a form of regenerating motor energy to achieve slowing and stopping. Deceleration can be described as controlled stopping. It is common to see a holding brake, electrically actuated when there is no motion at the motor. Once in normal running mode, these types of brakes should not be used so that the CEMF dynamic braking through resistors or back to the mains can be optimized.

Sizing of dynamic braking resistors usually is left to the drive manufacturer. However, it is easily calculated if you know the following:

- ☐ Duty cycle—braking times and how often
- ☐ Maximum speeds
- ☐ Power rating of motor and efficiency
- ☐ The dc bus voltage

Another form of braking is *dc injection braking*, sometimes called simply dc braking. A dc voltage is forced between two of the phases in an ac motor, causing a magnetic braking effect in the stator. This type of braking is commonly found in systems that can tolerate braking at frequencies of 2 to 3 Hz and lower. The reason for not applying this type of braking at higher frequencies is because the brake energy remains in the motor and could quickly cause overheating.

Why do drives stop running?

Sometimes an electronic ac or dc drive stops running. It has tripped or faulted, and when this happens, we get a domino effect. The electric motor stops running, and thus the load that it was driving has stopped. This might stop another part of this machine or other machines in an assembly line. This is how one little drive fault can cost a factory thousands of dollars in downtime. We have seen in earlier discussions why drives fault and how to troubleshoot drives. Drive faults happen for a reason—most often to protect the motor or itself from damage.

There is never a good time for a drive to fail, especially in the production environment. But there are optimum times such as at startup or when maintenance or machine overhaul work is being done. But most of the time the drive quits during peak demand. And when the drive quits running, the technicians spring into action. It is their job to find out why it stopped and get it back online.

Unfortunately, this troubleshooting process is not always clear, and to further complicate the picture, many times the problems and reasons why the machine goes down are unexplainable. Try telling the boss, after losing a full day of production time at a thousand dollars per lost hour, that the drive just started to run after you hit RESET. Simply clearing a drive fault, at the right time, will get the machine running again. You might lose credibility, but it might not be your fault. Electrical noise, intermittent disturbances, and other factors come into play with sensitive electronic devices.

Sometimes the unexplained drive occurrence remains unexplained—until the next occurrence. Eventually, if records are kept, a trend might develop, and corrective steps can be taken. Many electronic controllers now have battery-backed historical log functions for handling faults and diagnostics. When a fault occurs, the machine's controller logs the time, what the fault is, and how the machine was reset. This is better than keeping a written record at the machine, which usually gets dirty, lost, or forgotten. If the plant technicians can become knowledgeable about a machine and its controller, then the high-priced manufacturer's field service personnel won't have to be summoned. These trips can cost thousands of dollars, and many times simple solutions are found.

When it comes to downtime and time studies, good record-keeping is essential. Besides having personnel properly trained, good documentation (manuals and drawings at the drives and extra copies in an office), and a dependable source for support, knowing how, when, why, what, and who did what and when will make the drive installation a reliable installation. Bad data in equals bad data out (i.e., garbage in, garbage out). So often records on the plant floor are incomplete and misleading. Outside personnel can be hired to come in and do these analyses, but this can be expensive and unpredictable to have someone watch a machine. Disciplined machine operators and technicians can log good, pertinent data for later use when a problem occurs.

A sample drive downtime analysis log is shown in figure 10-7. This form should be taped to the drive enclosure's door or tucked within a pocket on the enclosure door so that it is readily available to the next operator who has to manually reset the drive. It is very helpful if the operator logs onto the sheet the time, the fault type, if the drive displays this, and whatever else happened prior to, during, or after the fault/reset. Of course, this format might have to be customized for a particular machine or purpose. Once data is collected, the manufacturer of the machine or controller can be summoned, and productive troubleshooting can commence.

So many times a user calls a manufacturer with incomplete information about what really happened, and the manufacturer has no clue as to what's wrong. This then frustrates both parties and nothing can be resolved. Having lots of good data available and doing some good, preliminary troubleshoooting (using the manual provided) will go a long way in getting the electronic drive up and running.

Drive training and uptime

Why do employees need training, retraining, and more training? Why do so many spare parts have to be inventoried at the plant to minimize downtime? Why aren't warranties of electronic drive products for longer periods? Why does field service for drives cost as much as it does from a manufacturer? The answer to all these questions: because drive technology is constantly changing. All are sensitive, expensive issues with industry today. Continually evaluating each to know where one stands with personnel and hardware, and not neglecting any one issue, will help keep a user of specialized, electronic equipment more secure and independent.

DRIVE DOWNTIME LOG

Drive number:
Drive name:
Drive supply voltage:
Drive current rating:
Motor data:

Incident	Date	Time	Fault	Solution	Initials	Remarks

■ **10-7** *A drive downtime log.*

Electronic drive training has never been more important. With so much digital drive equipment, there's good news and bad news. The onboard microprocessor makes for better diagnostics and troubleshooting because this information is available today where it wasn't a few years ago (with older drive technology). On the other hand, a drive's microprocessor requires a program and is sensitive to electrical disturbances. In addition, if the main CPU fails, how do you troubleshoot?

Employees change jobs and functions within the industry. An employee who is trained on a particular system might not be available later on, so other employees have to be trained and cross-trained. This is why technicians need to be trained on many different products and applications. The training doesn't end with a few days here and there; it must be ongoing. Technology is constantly changing and manufacturers change their designs to suit their needs, not necessarily the end user's. Therefore, the electronic drive technician must keep up with these changes.

Training is expensive; it could be argued whether it is really justified. The answer is in the question, "How important is it to keep these drives and machines in service?" If the machine operator can do some minimal troubleshooting, the machine might keep running for an additional couple of hours. At a thousand dollars an hour in downtime, a lot of training can be bought!

Spare parts inventories are one of those black holes that companies wish they didn't have. Say a machine goes down. You don't have a spare board. You call the manufacturer and order one. The manufacturer tells you that you should be stocking the boards, or that the boards you have in stock are obsolete. So they ship you two boards, one to use now and one for the stockroom. The probability is that the spare board will never be used. Thus, the spares inventory grows and grows. Technology changes and new electronic equipment is installed in the plant. More spares are placed into the storeroom. And rest assured, manufacturers do not take back unused spare parts.

One solution, if 24 hours of downtime can be tolerated, is to rely on the manufacturer of the machine to keep in their stock all spares available for overnight delivery. Most often they will have the parts, saving thousands of dollars in spares inventory which may never get used.

Warranty issues first arise when negotiating the purchase of an electronic product. One year from date of shipment is standard and covers parts and labor. When a board fails within the first year, it should be fully exchanged for a new one. Some manufacturers offer two-, three-, and even five-year warranties. And they should. Many electronic products have gotten fairly reliable, and there is no reason why longer warranties shouldn't be the norm. Also, many times with systems projects, electronic product sits in a receiving area for several weeks, even months, before it is installed. The warranty period should start at the time of installation. Warranties can be negotiated and longer warranty periods can be requested.

When all attempts fail to get the machine running with available spares and personnel, it is time to bring in the heavy hitters: the factory authorized field service engineers who get paid from the moment they leave their house until they return to it (portal to portal). Their cost per hour can range from $100 to $200 depending on the complexity and the rarity of the product. A company with a down machine will often pay big bucks to get back online (this is where having a local system integrator, who can provide service at half the hourly rate, is valuable). Having one's own facility personnel trained and keeping them well compensated can head off these expensive field service charges!

Closing remarks on drive integration

Integration can involve merely packaging the drive into a compact space on a machine or it can be as elaborate as attaching a high-speed network to a multidrive installation. If anything, the electronics, controls, and automation industries will continue to grow. New equipment will be developed and company needs to keep up with the technology will grow. Technological change is occurring at an exponential rate. One or two important developments can affect the entire industry. This change could mean that new products will be introduced, quickly making existing equipment obsolete. Good system integrators will be hard to find. Once a good working relationship is established between an end user and a system integrator, that integrator will practically become a full-time employee of the end user because of the many hours spent in their factory.

Old-style, mechanical machines will be replaced with high-tech, faster electronic models. Energy efficiency and reliability will be factors. Robots with their electronic drives will continue to be implemented, especially into hazardous applications. Drive controllers will become more powerful, and technicians will constantly have to train, be trained, and educate themselves in order to keep pace with technology. Like it or not, we are on this course and must learn to accept the technology rather than fight it. Which brings to mind the question, "Are we driving the drive technology—or is it driving us?"

11

Electronic drive applications

THROUGHOUT THIS BOOK WE HAVE DISCUSSED VARIOUS types of ac, dc, and specialty drives. We have looked at the hardware, software, and control concepts, along with electrical and motor basics. This knowledge is the basis for applying these drive products in industry and just about anywhere there is an electric motor. Many applications have traditionally been successful with one type of electronic drive or another. But today, as the different drive technologies tend to merge, many applications can be successful with the "user's choice" of drive product. Where a dc drive was the only solution in the past, an ac drive might perform well—perhaps even better and for less initial cost.

Whenever looking into applying a drive in a particular application, you will need to set some basic priorities. The particular needs of the user, location of drive, and performance required should determine the priorities. Following is a list of issues to consider:

1. What are the hardware costs? Compare ac to dc motor and drive package costs.
2. Will there be software costs? Some digital drives need to be programmed.
3. Will one drive type (ac, dc, or other) be easier to install than another? Is the motor existing and will it be reused?
4. Is the plant familiar with the drive of choice from a troubleshooting standpoint? Will new spare parts have to be purchased?
5. Performance. What are you trying to accomplish with the use of a drive? Which type will give the best results? (This should be the number-one consideration, but often is not!)
6. Environment. Can a dc motor be used? Can an ac motor be used? Where will the drive for the prospective motor be located?

7. Drive technology and integrity. Are both reliable and substantiated?
8. Drive supplier service, support, and experience. This is another major issue. When installing the drive, the drive manufacturer's experience in that particular application will be relied upon. Once the drive is installed, the drive manufacturer's support and service will be important.
9. Startup assistance. Can and will the drive supplier get the application running—quickly and proficiently?
10. Integration. Many times the drive is just one piece in the entire application's electronic make-up. Many times drive suppliers have to either supply additional, peripheral components or be able to interface with a variety of like electrical components.

Obviously, this list does not contain all application considerations. Every drive installation has its own possible pitfalls and idiosyncrasies. Use the aforementioned list as a starting point and guide. For additional reference, Table 11-1 shows a matrix of different drive types, their basic speed performance capabilities, and a few other notes.

■ Table 11-1 Comparison of drive types.

Drive type	Typical speed regulation	Typical speed range*	Remarks
dc drive (analog)	1–3%	300 : 1	Analog tachometer feedback
dc drive (digital)	0.01%	600 : 1	Digital encoder feedback
ac drive (open loop)	1–3%	20 : 1	Only CEMF feedback
ac flux vector drive	0.01%	1200 : 1	Digital encoder feedback
Servo drive	0.01%+	1200 : 1 +	Brushless motor w/encoder or resolver fdbk.

* Auxiliary motor cooling assumed whenever necessary

Beyond that, the rest of this chapter is dedicated to several individual applications and issues common to each. Again, these application notes are to be used as a guide. Every day, the technologies are being tested. There is also more than one choice for many installations. Carefully consider as many factors as possible for your own application. And always consider safety a top priority. Remember, we are dealing with high-voltage electrical products in the drives, and controlling motors that are usually moving heavy objects. People can get hurt and even killed!

Application: Extruder

Possible user: Plastic film manufacturer; film packager

Scope of application (figure 11-1): Extruders are made up of a die section, encased screw system, motor, and oftentimes, a gearbox. Extruders typically have a wide speed range. A plasticized material is forced through an extruder die while being heated to form a specific cross section or shape.

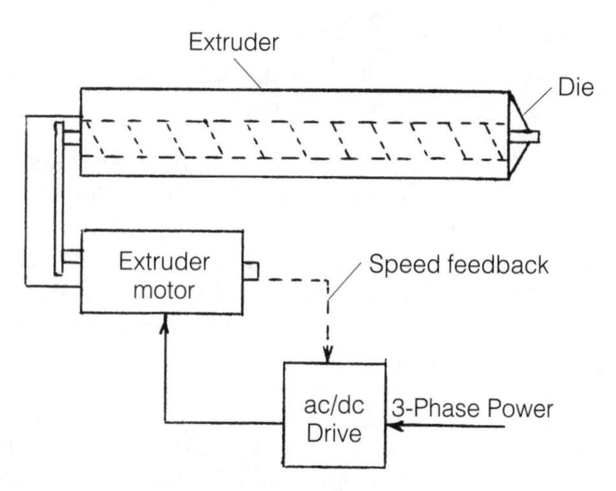

■ **11-1** *An extruder application.*

Specific concerns: Since the extruder has a wide speed range, serious attention must be given to the low speeds. Full torque is required at these low speeds. To compound matters, no noticeable pulsing or interruptions can be tolerated from the drive's output to the motor. Speed regulation has to be extremely tight. Any pulsing will have an effect on the material. It can mark the material, make it too thin or too thick, and an interruption can break the flow of material altogether. Smooth motor operation at low speed, full torque is mandatory. Also, the extruder drive might have to deliver 150 to 200% breakaway torque when the extruder is started and the heating temperature has not fully heated the material being pushed through the die. Another issue with extruders is that they are a nonregenerative application and cannot tolerate being reversed.

Drives commonly used: Traditionally, these applications have been done with nonregenerative dc drives and motors. The dc motor's higher level of rotor inertia helps the motor to ride through load changes and to run smooth at low speeds. However, ac flux vector drives have been able to perform well in these extruder applications. Instead of using rotor inertia to ride through load changes (the ac motor's is lower than the dc motor's), the flux vector drive overcomes this obstacle with fast response and full control of the flux angle for torque.

Benefits: An ac drive package with blower-cooled ac motor, feedback device, and ac vector drive is less expensive than the equivalent dc system. Also, an ac motor is better suited for explosive atmospheres, which are prevalent in many plastic film plants. Explosion-proof dc motors are very expensive, and they still require brush maintenance.

Application: Centrifugal fan

Possible user: Commercial user with supply and return fans on a heating and cooling system.

Scope of application: These applications are mainly for energy savings. As shown in figure 11-2, the fan curve for power required using outlet dampers versus an ac drive is documenting the energy savings. Dampers run full-open in a ventilation dust system, letting the variable-speed drive control the flow of air based upon pressure in the system. Figure 11-3 shows in block form an HVAC system utilizing a drive.

Specific concerns: Many times existing ac motors are used; therefore, physically locating the ac drive is an issue. Another issue is making sure that the dampers cannot ever be fully closed. The

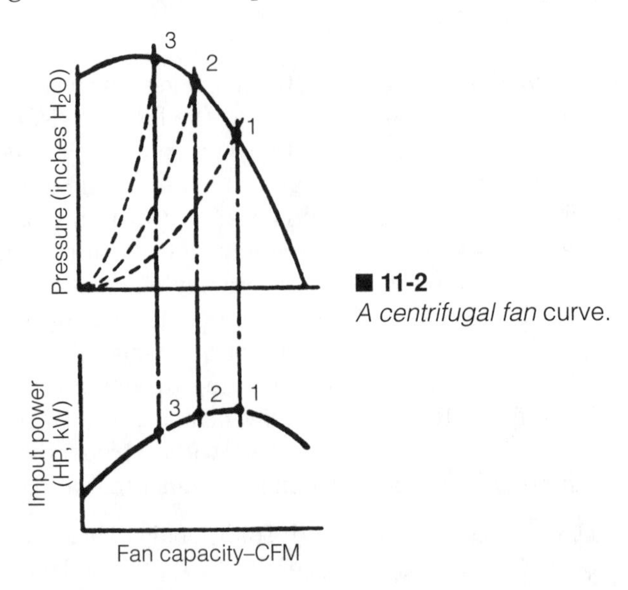

■ **11-2**
A centrifugal fan curve.

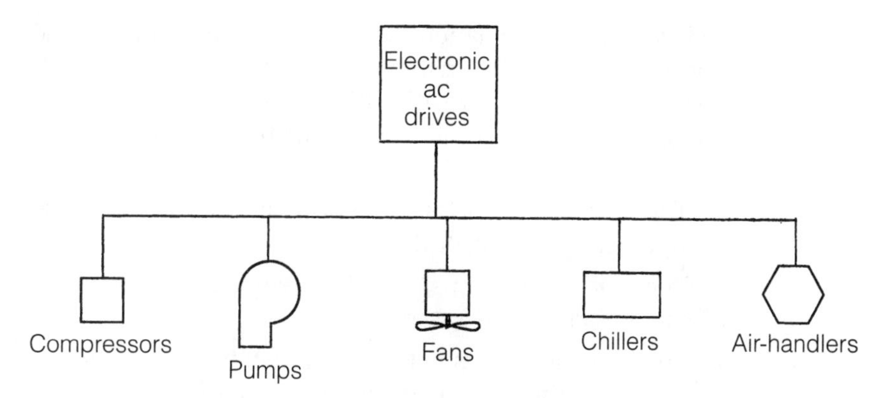

■ **11-3** *Heating, ventilating, and air-conditioning (HVAC) and drive applications.*

drive, running off of a pressure signal, could inadvertently provide full torque and damage ductwork. Frequently when the drive is introduced into office-building surroundings, noise becomes an issue—both audible and electrical. Sensitive computers can be affected if the drive is installed without consideration of the electrical system's loads and impedance. Filters might be in order. Also, there can be added noise from the fans and transistors. Paying attention to location up front is recommended.

Benefits: Obviously, energy savings will be the biggest benefit; however, another is that the soft starting of the motors will lengthen their life and save the rest of the mechanical system excessive wear, also. Many times the supply and return fans can be the same size motor and run simultaneously. Thus, one ac drive sized for two-motor operation and current can be used to operate two ac motors. This will further reduce initial costs.

Additional notes: Cooling towers, dryers, and other fan systems all move air or other gases. Air has different characteristics when cool versus when it is hot. Likewise, the gas has be treated differently when under pressure. In addition, horsepowers and speeds have specific relationships. These relationships are often referred to as the *affinity laws*. Attention must be given to these laws whenever applying a variable-speed device such as an electronic drive. A simple equation for calculating the horsepower of a simple fan or blower is as follows:

$$HP = \frac{\text{cubic feet per minute (CFM)} \times \text{pressure (psi)}}{33{,}000 \times \text{efficiency}}$$

This is a quick calculation for a system with normal temperature and pressure characteristics. However, horsepowers and motor speeds have distinct relationships in fan applications. For centrifugal, or variable-torque, situations the affinity laws can apply. They are as follows:

1. The flow in a system is a proportional relationship to speed, whether increasing or decreasing.
2. The speed in a given system increases by a square function as the pressure increases.
3. The speed increases in a given system by a cube function relative to the horsepower increases, and vice versa.

What the affinity laws mean is simply this: A fan's characteristics will follow these curves for flow, pressure, and horsepower. If a change is made to one or the other, then one must be willing to make the necessary changes to the rest of the system in order to get the performance desired. This can mean installing larger motors, changing base speeds, or changing gear ratios.

Application: Single centrifugal pump with bypass

Possible users: Water/wastewater facility; commercial and industrial facilities.

Scope of application: Most radial flow, or centrifugal, pumps have similar components and operate in much the same manner. The basic components, including an electric motor, a pump casing that houses the impeller, and associated seals, are shown in figure 11-4. The centrifugal pump generally moves high volumes of fluid at low pressures. A drive placed on a pump motor to reduce the motor speed, which, in turn, reduces the flow is now a common occurrence. Because the pump system must have the capability to always run, a line bypass method is usually installed with the drive (on ac systems only). This is shown in figure 11-5.

Drives commonly used: Most pump installations will use an ac motor, thereby making the drive an ac variable-frequency type. However, there might be some remote installations that use dc power, and therefore, a dc motor is in place.

Specific concerns: A partial vacuum is formed when the centrifugal pump impeller rotates, thus drawing more fluid into the pump. However, when the suction is broken, cavitation takes place. These load interruptions could trip the drive offline. Likewise, "hammer" in the piping system can exist, sometimes when a centrifugal pump is on a system where there is a lot of head pressure. This is not so much a problem for the drive as it is for the piping system itself.

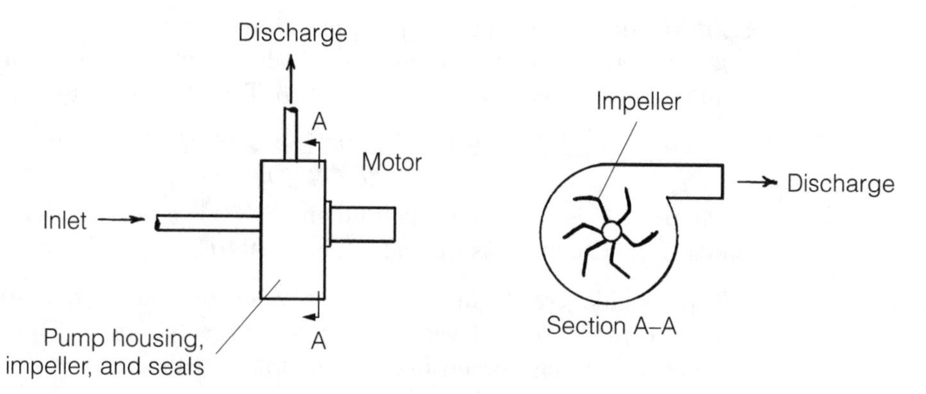

■ **11-4** *Common pump systems and their components.*

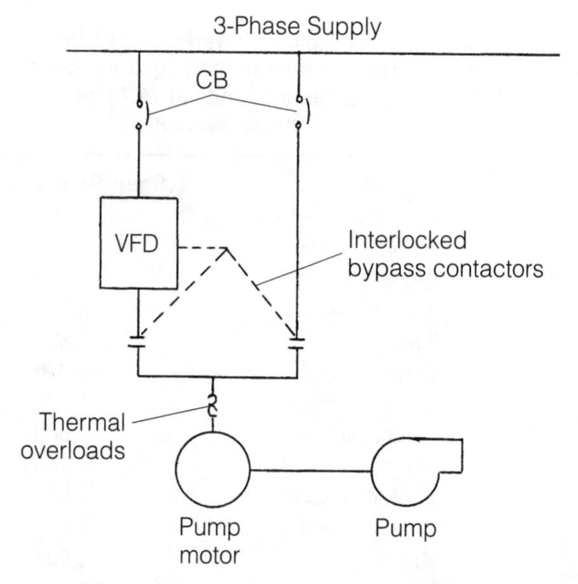

■ **11-5** *A drive and bypass single-line diagram.*

Benefits: The obvious benefit is the energy savings. Instead of using valves and vanes in the system, the ac drive controls the flow. Another benefit is the ability to run full-speed via the bypass system. Yet another benefit is soft acceleration and soft deceleration, which can help reduce hammer in a pump's piping system. The ac drive can also ride through many line fluctuations and through some pump cavitation instances.

Additional notes: In pumping applications it is necessary to size a given motor for a certain flow rate (gallons per minute, or GPM) and a certain pressure (or head, in feet). The equation is as follows:

$$HP = \frac{GPM \times head\ in\ feet \times specific\ gravity\ of\ the\ liquid}{3{,}960 \times efficiency\ of\ pump}$$

where GPM is gallons per minute and 3,960 is a horsepower constant. Also head pressure, in feet, is equal to 2.31 psig.

The specific gravity of water has a value of 1 at normal atmospheric pressure and warmer temperatures (70 to 80°F). This value will change based upon temperature and pressure. These are factors that have to be considered whenever sizing a drive and motor for a pumping application. Also, many other nonwater liquids have different specific gravities. Some common liquids and their specific gravities are shown in Table 11-2.

■ Table 11-2 Various specific gravities for common liquids (English system of units at 14.7 psia and 77 degrees F).

Liquid	Specific gravity
Acetone	0.787
Alcohol, ethyl	0.787
Alcohol, methyl	0.789
Alcohol, propyl	0.802
Ammonia	0.826
Benzene	0.876
Carbon tetrachloride	1.590
Castor oil	0.960
Ethylene glycol	1.100
Fuel oil, heavy	0.906
Fuel oil, medium	0.852
Gasoline	0.721
Glycerine	1.263
Kerosene	0.823
Linseed oil	0.930
Mercury	13.60
Propane	0.495
Sea water	1.030
Turpentine	0.870
Water	1.000

Application: Parallel pumps on variable frequency drives

Possible users: Water/wastewater facility; commercial and industrial facilities

Scope of application: To optimize the motor and drive equipment available for a variable-flow pump system. This might allow running multiple pump motors from one drive and also allow the ability to run full, constant speed in the event the drive fails. In addition, the flexibility of sequencing pumps based on demand and alternating pumps to equalize wear should be inherent with the drive control system.

Drives commonly used: The ac drives are used extensively in parallel pump systems. The dc motors cannot operate directly off the ac supply. Complex bypass schemes, contactors, and control logic is also a part of the package.

Specific concerns: Several configurations for parallel pump systems can be developed. However, there will be cost and complexity issues to address as more functionality is expected from a particular system. For instance, the simplex drive and motor system from figure 11-6 with its bypass scheme was fairly simple. By adding a motor and another drive to the system, we have now developed a double-simplex pump system. This is shown in figure 11-7. Control logic is required specific to when contactors close and open. A two-pump motor installation, shown in figure 11-8, is similar; however, this time there is one less drive but still two pump motors. Some installations use the drive to bring the first motor up to line speed (60 Hz) and, once matched, switch it over to the 60-Hz line and then pick up the second motor based on demand.

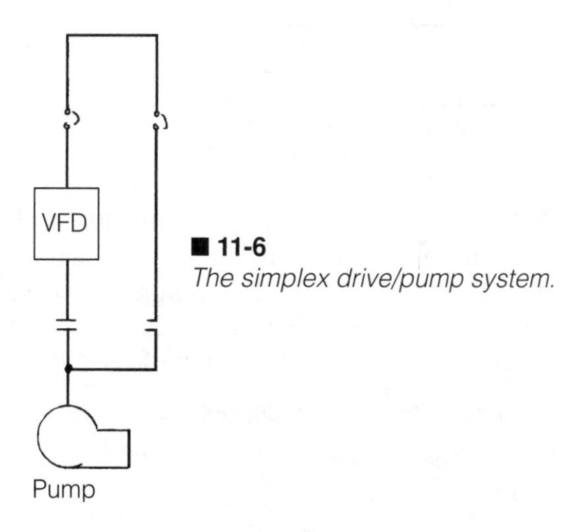

■ 11-6
The simplex drive/pump system.

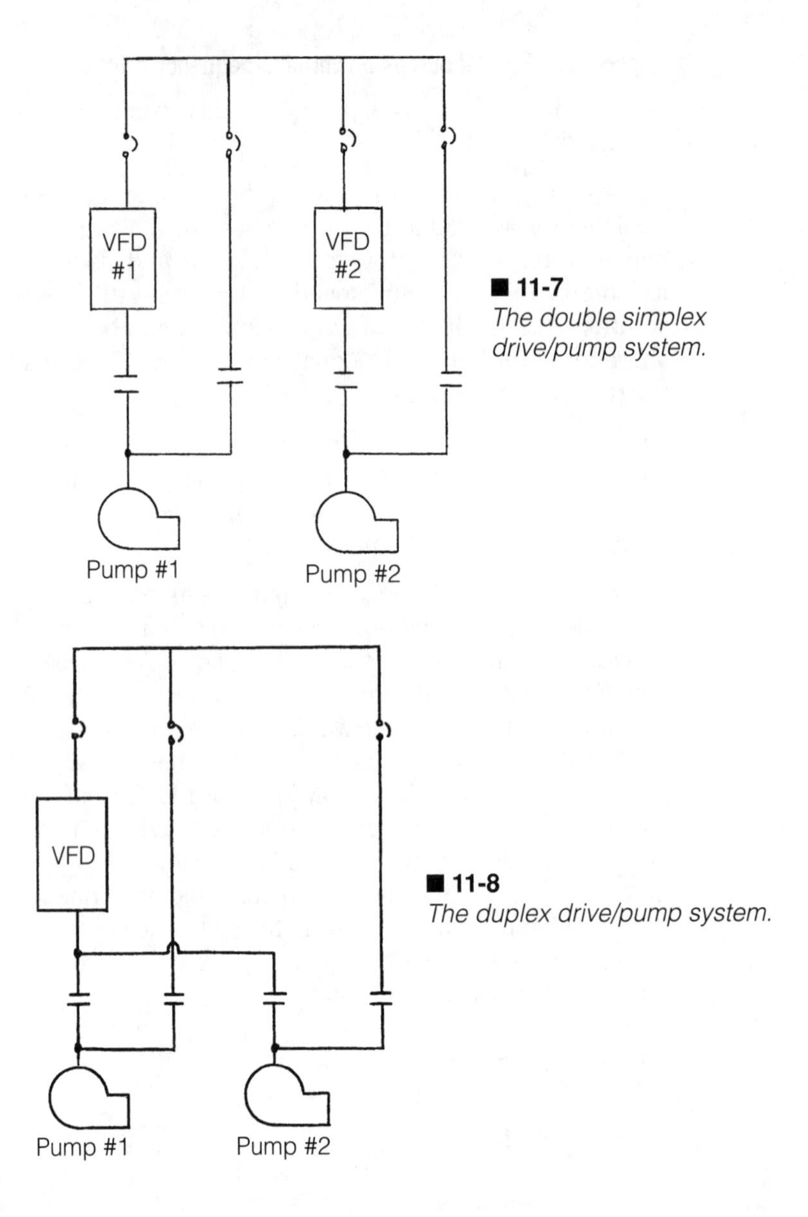

■ **11-7** *The double simplex drive/pump system.*

■ **11-8** *The duplex drive/pump system.*

Benefits: The electronic ac drives provide energy savings; the ability to select which pump motor to run, thus equalizing motor wear; and the ability to bring a pump motor up from a soft start to full speed and transfer it to the line. The bypass system allows for operation whenever the drive is out of service.

Application: Positive displacement (PD) pump

Possible users: Many industrial facilities (paper, petrochemical, metals, mining, etc.)

Scope of application: A positive displacement pump typically starts with a definite load. There might be a certain amount of head pressure to overcome upon starting, or the pump itself might be constructed such that the mechanics within need extra torque for starting (see figure 11-9). Positive displacement pumps include reciprocating types with piston, plunger, and screw pump versions. Positive displacement pumps typically move low volumes of fluid at high pressures. PD pumps often have to move solid particles and other nonfluid materials.

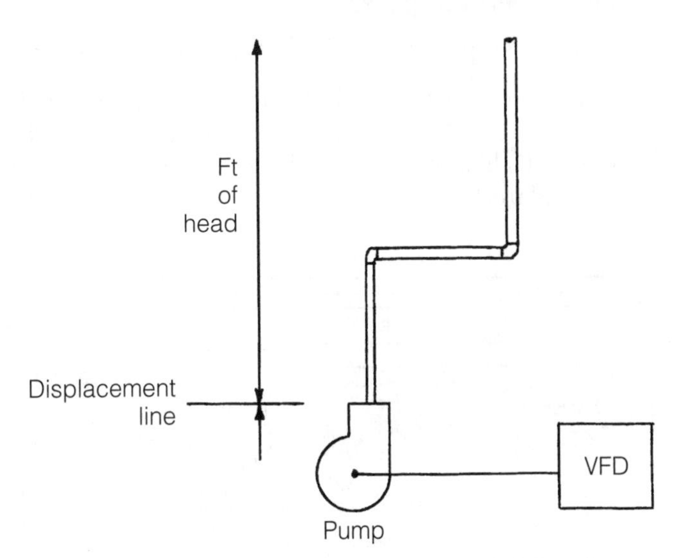

■ **11-9** *A PD pump having to overcome high head pressure to start.*

Drives commonly used: Because these types of pumps require high-starting torque, they are classified as constant-torque applications. Therefore, standard ac drive will have trouble starting and at low speeds. Instead, dc drives and high-performance, vector-type ac drives are used frequently.

Specific concerns: In order to run more efficiently, usually these types of pumps are in a heated element. A heated fluid will move more efficiently through a pipe system than will a cold fluid (especially with some solid particles present). This heating means that the motor and pump must be able to withstand this condition. The drive will be affected by this heating mainly when the heating isn't present. If the material is cold, it will take a lot more torque to start and to move. Also, whenever solids are present in a pump system, there is always the potential for a jam or blockage in the pump/piping system. This means that the drive must be appropriately set up to trip quickly on overload so as not to burst any seals.

Benefits: The energy savings on this type of pump system is not as dramatic as the centrifugal pump's. However, some savings are still incurred whenever the pump does not have to run full speed. The soft-starting drive and ability to deliver high-starting torques are both advantages. The ac motor is also more suitable to the harsh environments.

Application: Cooling tower

Possible user: Any facility that needs to cool recirculated water.

Scope of application: Based on the temperature of the water, the cooling tower fan will cycle on. This fan motor drives a fan blade that pulls high volumes or air through warm water, thus cooling it (see figure 11-10).

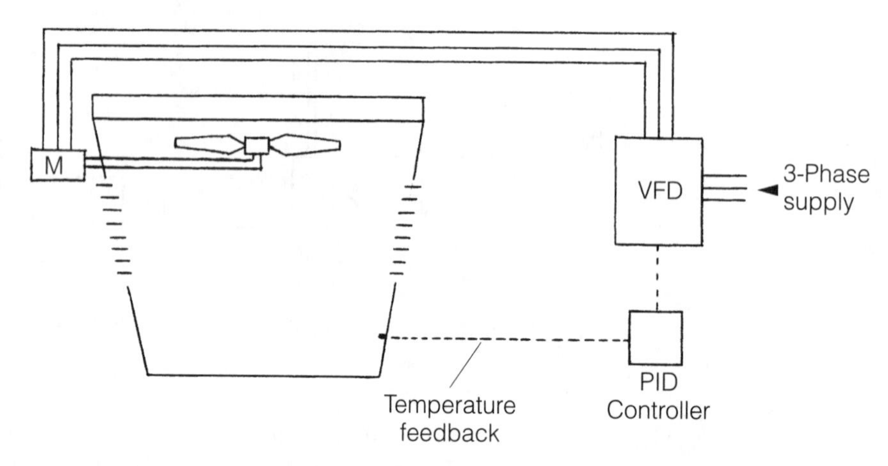

■ **11-10** *A cooling tower drive application.*

Drives commonly used: In the past many cooling tower fans utilized two-speed motors, variable-pitch fan blades, and dc motors. Today, with ac drives capable of two-stage ramps and able to provide dramatic energy savings, the choice is mainly to use an ac induction motor with a variable-frequency drive. The ac drives with built-in PI (proportional-integral) loop control for operation relative to a temperature setpoint are the best choice.

Specific concerns: Resonant frequencies are possible with such a fan installation. The fan motor, fan blade, motor, and coupling are usually substantial. Any vibration can be amplified to the point that bearings, shafts, and couplings can fail. To avoid this vibration, the ac drive has the ability to skip critical frequencies where vibration can exist. Additionally, whenever there is a condition where the fan motor and blade are commanded (or drive has tripped) to coast to a stop, the fan can keep rotating for several minutes. Most ac drives have the capability to catch the spinning load and reaccelerate the motor to its requested speed.

Benefits: By cycling motors in a multiple cooling tower scheme, even motor wear and eventual longer life can be achieved. Lighter loading will not require that all the cooling towers be operational. Varying the speed of each fan by a select volts-per-hertz pattern will save energy. Soft starts and the ability to pick up a spinning fan motor are other benefits.

Application: Roll grinding

Possible users: Steel and aluminum mills of all sizes

Scope of application: Rolling mills use several rolls in the process. Rolling steel, aluminum, or any other metal into thinner, finished sheet product is a process that is hard on the rolls themselves. They literally get beat up, chipped, and lose their finish. These work rolls and finishing rolls must come out of service and be re-ground to a finish worthy of getting them back into service (only to get beat up again). The drives that control the speed of the carriage, headstock, and grinding wheel (see figure 11-11) must be capable of reliability and performance to quickly return a roll to service with finishes sometimes as glossy as a mirror!

Drives commonly used: Traditionally, dc motors and dc drives have played the dominant role in roll grinding applications. Today's digital dc drives are capable of excellent speed and torque regulation. However, ac flux vector drives are also capable of the same speed and torque regulation needed in a roll grinding appli-

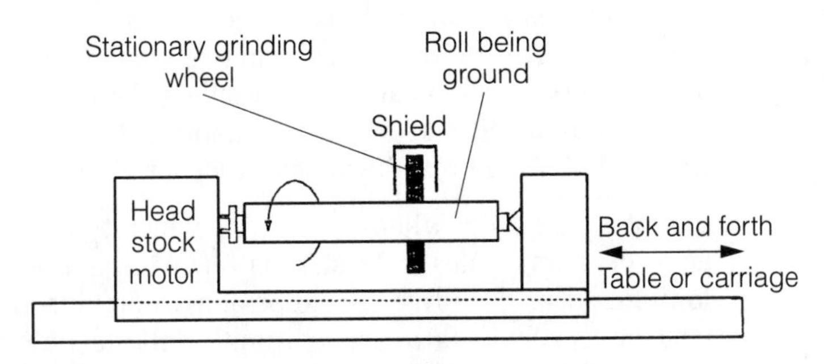

■ **11-11** *Roll grinder application.*

cation. Also, some high-performance and high-cost roll grinders are driven by coordinated servo amplifiers and servomotors.

Specific concerns: The grinding wheel is a high-inertia load. It requires 150 to 200% torque for starting. Likewise, it needs regeneration or dynamic braking to stop quickly. All the drives have to function in a coordinated manner. If one is out of synch with any other, then "chatter" marks can appear on the roll, and it will have to be ground again. Routing of feedback wiring is important. Since there is a traversing component to the grinder, the feedback wiring has to be routed through flexible conduit, a track system, or in a trough. This can potentially lead to broken wires or to interference from high-voltage wires.

Benefits: The use of ac, dc, or any other type of drive should give the operator very good speed control. This speed control, coupled with the individual drive's performance in torque regulation, means fewer passes should be required to grind a roll. Thus, more rolls can be ground in a day and the finishes should also be smoother.

Application: Belt conveyors

Possible users: Coal and mining facilities; food and beverage plants.

Scope of application: Conveyors, comprised of many rolls (some driven and some as idlers) and a continuous belt, are the primary vehicles to move material from one location to another. They have to run fast or slow, fully loaded and empty. Therefore, there can be many methods of running and controlling them. Many mechanical systems such as variable-pitch pulleys, fluid couplings, and even eddy current clutches have been used for years. Many conveyors are single-speed types utilizing a gearbox to get the low resultant speed at the belt.

Drives commonly used: Beyond the mechanical drives used, the electronic drives most often seen in this type of application are ac and dc. Servo systems are overkill unless the conveyor has to be matched in speed and position to another factory machine. See figure 11-12 for a breakdown of a conveyor system with drive.

Specific concerns: When considering which type of drive to use, ac or dc, many times it doesn't matter. What matters are the following issues: The drive should be able to provide variable speed control, usually a 10 to 1 speed range. Soft starting is a must because pieces can fall from the conveyor when started abruptly. If the conveyor is capable of overhauling (does it run uphill or down-

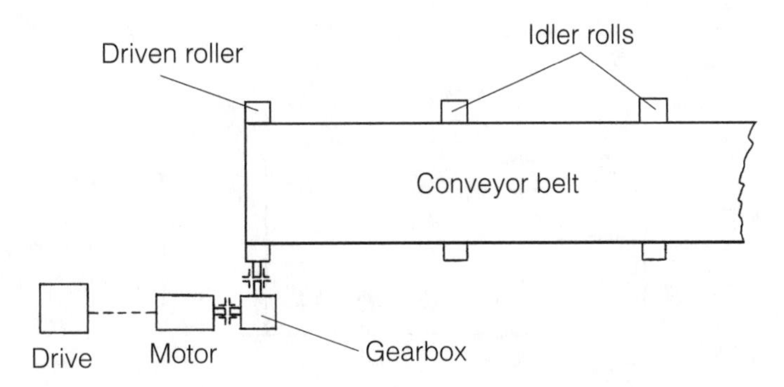

■ **11-12** *Belt conveyor application.*

hill), regeneration or braking must be considered. Lastly, if multiple rolls are to be driven, is one drive to run all the motors or will each motor run from its own drive? Once it is determined how the conveyor system will be powered and controlled, then the drive selection can be expedited.

Benefits: The ac and dc drives provide excellent speed regulation, and as long as the speeds are not too low, torque regulation should not be an issue. The inherent soft-starting capabilities of the ac drives are actually displacing many reduced-voltage starters these days. If the motor is inaccessible, a low-maintenance motor such as the ac induction type is excellent.

Application: Chain conveyors

Possible users: Automotive assembly plants; foundries; paint lines

Scope of application: Chain conveyors carry parts through factories and can stretch for several thousand feet. Load conditions will vary, too. Typically, multiple drives are incorporated on long chain-conveyor systems with heavy loads and many inclines and declines. Paint booths and dip tanks, foundries, and ovens are all recipients of chain-conveyor passengers.

Drives commonly used: The dc drives, again, have tradition on their side. With loads changing dramatically and moving these loads up, down, and around a factory, the dc motor is a good workhorse. Load sharing is very important between motors. Similarly, ac drives can be utilized on chain conveyors in hostile environments, where brush maintenance is undesirable, and where ac motors might already be in place. Both ac and dc drive arrangements for multiple motors on a chain conveyor are shown in figure 11-13.

Benefits: The inherent soft starting of electronic drives is kinder to the chain and the remaining drivetrain components. Multiple

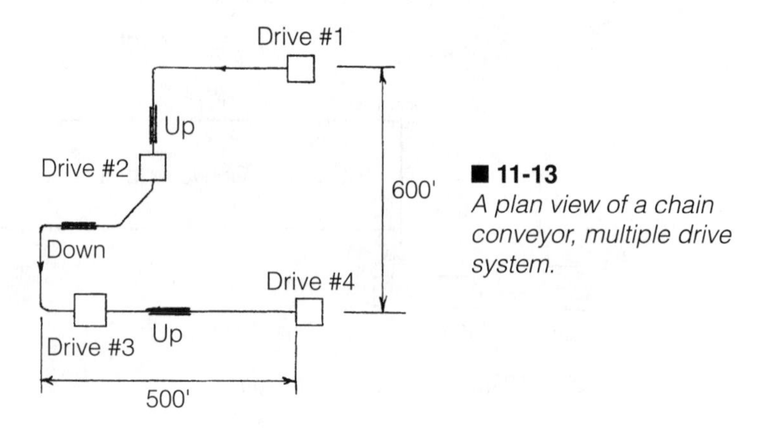

11-13 *A plan view of a chain conveyor, multiple drive system.*

drives and motors will reduce excessive chain tension, and should one or two drives fail, the other drives on the chain system can keep the conveyor running. Electronic drives are also good at load sharing, even during periods of high-to-low/low-to-high load changing.

Application: Overhead cranes and hoists

Possible users: Heavy metal manufacturers and all facilities with overhead cranes.

Scope of application: As shown in figure 11-14, the typical crane system is made up of a rail, a north/south motor, an east/west motor, and a hoist motor. The traversing motors have to run smoothly at creep, jog, and commanded speeds. Additionally, all the motors have to be fitted with some form of fail-safe brake.

Drives commonly used: The dc motors and drives have been used for years. Although ac drives are now being implemented, this is primarily a dc market.

Specific concerns: Since cranes are lifting some heavy object and transporting it to another location, safety and performance are paramount. The drive has to interface directly with holding brakes and fail-safe brakes. The hoist motor, lifting and lowering objects, must be capable of regeneration, as loads will tend to overhaul. Therefore, the dc drive and motor are good choices.

Benefits: The crane drives must interface directly with all the motor brakes. The ac drives are usually set up with specific software to accomplish this. With regenerative packages, loads will not overhaul. The operator has a wide speed range of control with the variable-speed drives.

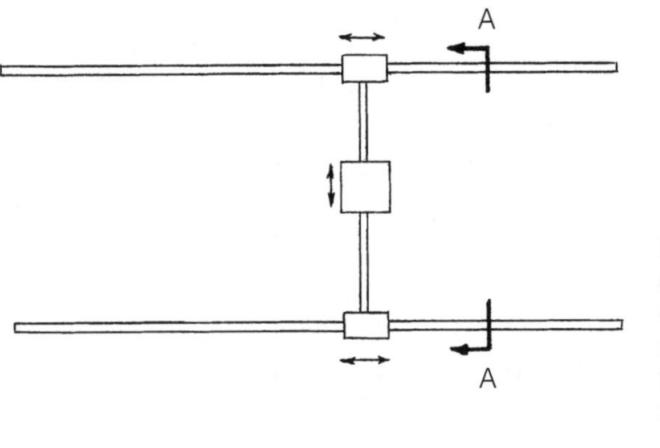

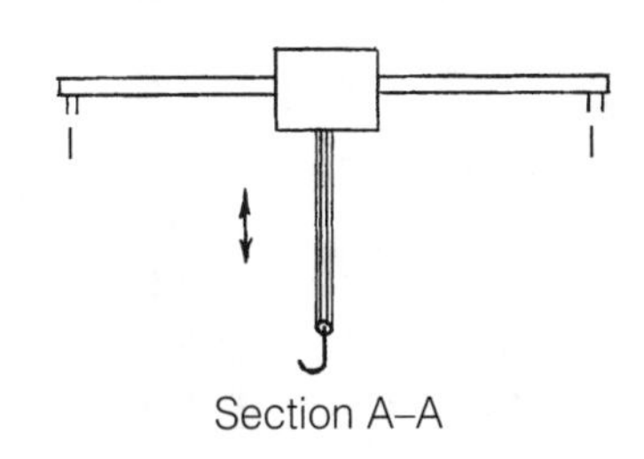

■ **11-14** *An overhead crane.*

Application: Mixers

Possible users: Paint and resin facilities; tire and rubber plants; chemical plants.

Scope of applications: This has to be defined per the individual mixer. As was shown with the possible user list with this application, mixers can differ and so, too, can their drive requirements. Driven loads can vary from light (liquid mixers) to medium (slurry and dough mixers) to heavy (solids and Banbury-type mixers). Others, such as powder and sand mixers or screw-type mixers have hard-to-start loads. Thus, the scope of the application is to drive the mixer motor, but it isn't always that simple. A typical mixer is similar to a blender and is shown in figure 11-15.

Drives commonly used: For the medium- and heavy-duty mixer applications, dc drives and motors are a good choice. Their ride-through capability during severe load changes and the 200%+ torque when necessary are good for mixers with hefty service factors. As for the ac drive, flux vector drives are an available alternative, and as long as they are sized for overload conditions in the mixer, they can be used. Volts-per-hertz ac drives are excellent choices for light-duty mixers.

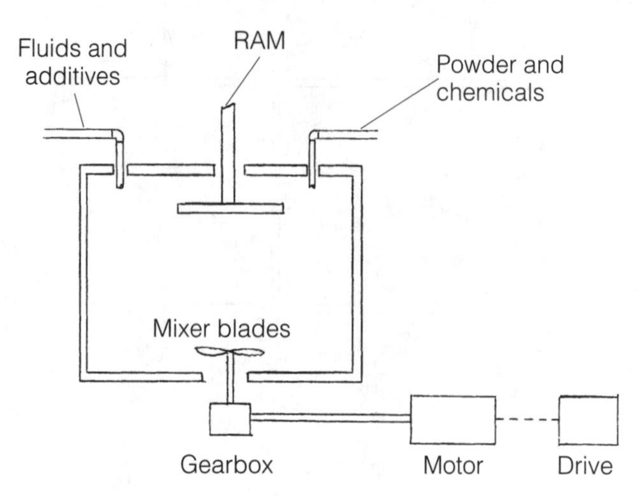

■ **11-15** *An industrial mixer.*

Specific concerns: Before selecting a drive, know the type of mixer and its duty cycle. Also, predict what possible "shock loads" to the mixer can occur, so the drive selected can be sized to ride-through.

Benefits: The ac drive, when applicable, will save wear and tear on the motor windings. Line-starting motors typically hit the windings with 600%+ inrush current. An electronic drive limits this to no more than 100%. The dc drive and motor can "churn" through tough mixing applications. Once running, the mixing operation can tolerate a hard piece of material, and the dc motor should keep right on running.

Application: Batch centrifuges

Possible users: Sugar manufacturers; pharmaceutical companies.

Scope of application: A batch centrifuge is made up of a basket in which material resides, a motor assembly, and controls. The basket is a high-inertia cylindrical component that must accelerate extremely fast to very high speeds in order to separate the liquid from the solids. The solids collect on the outer wall of the centrifuge basket and are later scraped off with a mechanical knife. The motor assembly commonly is mounted vertical to the basket, as shown in figure 11-16; however, horizontal mounts and belting arrangements are often used. The controls include a master controller and motor drive control. The profile of a typical speed-versus-time curve is shown in figure 11-17.

Drives commonly used: The dc drives have been used extensively because of their inherent ride-through capabilities, their regenera-

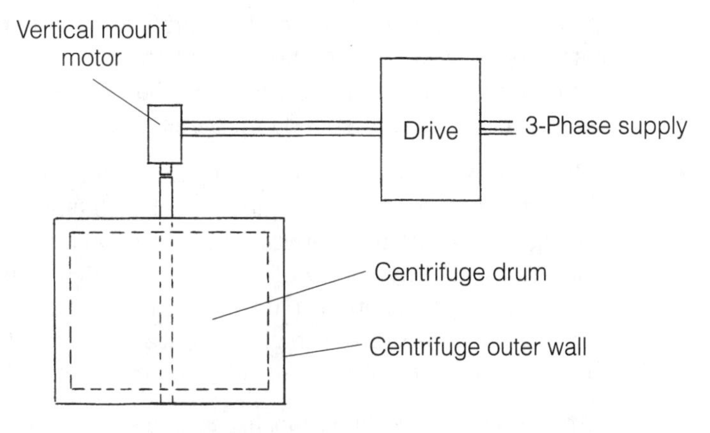

■ **11-16** *A batch centrifuge.*

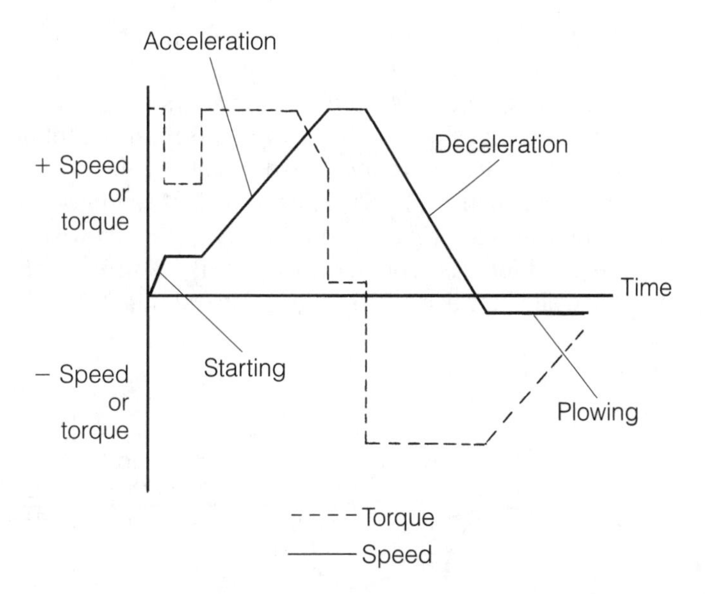

■ **11-17** *A batch centrifuge speed/torque versus time curve.*

tive power scheme, and their ability to develop high values of accelerating torque. However, ac drives with an extra power bridge to furnish regeneration have been used along with inherently regenerative current source inverters. Due to the high cycle rate of batch centrifuges, dynamic braking is not a good option for stopping.

Specific concerns: The drive has to accelerate in virtual current limit without tripping offline. The same is true when stopping the centrifuge, as any extra time in the cycle reduces the amount of batches run in a given day. The drive must be capable of reversing in order to discharge the solid material, and tight speed regulation at low speeds is required here. Since the centrifuge basket is spinning at extremely fast speeds, there has to be a built-in means of safely handling high-speed emergencies and basket runaways.

Benefits: An ac drive system is attractive because of the size of the induction motor: a smaller motor package is mounted on the centrifuge. Four-quadrant operation is necessary with this type of application, and any saving of energy in each cycle can add up over time. Like the regen ac package, a dc drive can regenerate, reverse, and accelerate in high-current situations. It has been used in these heavy-duty applications for years.

Application: Presses (drill, punch, and brick)

Possible users: Metal fabricators; brick manufacturers

Scope of application: In metal applications, a punch press is fed a sheet of metal, and the metal is either sheared to a length or a piece of the metal is knocked from the sheet. This is done by striking the metal sheet with a very high-pressure tool punch against a die. This pressure is developed usually by the interaction of a large flywheel (figure 11-18). Thus, the electric motor must be capable of high accelerating and running torques. Likewise, the drive must be sized for this type of duty cycle. Brick presses and drill presses have similar needs to that of the metal punch press.

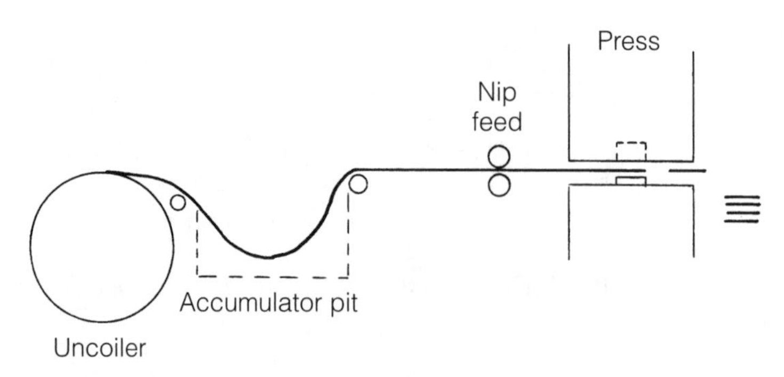

11-18 *A punch press.*

Drives commonly used: The dc drives and motors along with eddy current clutches have been used for years on these kinds of presses. The dc motor is robust and can handle the high inertias and high values of torque required. However, more and more ac drives, both flux vector and standard volts-per-hertz types, are being used.

Specific concerns: The built-up energy in the flywheel is the biggest concern. It must constantly be dealt with. This energy must be furnished to accelerate the flywheel and dealt with as the flywheel stops. However, once the motor has reached full speed, peak running torques are the most important consideration in the cycle.

Benefits: An ac motor is welcome in these ambient environments. Presses notoriously are in dirty, dusty conditions, and dc motors tend to have maintenance problems. A drive, whether ac or dc, will give consistent press action as long as the peripheral equipment is working well. Additionally, drives interact well with press feeder drives and out-feed equipment.

Application: Lead screws

Possible user: Any robot, xy stage, or precision machine tool

Scope of application: A lead screw converts rotary motion into linear motion (see figure 11-19). This linear motion normally has to be controlled to very exacting tolerances. Most positioning-type applications will incorporate high-precision lead screws and recirculating ball bearing nuts to achieve high accuracies.

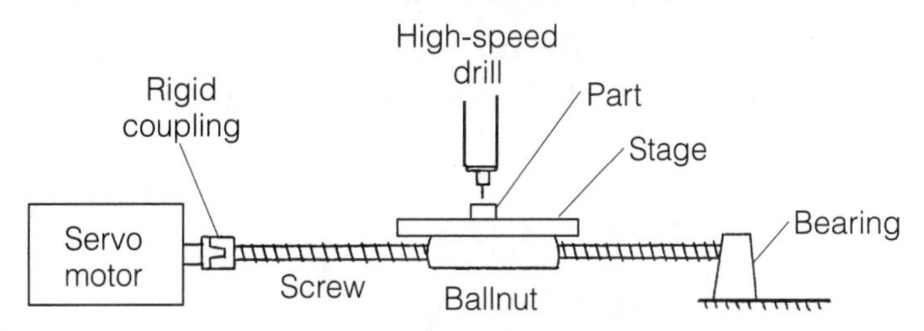

■ **11-19** *A lead screw, or ball screw, application.*

Drives commonly used: Servo drives and servomotors with encoder or resolver feedback will be able to hold positioning tolerances down to a fraction of a micron (0.000040"). Stepper motors and closed-loop stepper controllers can also perform well in lead screw applications.

Specific concerns: Since positioning accuracy is the goal in these types of applications, many external factors can affect performance of the drive control. A sound mechanical system is paramount. A flexible coupling should be used between the motor and lead screw to provide a certain amount of damping. This will also help with premature bearing failure at the feedback device due to misalignment. Hysteresis, or the motor's tendency to resist a change in direction, is another concern when repeated reversals are required.

Benefits: The lead screw (sometimes called a ball screw because of its ball bearings and ball nut assembly) is an excellent choice for controlling position in compact and confined spaces. High-precision screws with preloaded ball nuts are available and with today's positioning controls, accuracies well below a micron can be achieved.

Application: Glass furnace, shuttle, and quench line

Possible user: Glass windshield manufacturer

Scope of application (figure 11-20): Imagine the following scenario: A method is required to exactly match the line speed of a conveyor carrying windshields with that of a shuttle that has vacuum grippers to transfer the windshields to a second conveyor. This means that a photocell, triggered when sensing the leading edge of the windshield, needs to send an output to a controller. Meanwhile, the conveyor line is still traveling at a fixed speed. Also, once the shuttle picks up the glass, it must accelerate to the next conveyor, decelerate, and match the second conveyor's speed. The shuttle will have to perform this task several times per hour.

Specific concerns: How to time the arrival of the shuttle, precisely match the speed of the conveyor, and pick up the glass windshield

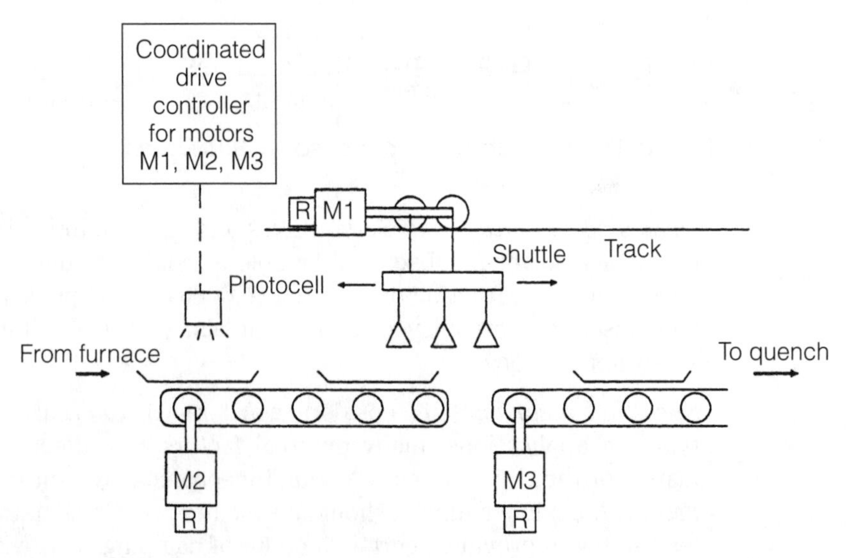

■ **11-20** *A glass furnace and shuttle quench line.*

without any damage to the glass piece. All this prompted by a photocell signal.

Drives commonly used: Servo positioning drives and a coordinated controller are used. The drives control both the conveyor and the shuttle, so speed matching is easy. The problem is to detect the edge of the glass and match the shuttle speed to the glass speed on the conveyor.

A mode on the axis controller card is used to achieve the capability of interrupting the program with the photocell input. When the photocell is triggered by the edge of the glass, the axis controller card stores the position of the glass on the conveyor (within 1 ms) so the shuttle can pick it up. It does this by reading the resolver count off the servomotor.

Up to 10 values (position, absolute, etc.) can automatically be stored in the first in, first out (FIFO) stack without program intervention. The data can be read out of the axis's FIFO as needed. The axis controller card also has three direct inputs into its processor. These are mainly used for ends of travel and the home switch. In a conveyor line such as this, the end of travel function is useless. Therefore, the photocell output goes directly to the axis card, thereby guaranteeing a maximum of 1-ms update time.

Benefits: By cutting down the time required to determine motor shaft position, the customer is able to run the conveyor line at the fastest speeds possible. This function also ensures that the glass windshield would not be marked or damaged.

Application: Winders, including centerwinds, unwinds, rewinds, and surface winders; payoff reels; and coilers

Possible users: Paper, film, foil, metal, wire, and cable manufacturers

Scope of applications: To unwind a material, do some work to that material, and then wind it back up on a winder. This can involve many distinct functions, but this discussion is only concerned with the unwind and winding aspects for drive control. (A later application discussion will cover an entire web line.) As shown in figures 11-21 and 11-22, the unwinder and winder functions are similar from industry to industry. The most common types of winders and unwinds are the surface winder type and the center winder type (figures 11-23 and 11-24). The type of winder will dramatically affect the size and type of electronic drive selected.

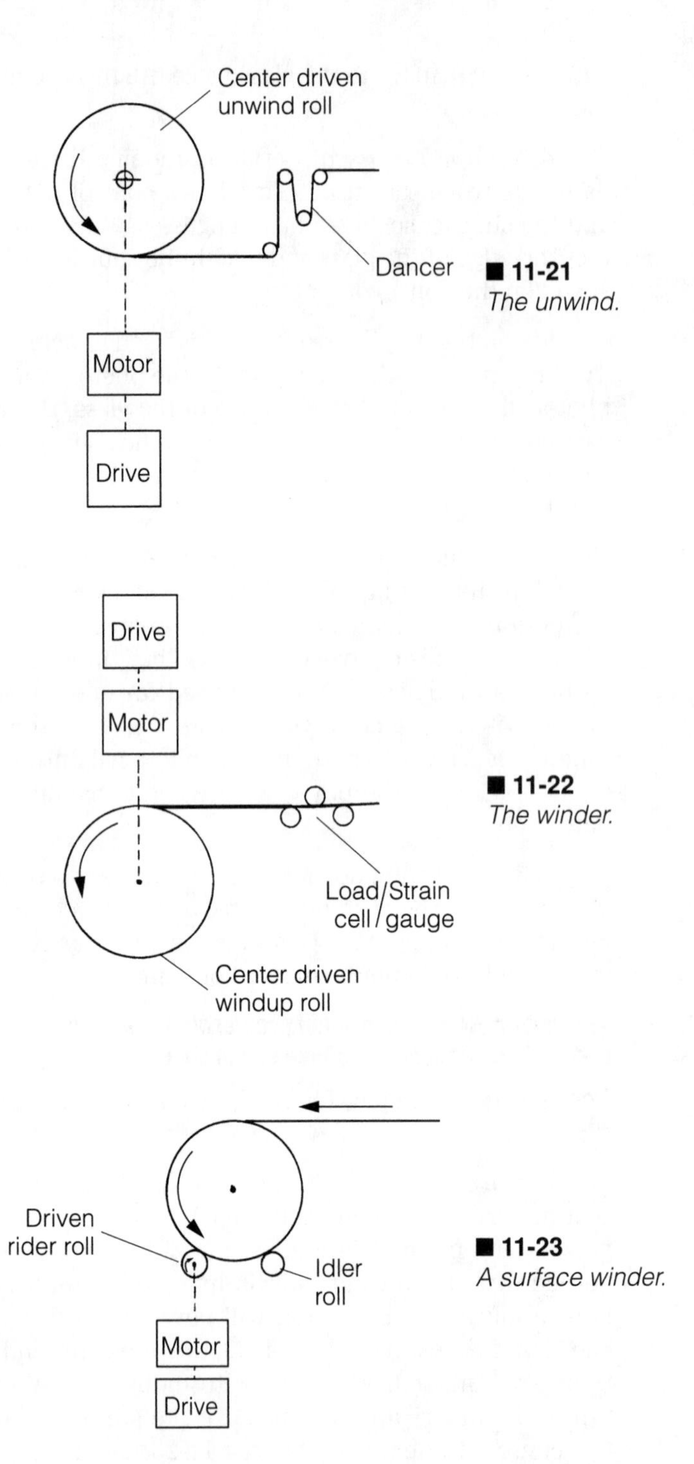

■ 11-21
The unwind.

■ 11-22
The winder.

■ 11-23
A surface winder.

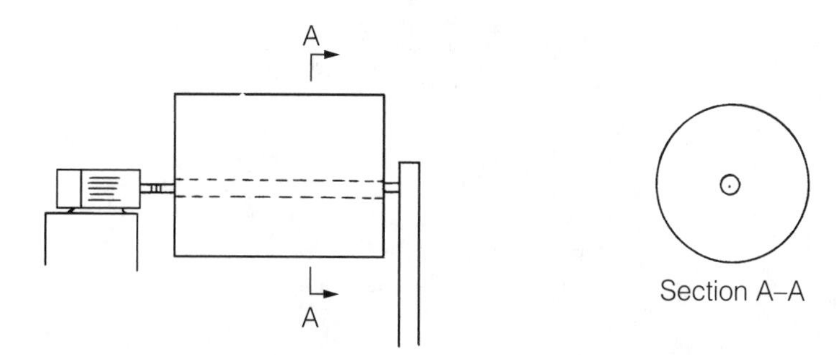

■ **11-24** *A center winder.*

Drives commonly used: Surface winders are much easier to control, and therefore, ac or dc nonregenerative-type drives can be used. System friction will usually stop a line in plenty of time; however, dynamic braking resistors should be considered for safety and emergency stops (this is even recommended for regenerative installations). On centerwind-controlled winders and unwinds, the buildup ratio on the roll is a major factor. This buildup ratio is shown in the calculations for a given horsepower (figure 11-25). The motor and drive have to be sized to handle the worst torque condition or the heaviest possible load. In addition, holdback torque is required for an unwind, and wind-up rolls have to be prevented from overhauling. Thus, regen drives are required.

Specific concerns: As mentioned previously, regeneration is very important to centerwind-controlled drives. Stopping, overhauling, and holdback for tension control is a concern. A fine line exists between breaking a web and applying the proper tension. Metals, wire, and cable are much more forgiving in these types of applications.

Benefits: Using drives allows for tremendous speed and torque regulation throughout the overall speed range. As a roll of material unwinds in center-controlled unwind mode, a consistent line speed must be maintained as the load changes. The drive response to these changes will have a dramatic effect on the material's journey through the rest of the machine. Installing ac drives on surface winders is now a common approach.

Application: Tying multiple drives to a common bus

Possible users: Any web system having unwind and rewind needs.

For a web tensioning system with a winder roll, the build-up calculations are as follows:

Givens: Tension = 2 pounds per lineal inch (PLI)
Line speed, or FPM = 1000 FPM
Web width = 78 inches
Winder core diameter = 8 inches
Maximum winder diameter = 40 inches

Therefore:

Horsepower = F × FPM / 33000 = (2 pli × 78") × (1000 fpm) / 33000

Horsepower = 4.73 × 5 (Build ratio of 40/8) × friction

Horse power = 4.73 × 5 × 1.1

Total winder horsepower needed is 26.015 HP

Therefore, use a 30-HP motor.

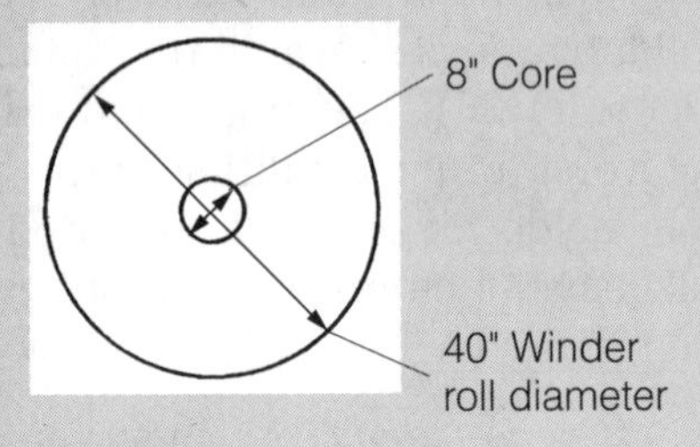

■ **11-25** *Typical winder buildup calculations.*

Scope of application: As shown in figure 11-26, there is energy needed and energy provided in a system such as this. When one drive is motoring, another can be regenerating. Thus, energy is conserved.

Drives commonly used: This approach for handling holdback and tension control in a continuous process is mainly suited for ac drives. Because the ac drive has the link section with a dc voltage ahead of the inverter, this section can be tied into and shared among drives. In this way, power can be delivered to the drive that needs it, and whenever a drive has to divert power from the motor, it can get back onto the dc bus.

Specific concerns: This approach will work as long as there is a balance in power need and power return. If the power balance sways too much in one direction, drive undervoltage or overvoltage faults can occur.

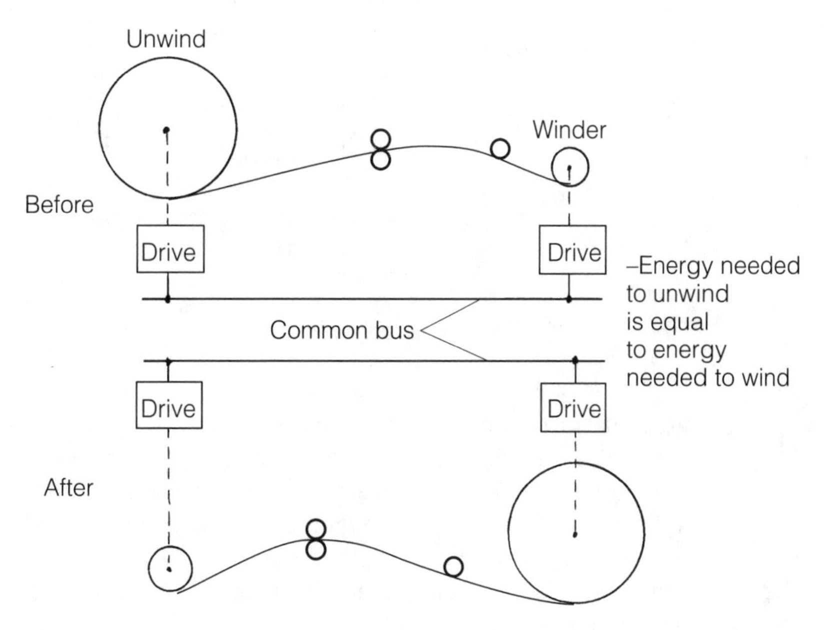

■ **11-26** *Common bus scheme—before and after with winders and unwinds.*

Benefits: The major benefit is that ac motors can be used in traditional dc regenerative applications. For those drive suppliers who do this routinely, real savings in hardware can be achieved.

Application: Compressor

Possible user: HVAC and refrigeration manufacturers

Scope of application: A typical refrigeration system consists of four major components: a condenser, evaporator, heat exchanger, and compressor. The compressor drives the whole system (figure 11-27). A refrigerant is changed to a gas under pressure. It exchanges heat that it picks up along the way, condenses back into a liquid, and the cycle starts over. The compressor is actually a motorized pump that initiates the flow and pressure. It can be controlled by an electronic drive.

Drives commonly used: The ac variable-frequency drives are a good choice for many reasons. First of all, the loading varies in the refrigeration system depending on whether it is summer or winter. This reduction of load calls for an energy-saving device such as a drive.

Benefits: The compressor will run more efficiently off of the drive. The ac drive will also be easier on the motor and compressor with

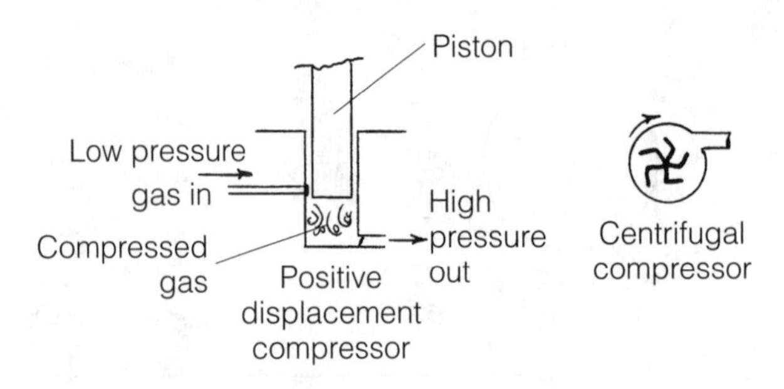

■ **11-27** *Compressor mechanics.*

its soft-starting capability. A benefit yet to be fully explored involves new refrigerants now coming on to the market. The EPA has banned certain Freon refrigerants, and the allowable types require more flow in order to gain back the capacity needed in the system for proper heat exchange. A variable-frequency drive can run 10 to 20% over base speed, thus possibly getting the extra capacity needed.

Application: Typical multiple-section web control system

Possible users: Paper, film, foil packagers and converters

Scope of application: These applications usually require both fine speed regulation and synchronization between the various sections in the web line. As shown in figure 11-28, a master controller is ideally suited for both. However, true distributed control can mean that each drive acts as its own closed-loop controller for the motor it is controlling.

The *web* is nothing more than the material (paper, plastic, foil, etc.) being pulled or drawn through the line for treatment or converting. It can be thick or thin, wide or narrow, and might or might not stretch. A *section* is defined as a part of the overall converting line with its own specific purpose and typically its own motor. A *tension zone* (figure 11-29) is an area, usually between two driven sections of the web line, where the material is subject to elements (nipped rolls, festoon, etc.) that can affect the tension on it at that part of the machine. There can be several tension zones in a long web line.

Tension is usually described in pounds per lineal inch (pli) of the material's width. Line speed is usually described in feet per minute (fpm) or inches per second (ips). Maintaining a consistent tension on the web (without breaking or deforming) in various sections is most important.

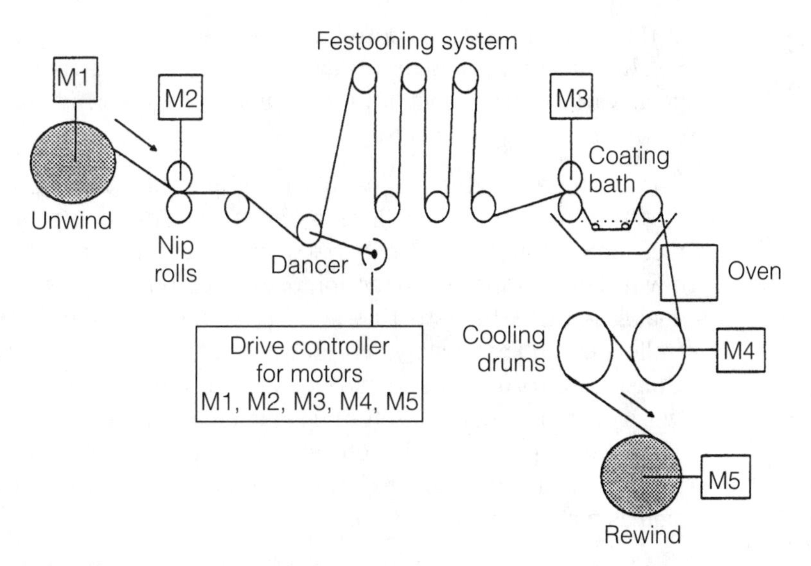

■ **11-28** *A typical web line.*

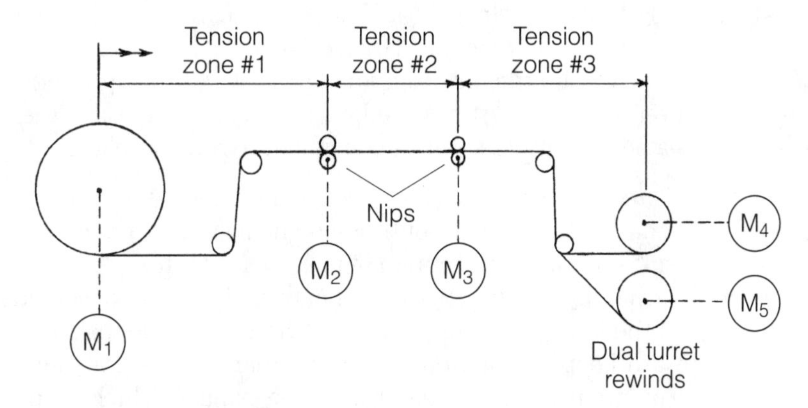

■ **11-29** *Tension zones in a web line.*

Specific concerns: How to maintain a constant tension on the material without breaking the web or allowing the web material to droop too much is the challenge! The drives have energy that must be accounted for all through the web system. That is why regenerative drives are preferred. In addition, from unwinders to rewinders, all sections of the machine have to be synchronized.

Drives commonly used: The unwind should be driven by a regenerative dc drive (regen for holdback) or servo drive with a common bus configuration or a dynamic braking package (which is not efficient). The material can also be "pulled" from a nondriven unwind by a set of nipped rolls, one of which should be driven. Any drives in this system must be coordinated with each other in some manner. Coating drives and dryers can be nonregen drives and even can be ac variable-frequency type. This makes for a consistent coating on the material and good drying.

Prior to being coated, the web tension must be held constant. A method for doing this utilizes a dancer. Dancer setup involves a mechanical arm that actually rides on the material, moving up and down. This up-and-down motion pivots at a point monitored by an analog sensor. The 0- to 10-V analog signal can be sent to the controller so that the upstream or downstream drives and motors can adjust their torque or speed to maintain the proper setting. Other types of transducers are often used such as LVDTs and load cells. They often appear in the web system after an unwind to maintain tension off the roll, and before a rewinder to again maintain tension as material is wound up.

Some processing lines even bypass the rewind roll at this point and move on to another section for more converting or treatment. Cut to length, slitters, stackers, and other applications can be incorporated further downstream in a large web converting line. Many argue that dc drives are the only solution for these types of applications. However, many ac systems are being installed every day. Even coordinated servomotors can be configured to run a web line.

Benefits: The drives can work in conjunction with a programmable controller and other controls. In one instance, a master controller can coordinate the motion of many slave motors. A controller with 1 ms or better update times is extremely useful in web lines. Also, as the line speed changes, the digital control loops with feed-forward keep the null position of the dancer in a constant position. This reduces the actual size of the dancer, which helps to keep from deforming the web.

In most web lines, however, the drive of choice is still the dc drive and dc motor. This is because of the dc drive's ability to regenerate energy back onto the supply line. Thus, tension control can be easily maintained, and with digital dc drives, synchronization of multiple drives has become easier. Also, the dc motor's high rotor inertia allows it to ride through load changes in the web line. This ride-through capability allows for tight speed regulation between drive sections.

Application: Box palletizer

Possible user: Any manufacturer whose product is boxed, stacked, and shipped.

Scope of application: Suppose a manufacturer of particleboard furniture needs a means of stacking individual boxed pieces as they come off the production line. In this particular case, the application calls for fast movement (for increased piece count) and accurate positioning. One problem is that every box that comes off of the feed conveyor has a different size and weight. In most applications, multiple pieces would be similar in size and weight.

The customer requires that the motion controller be simple and inexpensive. Only two axes of motion have to be controlled, but both have to be coordinated. The customer also requests that the controller be such that his personnel could write the actual motion programs.

Specific concerns: Different sizes and weights of boxes. Many times a teach type system is available with servo controls. However, when physical coordinates change due to different products, the motion program can be written such that calculations can be made by the controller's microprocessor, and the corresponding position commands can be executed. Many users want a servo system that they can understand, program easily, and troubleshoot.

Drives commonly used: Since two servomotors had to be coordinated, a servo controller with a common bus architecture is selected. With its built-in power supply, input-output capability, diagnostics, and troubleshooting, a multiple axis servo controller is a logical choice. This choice minimizes components. In addition to the flexible automation controller, all that is required are two servomotors, two PWM drives, and the necessary cables. See figure 11-30.

All the motion functions, I/O functions, arithmetic calculations, and screen commands can be entered by the customer via a hand-held programmer and the offline development software so that different size and weight boxes can be handled upon setup. The operator simply enters the box dimensions into the custom screens of the program, and the controller automatically calculates the position to move either the Y or Z axis in order to pick up or drop off a box. The vacuum grippers that secure the box are triggered on and off by the input/output functions of the controller.

Benefits: This type of control makes it possible to automate a portion of the manufacturing line. Previously, the boxes had to be handled physically by an employee. Sizing the servomotors for the heaviest box also allows tremendous flexibility. In the future, more servomotors and drives can easily be added to the system simply by adding plug-in cards. This is typical in the servo industry.

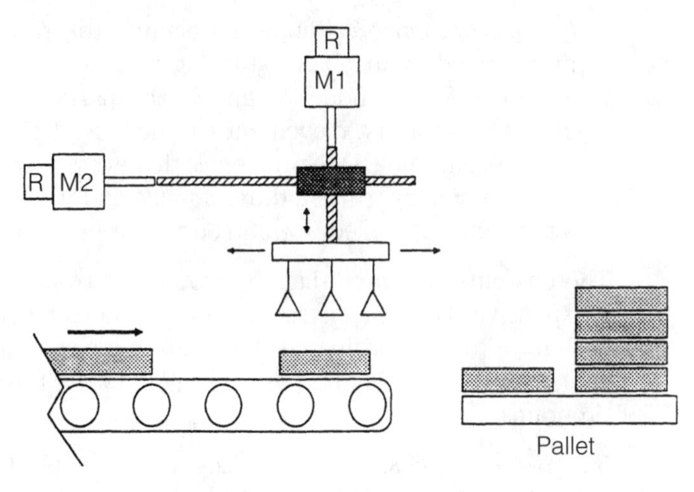

■ **11-30** *A box palletizer application.*

Many applications end up having to utilize an existing special motor or have to incorporate a special control algorithm. Often the special motor is a real challenge. Wound rotor motors, two-speed motors, and synchronous motors are often considered as recipients of an ac drive's output. This is possible in most instances; however, certain accommodations must be made. For instance, a two-speed motor has two windings: a high-speed winding and a low-speed winding. To maximize the usefulness of an ac drive, the motor should be set up to only run on the high-speed winding, thus allowing for a volts-per-hertz curve all the way from zero to full speed. Likewise, a wound rotor motor has to have its slip rings tied together in order to make it accept an ac drive as a controller. These considerations will come up from time to time as more existing motors are reused.

Other application considerations are customizing the drive controller to the motor and application. S-curves and custom volts-per-hertz patterns are just a couple of possibilities. Shown in figure 11-31, the S-curve is commonly used to provide a soft start to any ac motor. This might also be a requirement of the application.

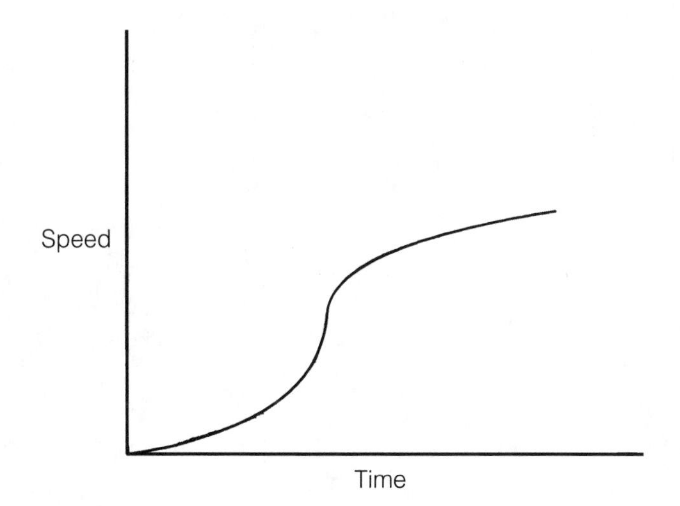

■ **11-31** *An S-curve for soft starting.*

12 Harmonics, power factor, and other electrical line phenomena

BECAUSE ELECTRONIC DRIVES ARE POWER-CONVERTING devices and because most now carry a microprocessor on board, electrical line disturbances are of keen interest to many. The electric utility, the end user, the drive manufacturer, consultants, and many others all want to know everything they can about harmonic distortion, electromagnetic interference (EMI), radio-frequency interference (RFI), and any other line disturbances. The need to know is precipitated by the need to avoid or correct the problem. The aforementioned electrical line ailments are actually distinct, yet often confused with one another. An understanding of power factor must enter into the equation so as to not compound electrical problems in the plant or building. This chapter will provide an overview of all these phenomena and help separate them from one another.

Drive harmonics

The topic of harmonics and harmonic distortion is very broad and complicated. Many individuals have not taken the time to understand the subject but claim to know what harmonics are and how to handle them. Talk to an electrical engineer or someone on the plant floor, and he or she will have a theory on what harmonics distortion is. Harmony is defined as, "to be in agreement with or to blend with." Therefore, harmonic distortion means that something is causing our system "to be out of agreement." Why does harmonic distortion occur and how should it be handled are the most often asked questions. Nonlinear loads and nonsinusoidal waveforms, items typically associated with electronic drives, have ushered us into a new realm—harmonics.

The harmonic is a sine wave-based component of a greater periodic wave having a frequency that is an integral multiple of the fundamental frequency. In simpler terms, we start with a pure sine wave and evolve into a nonsinusoidal condition. This is due in part because of the nonlinear loads that are produced by static power conversion (ac and dc drives), arc furnace devices, some rotating machinery, and so on. These nonlinear loads create havoc with highly sensitive communications equipment and microprocessor-based equipment. If all loads were linear, we wouldn't have the IEEE 519-1992 standard to use as a guideline when dealing with harmonic distortion.

Fourier analysis is the term given to the study and evaluation of nonsinusoidal waveforms. In 1826, Baron Jean Fourier, a mathematician, developed a series of formulas to work with these types of waveforms. There must always be a fundamental component that is the first term in the sine and cosine series. This is the minimum frequency required to represent a particular waveform. From here, there are integer multiples of the fundamental. The component called the third harmonic is three times the frequency of the fundamental frequency, the fourth is four times, and so on. In a six-pulse system, typical of a drive's converter, the real harmonics of concern are the odd-numbered harmonics not divisible by 3. The most critical harmonics to filter usually are the fifth and the seventh. In a 12-pulse system, the lowest producing harmonic is the 11th. What is created from these unwanted waves is the disputed issue among engineers, technicians, and anyone who blames harmonic distortion for the electrical problems in their facility.

Harmonic distortion is often referred to as electrical noise. Whether called garbage, hash, or trash on a given electrical line, it is unwanted and undesirable because of the complications it creates with computers and other sensitive equipment now used in many process control environments. Today, more is understood about what causes the distortion and what can be done to correct the problems. Yet manufacturers of power-converting devices are always looking for new ways to solve the problem—while possibly creating a new set of problems. For now, however, the drive industry is working collectively to deal with harmonic distortion. This cooperation is based more on competition than anything else; if your drive causes severe harmonic distortion, your drive will not be accepted in that facility.

When static power converters convert ac power into dc power and vice versa, a disturbance is created both to the incoming supply power and to the output of the converter. This is due to the waveform being distorted from its original sinusoidal state. Depending on its severity, this distortion can have detrimental effects on other electronic devices. Studies have provided industry standards that we can apply to at least minimize or predict the harmonic content of a given system; then we can take corrective measures. One such standard is the IEEE Standard 519, finally revised in 1992. In its new form, the standard has become more than a guideline; it now makes recommendations. This means that the user with the harmonic problem now has some options. To get a copy of the current IEEE 519-1992 standard, contact:

Institute of Electrical and Electronics Engineers, Inc. (IEEE)
Standards Group
445 Hoes Lane
Piscataway, NJ 08854
phone: (908) 562-3803
fax: (908) 981-9667 or (800) 678-IEEE

Prior to this, a facility just had to live with harmonic distortion problems; this is basically how the rumors about harmonics causing terrible electrical catastrophes got started. It is important to note here that there is no one single solution to a harmonic distortion problem, and it is not practical to think that merely meeting an IEEE standard for allowable distortion levels will prevent your plant from having problems. This issue of harmonic distortion definitely requires some study and usually some expense for correction. A cost-versus-performance curve is shown in figure 12-1. In order to meet certain percentages of allowable distortion, more cost is incurred as the percentage lowers.

When discussing percentages of distortion and allowable levels, there are a few terms we need to define. TDD, or total demand distortion, is a value derived from the total RMS (root-mean-square) harmonic current distortion. It is typically expressed in a percentage of maximum demand load current. THD, or total harmonic distortion, is the most-often-referred-to level and is relevant to both voltage and current distortion. It is sometimes called the distortion factor and is expressed as DF, in a percentage of the fundamental. It is the ratio of the RMS of the harmonic content to the RMS value of the fundamental. Thus, in mathematical form:

$$DF = \frac{\textit{sum of the squares} \atop \textit{of the amplitudes of all harmonics}}{\textit{square of the amplitude of the fundamental}} \times 100\%$$

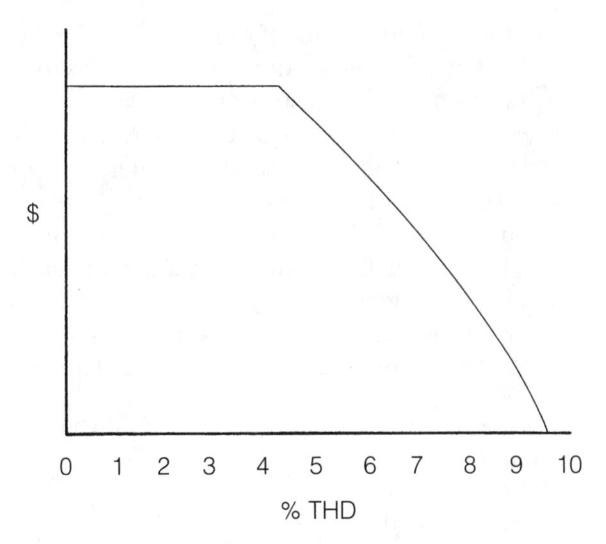

■ **12-1** *The relative costs for reducing total harmonic distortion (THD) in a system.*

However, most often when an allowable percentage has been met, something within the facility changes. New electrical equipment is added year after year, and before we know it, there's a major harmonics problem. As long as someone is monitoring these installations, one after the other, and by specifying that new equipment *cannot* contribute any more nonlinear loads, then the distortion levels can be kept in check. Another method, called harmonic analysis, involves performing checks after equipment is installed. This procedure is discussed at length below.

An initial analysis that can be used as a shortcut method for determining harmonic distortion whenever electronic drives are installed is called the *short circuit method*. By taking the available short circuit capacity (in MVA) of the bus where the drive will attach, dividing by the drive's rating (in MW), and multiplying that value by 100%, we get a very good indication of what amount of distortion to expect. However, a full and complete harmonic analysis by qualified individuals is the most accurate.

The harmonic analysis should neither be oversimplified, nor taken so seriously that all effort goes into looking for every minute load contributing to the nonlinear load total. The predominant loads will be apparent. In harmonic analysis, many entities play a role. The electric utility and the telephone company both have an interest. The telephone company is interested because of the TIF, or telephone influence factor. This is a complicated formula of weighted and nonweighted values regarding sine wave components used to place a working value on telephone interference due to distortions to the current and voltage waveforms. The factory or plant, obviously, has a lot to say, especially concerning their other sensitive pieces of equipment. The supplier of the phase-controlled converters or rectifiers has much to offer also. Therefore, it is important to include all parties, along with someone who understands the issue of harmonics and can act as the consultant, or even the mediator in solving problems and disputes. One fact is clear, there are definite misconceptions of what harmonics are.

Linear and nonlinear loads

As previously mentioned, in any installation with electrical supply, there are linear loads and sometimes nonlinear loads. A linear load can be defined as a predictable sinusoidal waveform generated by an electrical load. Examples of this include a facility's lighting system and its other resistor and inductor loads. The load is termed *predictable* because a relationship exists between the voltage and the current whereby the sine wave is "cleaner" and smoother. However, with the advent of rectifier circuits and power conversion devices also came nonlinear loads.

Other nonlinear loads have also been introduced over time for one energy-saving reason or another. Lighting ballasts, which help to save energy in lighting systems, exhibit a certain hysteresis, which contributes to magnetic saturation, causing the load to be nonlinear and the wave to be nonsinusoidal. Other nonlinear devices include metal oxide varistors (MOVs) and electrical heating equipment. With these nonlinear loads come waveforms, which are no longer nice, clean sine waves. The waves now contain many portions of other waves, making the resulting wave notchy and distorted. The challenge now becomes what problems does this condition cause and what can be done to correct the situation.

Since static power converters are used frequently and since they represent the largest contributor of harmonic distortion, there has been a significant rise in the awareness of harmonic distortion. Variable-frequency drives, UPS systems, and electrical heating equipment all convert ac to dc or dc to ac, and by doing so, create changes to the sinusoidal supply. This can cause definite interference problems with communication and computerized equipment prevalent in the industrial facility of today. A typical electrical circuit in a factory can have motors, drives, sensitive computers, lighting, and other peripheral equipment on it, as illustrated in figure 12-2. This single-line diagram of a typical factory system also shows various linear and nonlinear loads. Every factory is different and every circuit within the factory is different. Therefore, a complete analysis must be made of the system in order to properly identify the magnitude of the harmonic distortion and the corrective action required.

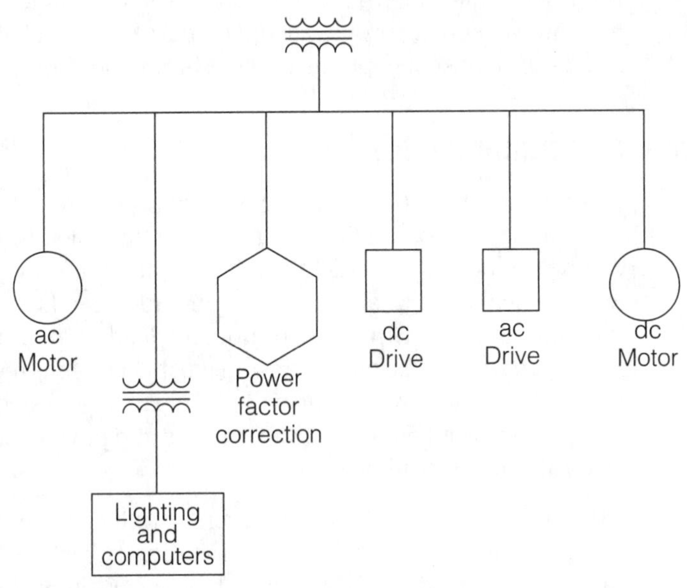

■ **12-2** *A single-line diagram of typical plant loads.*

Starting with a single-line diagram of all the loads and attributing values to each is the first step. From here we can begin to determine what kind of filtering, if any, is required and predict what might happen if other loads are added to the system. A *tuned filter*, as shown in figure 12-3, is a specific combination of inductors, capacitors, and resistors so configured to provide a value of impedance to handle one or more unwanted harmonic frequencies.

The tuned filter has some costs associated with it and might not be effective if the plant's loads change dramatically down the road. It might be simpler and more cost-effective to place a filter ahead of the only computer in the circuit, rather than install a more elaborate filter upstream. There is a relationship to the amount of harmonic filtering to the apparent costs involved. This has to be a consideration at some point, along with predicting load changes on the particular system in the future as new pieces of equipment are introduced.

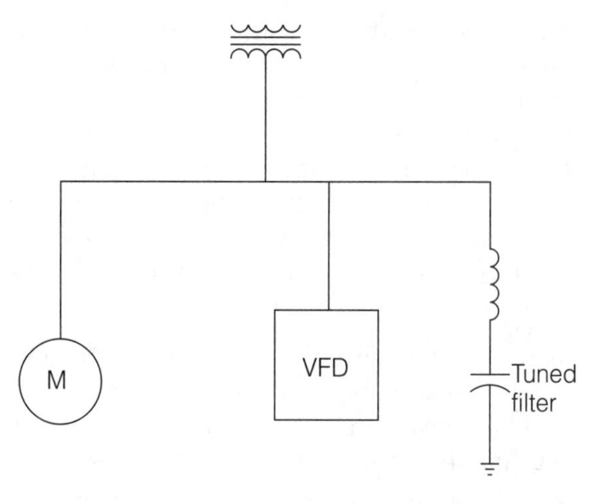

■ 12-3 *A tuned filter for a specific drive application.*

For any harmonic analysis, the point of common coupling (PCC) must be selected. This is an important location in the electrical system because it is the point where the harmonic distortion measurement will be made. As is shown in figure 12-4, this point is usually at the secondary of a transformer where many parallel loads of the same electrical system come together to connect to the main power supply.

Regarding transformers in general, a couple of comments: First, many people believe that isolation transformers eliminate harmonics. This is not so. Harmonic currents pass through a transformer. Voltage distortion can be affected by the impedance of the

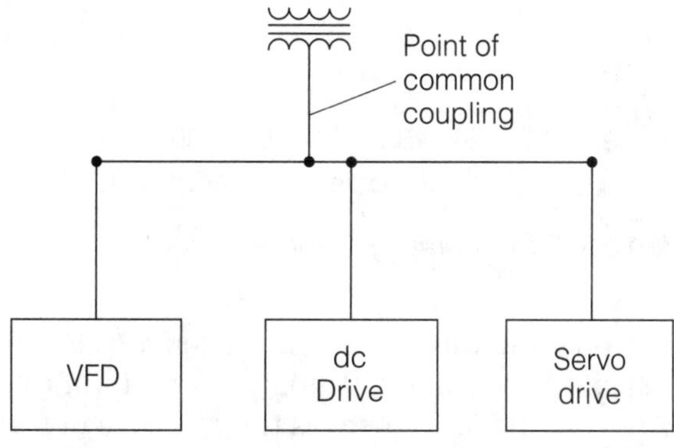

■ 12-4 *The point of common coupling (PCC).*

transformer, but this can also be accomplished with a less expensive line reactor, providing it has an equal impedance value. Second, transformers are designed to operate at 60 Hz. The harmonics tend to be present at higher frequencies, creating losses in the transformers in the form of heat. A transformer can overheat if subjected to currents containing high levels of harmonics. This has given rise to what is known as the K-factor element in transformer sizing in converter/inverter applications. These issues were discussed in a previous chapter.

Likewise, other electrical equipment, even power cabling, is often derated because of harmonic content to the waveform. A certain amount of heat can be caused by harmonic distortion. This value, while not dramatic, is still present. Figure 12-5 shows a cable-versus-harmonics derating chart for use on six-pulse systems. When designing an electrical system where converting equipment will be used, attention should be given to this matter.

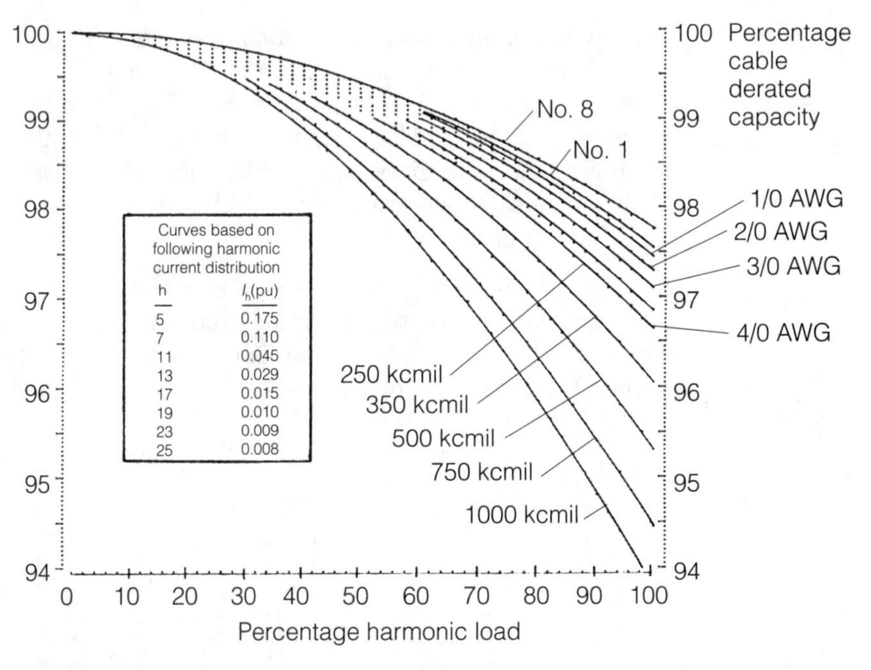

■ **12-5** *Cable derating graph.* IEEE

Not only will this power conversion affect the supply power system, it will also affect the output waveform's shape, most often to a motor. The occurrence is the presence of higher frequency disturbances that now tag along with the useful voltage and current.

Figure 12-6 shows voltage notches in a converted sine wave's current. Also shown are some equations by which the notch area can be calculated. These notches are called *commutation notches*, and their depth is of great significance. Also, the frequency with which these notches are seen is important.

Basically, since controllers are changing sine wave ac into pulses, and this is being accomplished by delayed conduction, harmonics are produced. Solid-state devices are the culprits in these rectifier systems, with SCRs and other thyristors being exceptionally notorious for high amounts of distortion because of their intermittent conduction. They appear as a short circuit to the system because they are not dissipating any energy when they conduct. Good for efficiency but bad for harmonics and power factor.

The effects of harmonic distortion in an output waveform from an electronic drive to an electric motor are well disputed. For instance, the drive manufacturer states that the wave is fine, while the motor manufacturer states that the wave is damaging. It all depends on whose side you are on. In an electric motor are windings that exhibit similar characteristics to the induction in a transformer. Therefore, should harmonics be present in the output to a motor, excessive heating can occur. This heating is lost power, and there can be more audible noise as well. Obviously, knowing ahead of time what levels of harmonics might be present can aid in selecting proper-size equipment for a desired performance.

Distortion limits are classified by the depth of the commutation notch, the area of that notch, and the THD, or total harmonic distortion, in terms of voltage. Now, with the IEEE 519-1992 standard, limits are also being set on current distortion. Tables 12-1 and 12-2 show recommended distortion limits for both voltage and current for different types of systems. What all this means is that manufacturers of power-converting equipment will have to provide filtering where required, causing system costs to go up. The harmonics themselves are not necessarily harmful—except to other sensitive electronic equipment on the same electrical system.

In summation, harmonics are blamed for a variety of problems in the plant. Blown fuses, burnt up motors, hot or burned connectors—all point to a potential in-house harmonics concern. It is worthwhile to make an analysis to determine whether or not harmonics are present and to what extent, rather than assume that they are the cause of the problem and spend time and money wastefully.

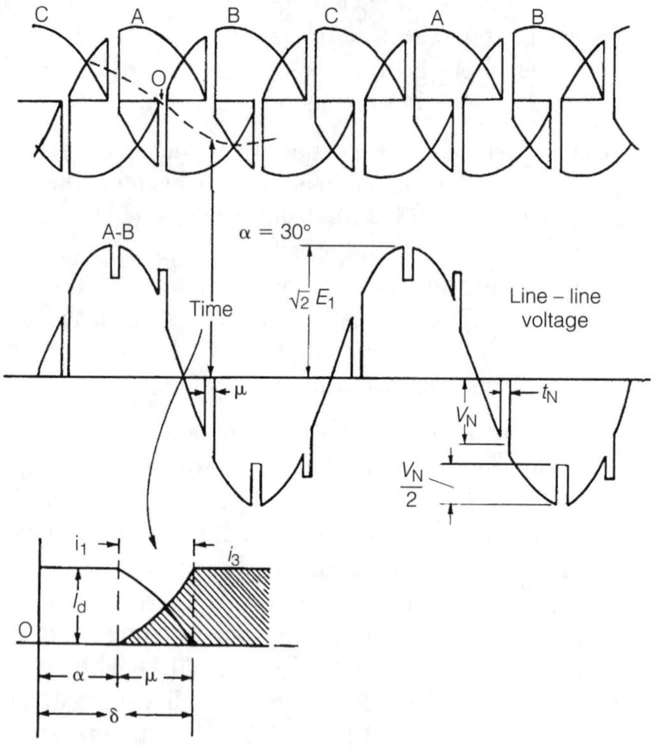

NOTE: the two other phases are similar to A-B. The width of the notches is exaggerated and ringing is omitted for clarity.

The notch area is calculated as follows:

$$V_N = \frac{L_L e}{L_L + L_t + L_S}$$

$$t_N = \frac{2(L_L + L_t + L_S)\, I_d}{e}$$

$$A_N = V_N\, t_N$$

where

- V_N = notch depth, in volts (line-to-line), of the deeper notch of the group
- t_N = width of notch, in microseconds
- I_d = converter dc current, in amperes
- e = instantaneous voltage (line-to-line) just prior to the notch on the lines to be commutated
- L = inductance, in Henry's, per phase
- A_N = notch area, in volt-microseconds

■ **12-6** *Voltage notches and area calculations.* IEEE

■ Table 12-1 Basis for harmonic current limits.

SCR at PCC	Maximum individual frequency voltage harmonic (%)	Related assumption
10	2.5–3.0	Dedicated system
20	2.0–2.5	1–2 large customers
50	1.0–1.5	A few relatively large customers
100	0.5–1.0	5–20 medium size customers
1,000	0.05–0.10	Many small customers

■ Table 12-2 Current distortion limits for general distribution systems (120 V through 69,000 V).

Maximum harmonic current distortion in percent of I_L						
Individual harmonic order (odd harmonics)						
I_{sc}/I_L	<11	$11 \leq h < 17$	$17 \leq h < 23$	$23 \leq h < 35$	$35 \leq h$	TDD
<20*	4.0	2.0	1.5	0.6	0.3	5.0
20<50	7.0	3.5	2.5	1.0	0.5	8.0
50<100	10.0	4.5	4.0	1.5	0.7	12.0
100<1,000	12.0	5.5	5.0	2.0	1.0	15.0
>1,000	15.0	7.0	6.0	2.5	1.4	20.0

Even harmonics are limited to 25% of the odd harmonic limits above.

Current distortions that result in a dc offset, e.g., half-wave converters, are not allowed.

*All power generation equipment is limited to these values of current distortion, regardless of actual I_{sc}/I_L.

where

I_{sc} = maximum short-circuit current at PCC.

I_L = maximum demand load current (fundamental frequency component) at PCC.

Data and equipment required to measure harmonic distortion

When considering the installation of a device that can convert ac to dc and vice versa, the area of the plant where the device is to be located should be looked over. Analyze all the equipment on the same circuit. Calculate the total linear and nonlinear loads. Look at the impedances of transformers and reactors in the circuit. Additionally, obtain the short circuit current data from the utility. Predicted harmonic distortion is easier to correct than after equipment is installed.

If it seems as though the entire plant could be affected, then all inductive and capacitive elements will have to be factored and attributed with some value of impedance. Any power factor correction systems must be accounted for because typically, when affected by resonant oscillating currents, a power factor capacitor can actually make the power factor worse in the system. It should be made known what value of power factor must be held. Also, if the capacitors are switched, it must be made known when the switching occurs.

If symptoms of a harmonics problem seem to exist, it is best to first ascertain whether or not any rectifying equipment is in the plant. Then it is necessary to list and document step by step all the electrical components on a particular circuit. This "plant map" can be invaluable in totaling loads and impedance values; also, it will indicate where sensitive equipment is in the plant and allow for planned filtering, if required. Concurrently, the utility should be able to provide data relative to the source of power and the available short circuit current, along with known impedances for upstream circuitry. After all, the power companies are partially driving the requirements of harmonic filtering and power factor correction.

When testing for harmonic distortion levels, certain test equipment is required. Along with the proper test equipment, the method of testing must be determined. Necessary equipment includes a current transducer, voltage leads, and some type of device to display and analyze the waveform. An oscilloscope, spectrum analyzer, harmonic analyzer, or distortion analyzer will also be required. These devices allow the user to take measurements of electrical wiring to determine harmonic distortion levels. They can display in their readings waveform shapes or graphs of a select harmonic. All this data can then be downloaded to a computer for further compilation. The digital analysis can be performed using either the Fast Fourier Transform technique or by using a digital filter. Both methods allow for the collection of spectrum data which can be analyzed. This data can be compared to allowable limits and acceptable values.

Once the data is gathered, it is necessary to display its results in some form. One method is by using a histogram, which will show the distribution of harmonic occurrences and the THD for the fundamental. A typical harmonic histogram is shown in figure 12-7. The data will help to determine whether corrective action is necessary. This corrective action can mean tuned harmonic filters or inductors have to be placed in the circuit. This can be expensive, so a good analysis by reputable sources is recommended.

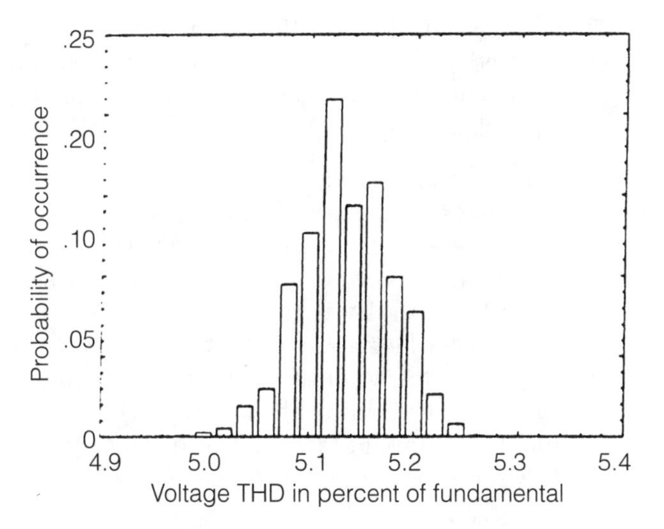

■ **12-7** *Harmonics distribution histogram.* IEEE

Radio frequency interference (RFI)

Contrary to what many think, radio frequency interference (RFI) is not harmonic distortion. Harmonics is the distortion of the sine wave due to power rectification found, for example, with variable-speed drives. Radio frequency interference is different. It is interference to frequencies above the audible by fast switching devices, similar to those used in variable-frequency drives. This is where the two are often confused. There are filters for shielding the input and output of a VFD so as to minimize the possibilities of interference.

Radio broadcasting frequencies are typically above 150 kHz and below 100 MHz (megahertz). Whenever current or voltage waveforms are nonsinusoidal, there is always the possibility of radio frequency interference. Newer switching technology in converter and inverter devices has started to infringe on the high-frequency domain of the radio waves. By the laws of the Federal Communications Commission (FCC), interfering with these signals is forbidden. There are new standards being discussed every day, as frequencies of inverters can be seen going higher and higher.

Switching frequencies, or carrier frequencies in drives, have reached values of 15 kHz with present-day drive technology. Since drives utilize the switching capability of different transistors, they have come under scrutiny. When current increases so quickly, as is the case with today's power semiconductors, noise can be detected at a radio receiver. Distance from the emitting device, shielding, and filtering are all factors relating to the existence of RFI. Other factors impacting RFI are the horsepower and current rating of the drive, the output and switch frequency, the impedance of the ac supply, and how well shielded the power modules are.

Filtering by means of inductors and capacitors is presently the best way to suppress RFI emissions. In general, the relationship between RFI and drive equipment is still ambiguous. Emissions and occurrences are rarely proven, as they are rarely considered. Enforcing the adherence to the laws and standards has not been paramount. As the occurrences increase, so too will the need to police the issue. One step is to clearly specify that the drive manufacturer must adhere to certain levels as set by the FCC. Then these levels must be measured after drive installation to see that they are met. By enforcing the issue, proper attention to the matter will be given.

Power factor

Power factor is another often misunderstood topic. Similar to harmonic distortion, it is more predictable whenever sine wave power is being monitored, but introduce nonsinusoidal waveforms and—watch out! Power factor is the ratio of the instantaneous, or active, power and the anticipated power. When power is being analyzed as a pure sine wave, all is predictable and understandable. When we deal with nonsinusoidal power, as is the case with drives, the analysis takes on new meaning. Normally, leading and lagging power factor assumes capacitance and inductance are in the system. This, again, is only valid when sine wave power is evident for voltage and current.

Power factor is actually made up of a displacement factor and a harmonic factor. As seen in an earlier chapter, power in the equation $P = EI$ is the product of voltage (E) and current (I). Voltage and current waveforms in phase means that their respective zero crossing points are the same. This condition allows for all the available power to be used as productive power. However, when the current gets out of phase from the voltage it is now termed *lagging*. The net result is that for this half-cycle, the power is not productive. This is the displacement power factor. The harmonic power factor is of a lesser degree, as it basically is the effect of the wave distortion in the same phase sequencing as previously described. The relationship between displacement power factor and distortion is best illustrated in figure 12-8. This relationship is common whenever a converter (such as an electronic drive) is in the system.

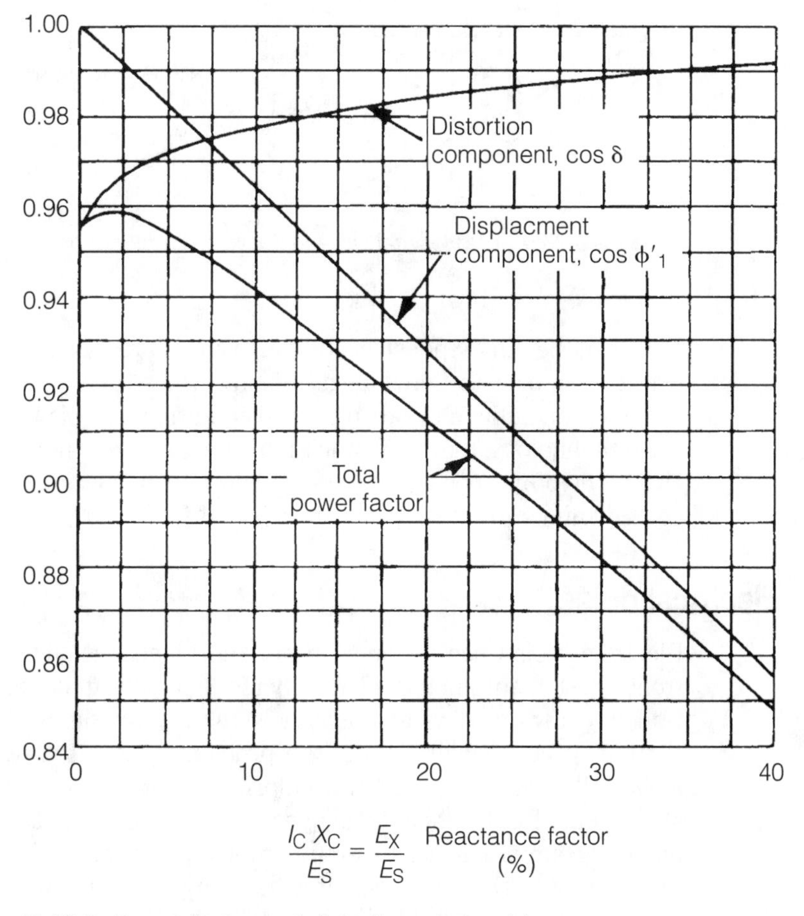

■ **12-8** *Power factor and distortion relationship.* IEEE

Power factor is expressed in terms of VARs, or K-VARs, which is a thousand VARs. VARs are volt-amperes reactive. This is the reactive power, a product of current RMS (a peak average) and voltage RMS. Power factor is proportionately equal to the watts divided by the volt-amps. A diagram commonly used to illustrate power factor is the power factor triangle. It is shown in figure 12-9.

Power factor is measured as leading or lagging conditions. Power factor correction comes into play when a predetermined value for a system must be met. Often the utilities demand that certain levels be maintained. A standard circuit for power factor correction can utilize capacitors and inductors to provide a tuned circuit, placed between the supply and the converting/inverting device.

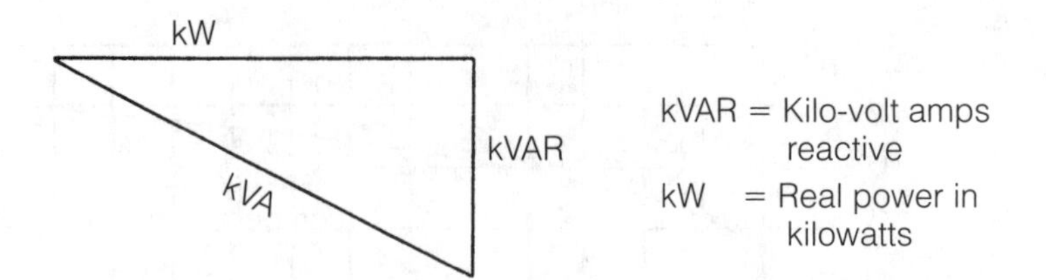

■ **12-9** *Power factor triangle—kW to kVAR relationship.*

This can be an expensive solution. When applying electronic drives, it is best to know ahead of time what their power factor value will be under all load and speed conditions. With present-day drive technology, a drive installation where there is none actually can be helpful, because a system's bad power factor can be discovered and corrected.

Electrical noise

Electrical noise has become one of the biggest challenges confronting the automated plant's personnel. Often it cannot be traced to any one device or any one source. Many times it is intermittent, which makes finding the problem even more of a challenge. Electrical noise is the culprit in many electronic plant shutdowns. The key here is that years ago machines and processes had few components that were microprocessor-based. The microprocessor and the low levels of voltage in the circuitry surrounding it are very susceptible to electrical noise.

By now, we are all pretty much aware that it is important to save and back up our files when using a computer. We have been trained to do this because all data could be lost in the event that a power fluctuation occurs. With electronic data traveling bit by bit across a conductor at low levels of current, it doesn't take a whole lot of higher level energy to disturb one bit or destroy it. This is basically what electrical noise will do.

There are many ways to describe, find, and handle electrical noise. When discussing noise, we are mostly concerned with noise emissions and noise immunity—where does it come from and how can the control system work adequately if noise is present. *Noise emission* is defined as electromagnetic energy emitted from a device. *Immunity* is the ability of a device to withstand electromagnetic disturbances.

In looking at noise emissions, we find that most electrical noise is man-made. There are natural emissions from the voltage coming off of all electrical components. This is sometimes referred to as *thermal noise emission*. The good news is that this type very rarely affects the electrical system. Another lesser type is the natural, atmospheric type noise mainly attributed to lightning storms. The last type is the man-made noise, and because it is created by humans, it can be controlled by humans. This type is made up mainly of radiated, conductive, and inductive noise emissions.

With radiated noise, we are concerned with that kind of emission that travels through the air. Conductive noise, on the other hand, is actually conducted over the wire. Being able to pinpoint where the noise is originating will obviously help eliminate the problem. Identifying all the circuits, isolating them, and locating the point of common coupling is necessary. This is sometimes dependent on how two, and sometimes more, electrical circuits have been incorporated into the system. Referred to as *electrical coupling*, this is the common point where the circuits meet.

Often this type of noise emission is inductive, although it can sometimes be capacitive. Inductive emissions occur when the magnetic field around a live cable affects another cable. This phenomenon is also referred to as an *eddy current*. Capacitive noise usually occurs when two electrical circuits share a common ground.

There are many different types of noise-related phenomena. Being able to distinguish one from the other and taking the proper steps to avoid each is important. These types include harmonic distortion, electromagnetic interference (EMI), ground loops and floating ground situations, power fluctuations, and radio frequency interference (RFI). Complete power outages and brownouts can also be thrown into the category of noise because an interruption of service is an issue. If the interruption is short (a cycle or less) then maybe the sensitive equipment will keep running. If it is longer, relays, most computerized equipment, and other electronic devices will shut down, thereby requiring resets of the equipment throughout the plant. This can be costly. Each has its own cause and each exhibits its own effect.

EMI and RFI

EMI, or electromagnetic interference, is not easy to pinpoint and usually does not occur that frequently. This interference occurs when a sensitive device, such as a computer, receives the EMI and trips, faults, or starts acting in an odd manner. Electromagnetic fields can exist around high-voltage, high-current carrying devices such as arc welders, large horsepower motors, and electric furnaces. This type of interference is best addressed when locating equipment within the plant. If not, loss of data, bad data, and false contact closures can be the result.

RFI, or radio frequency interference, acts much like EMI in problem but not in source. RFI is the presence of high-frequency waves in the general vicinity. These high-frequency signals can interfere with data transmissions and can create erroneous data at a microprocessor-based piece of equipment, whether a personal computer or a computer-controlled machine. The common sources of RFI are variable-speed drives with high carrier frequencies, wireless phones and microphones, walkie talkies, and so on.

The best solution is to avoid locating any high-frequency devices near sensitive pieces of equipment. If this is unavoidable, then a filter at the source can suppress the interference.

Grounding

Ground loops and poor grounding are often sources of electrical noise. A ground circuit is supposed to route higher levels of current to ground, rather than destroy valuable electronic components and shock humans. Unfortunately, this is also another route for low-level electrical disturbances to travel. This ground route can both send unwanted disturbances out to other sensitive devices and receive unwanted electrical noise. A ground loop exists when there is potential, or EMF, between pieces of equipment whose grounds are connected in the same system.

A typical ground loop condition is shown in figure 12-10. The best scenario is to ground each separate piece of equipment directly to its own ground. The trick is not to create a loop condition. This loop will allow unwanted signals from various sources to enter into a completely separate circuit—in many cases, a circuit whose equipment is very sensitive.

When done solidly and properly, grounding is a very good practice. However, grounds can become loose over time in environments where vibration and activity around the connection occur. Also, grounds can deteriorate over time, especially in corrosive atmospheres. Checking and inspecting ground wires and all wires in a circuit is good practice. Making sure the connections are constantly

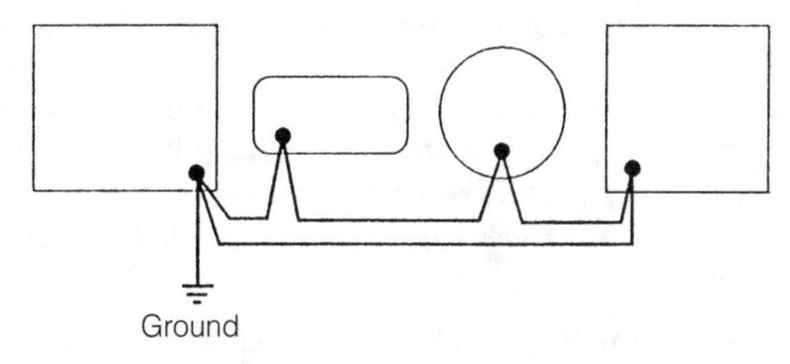

■ **12-10** *Ground loops.*

solid can save damage to devices that normally expect low levels of voltage and can also eliminate disturbances due to ground loops.

Power fluctuations

Since most electronic equipment desires good, clean electrical sine wave energy in a consistent form, we can categorize many power fluctuation phenomena into this section. Alternating current (ac) is nice and convenient, but it is also unforgiving if the supply is disturbed on its way to its recipient. The recipient in the automated factory is usually a computerized piece of equipment. That piece of equipment might have a host of protective devices on the incoming power side, including fusing, noise suppression filters, and voltage-matching transformers.

Even with some or all of this protection, problems can still occur. Problems discussed in this section involve complete power outages, brownouts, sags in voltage or current, oscillations (possibly due to harmonics), and power surges. Each has its consequence and each has its protective circuit—both at some cost and risk to the piece of equipment being protected.

A power outage can have many definitions. Did the lights flicker on and off? If so, is this is outage? Maybe yes, and maybe no. If the lights were off for a cycle or two, this is not a severe outage, or complete loss of power. If the lights and many other electrically driven items went off for several seconds, then this is an outage. Better call the utility and find out what happened. Lightning storms are a good culprit in the summertime. Some sensitive equipment, incorporating capacitor networks, can ride through a couple of cycles without power. Other more critical, sensitive pieces of equipment have to be placed on a circuit with a UPS, or uninterruptible power source, system so as to continue running throughout the outage.

Sometimes the outages occur at the same time over the period of a given day. This can be an excellent clue in determining whether or not the occurrence is external to the plant or internal. A chart recorder placed on one of the supply lines will record when and what happened. With this information a determination can be made how to solve the problem, or whether to just live with the outage.

Sags in voltage to a piece of equipment can also appear to be an outage of sorts. The end result is the same—the equipment trips and needs to be reset. The cause of the sag is just as important. Is there a large motor in the plant that, when it is started, demands so much current that the overall system sags? Or is there a similar current-needy device somewhere near the plant? If so, it should be isolated from all other pieces of equipment somehow. Some pieces of equipment can handle lower levels of actual voltage from the nominal level for extended periods of time. This should be prequalified with the supplier of the piece of equipment when these conditions are possible. A brownout is actually a low incoming voltage level over an extended period of time.

Dealing with electrical disturbances

Hash, crosstalk, garbage, harmonic distortion, oscillations, ringing, and notches to the electrical system's waveform are the common terms, and they are both unsightly and unwanted. They are unsightly because no one wants to see them on an oscilloscope, the device that normally picks them up. From an electrical standpoint, they are unwanted. Sensitive pieces of equipment need clean, smooth low-level waves coming into their circuitry. If this condition does not exist, erroneous data can be produced, software programs can stop executing, and damage can even occur to componentry. There are many causes of oscillations as well as several solutions. Filters, reactors, and capacitor networks can smooth out the waveform on the incoming line, but the primary focus should be to find the cause and stop the oscillations altogether.

Power surges and voltage spikes are types of power fluctuations that are probably the most unwanted. They are truly undesirable because they can destroy electrical components without warning. The best protection for these kinds of fluctuations is fusing or circuit breakers. If fuses are blowing often and randomly, it might be critical to source the problem rather than just replace fuses each time. All it will take is one instance, with all conditions right, to have the surge get past the fuse and take out an important piece of equipment. In addition, if there are spikes present in the system, transformers and power supplies will be stressed often, which can lead to premature failure of these components.

The key to dealing with electrical noise is to prevent it. Diagnosing noise and its source later is tough enough. If good wiring practices are followed and attention is given to what types of equipment are to be used together, some noise problems can be addressed from the start. With so many retrofit projects happening in the plants today and with engineers and designers trying to anticipate problem areas, it is still likely that electrical noise will be present in the system. Being able to test for it and getting the parties responsible are two important steps to solving the problem. There is always a solution, but at what cost and who should pay?

Following are some steps to take when installing new equipment or retrofitting old machinery in a plant with new. First, know the electrical characteristics of the new pieces of equipment. This means find out what types of power conversion devices are used, how fast they switch, and so on. Suppliers of this type of equipment are used to these types of questions and must answer honestly or else they will have to fix the problem later. Secondly, make a single-line drawing of all the equipment on the electrical system in question. Show both new and old equipment. List all the sensitive pieces of equipment on the circuit and find out from the operators which ones are the most sensitive. List any filters already in place. When doing this evaluation, also remember that some equipment can be a receiver as well as a sender when it comes to radio frequency interference and even electromagnetic interference.

Next, the analysis must take into account the plant's incoming ac supply. If we're going to connect up to this source, it will be nice to know what we're dealing with. Is the supply clean, or is there already a present disturbance? Is the supply stiff, is the short circuit current level high or low? In other words, can the supply system handle your loading and possible fluctuations? Having a handle on what's coming in can help pinpoint where problems might reside later. Likewise, the plant's grounding system must be measured. See if the grounds present are proper and if any show signs of deterioration. This analysis will also aid in the grounding of the new equipment due to be installed, anyway. Again, heading off these types of grounding problems will be well worth it.

After you are satisfied that all the bases are covered, it's time to install the equipment and apply power to it for the first time. But before running the equipment in the production mode, first test at various key locations for the presence of electrical noise. Check in and out of electrical enclosures, control panels, even at the sensitive computer stations. Many electrical noise problems will surface right away. Some might not. Be prepared to test the supply and outgoing lines to and from equipment. Have certain test meters and recorders handy. After all, there will be probably only a small window of time to get the equipment back on line into the production scheme.

Shielding

Shielding is the most common practice of protecting signal wire from electrical noise. As discussed earlier in this book, there are several techniques that can be used when routing signal wire and the type of cable actually used. Coaxial cable provides an ample degree of shielding due to its composition. The outer casing protects the inner signal conductor. Coaxial cabling is a safe choice when electrical noise is expected in a system, but not at too high levels.

Another common type of signal wiring is shielded wire. This should be used for any speed reference signal into an electronic drive. In figure 12-11, a three-conductor wire is encapsulated with a shield. This is tied to ground at one end only, which helps keep unwanted noise from getting through to the wires. Another method that helps minimize these intrusions of noise is to use twisted wire. Shielded and twisted wire provide an even better degree of protection.

Another form of shielding is actually installing physical barriers within an enclosure housing multiple electronic components. Often these barriers only have to be sheets of metal that separate compartments inside the enclosure. The shielding should prevent emissions, some natural and some man-made, from getting to the sensitive devices within the enclosure. As suggested, it might be worthwhile to predict this situation ahead of time. Quite possibly the sensitive component or the noise-emitting component can be located from the other external to the enclosure.

The reference signals entering into an electronic drive should be isolated. This will keep line voltage to ground potentials from occurring, which could disrupt the incoming signal. It is important to fully know where a signal is originating. Many times the source has its dc common tied to earth ground. This source, if a noise emitter, will often pass the noise to the signal receiver via the signal. That is why it is common to completely isolate the incoming process signal to a controller. This can be accomplished by solid-state relays or optical isolation.

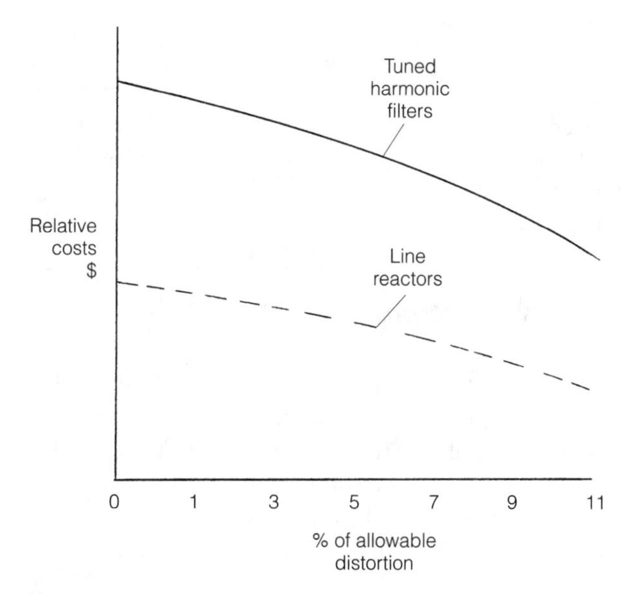

■ **12-11** *Shielding.*

Optical isolation

We have all heard the term *isolation*, but what is it and how is it incorporated? In a basic sense, isolation involves completely separating two components in an electrical system so that unwanted noise or signals cannot pass between. One common means of achieving this is by optical isolation. Since certain diodes can emit infrared light, it has become an electronics industry practice to use that capability to isolate expensive electrical components. By using the light emitter (sometimes called a phototransmitter) in conjunction with a light receptor, no physical connection has to be made. In this way, logic commons in low voltage control circuitry are fully separated, or isolated.

This is a safe and effective means of enabling or disabling a circuit. However, when interfacing the logic of one manufacturer's product with another, make certain compatible isolation techniques are incorporated. Doing so can avoid problems of incompatibility at the startup of the equipment.

With LEDs, or light emitting diodes, it has become a practice in many low-voltage control systems to use optical isolation. Basically, as the electrical current passes through the diode (LED), the light is emitted (see figure 12-12). A light receptor located adjacent to and dedicated just to that particular LED receives the light transmission and keeps the circuit flowing. The attractiveness of optical isolation is that there is built-in circuit protection. No ac-

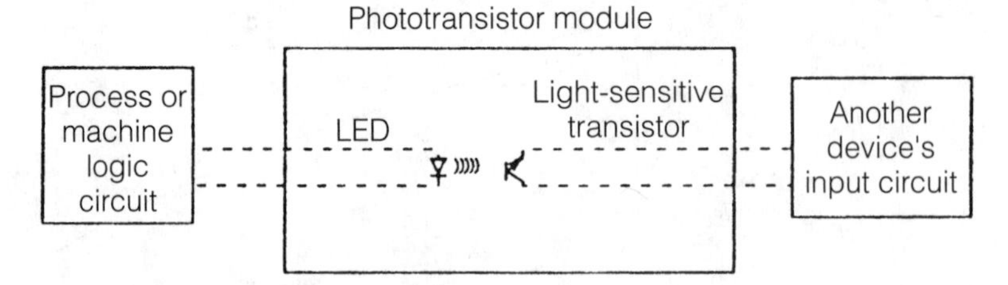

■ **12-12** *Optical isolation.*

tual electrical current flows through this isolation point. Since control equipment is usually expensive and board level components are not easily replaced (it's now customary to discard the entire board, rather than troubleshoot it and replace one or two components), protecting the low-level voltage components from all possible spikes and surges is extremely important.

Signal conditioners and filters

Many times the electrical noise in the circuit cannot be traced to a source and, therefore, eliminated. Fortunately, dedicated modules exist that can be installed in a circuit and provide the necessary isolation or even filtering so that we can go on to the next problem. Electrical noise and sensitive control/computer equipment don't mix very well. Thus, a new business of filters and signal conditioners has arisen.

Many filtering packages can be purchased off the shelf and installed quickly to eliminate noise. Often these filter networks are a resistor and capacitor in series parallel to the load being filtered. Looking at the resistive-capacitive filter scheme in figure 12-13, the values for the capacitor and the resistor are selected for the amount of filtering desired, the load, and the actual noise predicted. Usually this solution is the least expensive and will work as long as no one inadvertently removes the filter.

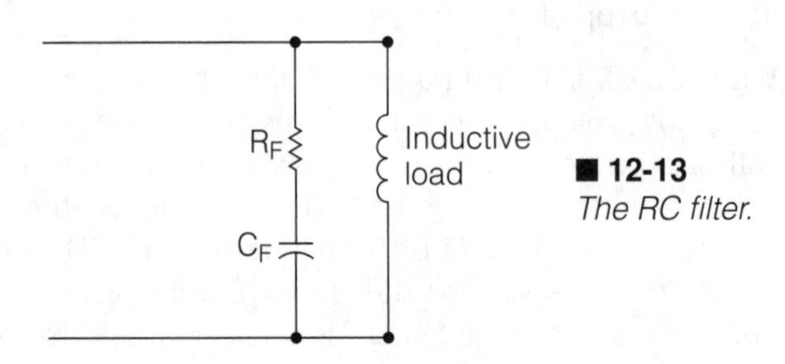

■ **12-13** *The RC filter.*

Sometimes a single capacitor can provide adequate filtering and dampening in a control circuit, although too much capacitance can make a system very nonresponsive and sluggish. This should be avoided. Another solution is to install metal oxide varistors, MOVs in the circuit to accomplish the same result.

These solutions are always worth the attempt. At worst, it won't filter enough and another approach will have to be considered.

13

Electronic drive technology's future

OVER THE PAST 15 YEARS THE DRIVE INDUSTRY HAS GROWN dramatically. Just when the market appears as though it is saturated with electronic drive product, another use or application for the product emerges. Industry and building continues to expand, and drives are now as commonplace as the personal computer. Energy efficiency, better speed and process control, and electronic communications are all considered with old and new installations. Electronic drive products give the user all of these as part of the package.

Whether the drive is ac, dc, or another type, manufacturers of these products are making them much more affordable, powerful, and reliable. All this makes them easier and more practical to implement. It is probably a safe statement to say that any electronic drive today has the capability to improve the control of any motion process. Improving a process by speed or torque control of an electric motor or by allowing the user to have access to information, not obtainable in the past, are those capabilities. But electronic drives now provide so many more advantages.

Any expense, whether for personal or business use, has to be justified. Spending money and keeping costs down are issues dear to everyone's heart. But whenever a product can provide the same functions that, in the past, were luxury functions (which always had additional cost associated with them), the purchase of the product is easily justified. Electronic drives have gotten to this point. In addition to being speed controllers for electric motors and improving a machine's performance or a process's output, the electronic drive acts as a motor protector, diagnostic tool, and energy saver.

As we have seen throughout this book, although many drives are similar, each type of drive has certain features that make it different than the other types. The differences are mainly in the power circuitry and the voltage types. The ac and dc drives have their niches, as do servos, steppers, and other drives. Some are not available in all sizes and some don't provide as much energy savings as others. Yet all drives today have some kind of microprocessor content and some form of communicating with external equipment, which makes them similar. They also all have to control an electric motor of some kind, and, therefore, issues of torque, horsepower, and power transmission unify them. In the future, drives will begin to look even more alike.

There is activity within the drive industry and its technology to make drives more configurable. *Configurable* here means that a drive should be able to run any electric motor, ac or dc, or otherwise. This makes sense because all electronic drives have to start with either three-phase or single-phase ac power as their input. From here, that voltage is converted into dc anyway. At this point, the dc voltage can be used to operate and control a dc motor, or it can be inverted back into alternating current to run an ac motor.

In the past, the drawback to a voltage-configurable drive has been cost. If a drive is going to be marketed as configurable, it would require a second power bridge for dc to ac inverting. In a competitive industry, this is a dramatic cost burden to any manufacturer's drive. However, with new control techniques and costs coming down dramatically, an ac/dc configurable drive is probably just around the corner.

Another type of configurable drive now coming onto the market is the open-loop/closed-loop flux vector drive. As discussed in an earlier chapter, a closed-loop version of an ac drive, along with the algorithms to correct quicker to speed and torque errors, is called the flux vector drive. This is a closed-loop drive because it has to have some type of feedback, usually an encoder or pulse generator, in order to perform its algorithms at all. This means that a module has to be included to decode the pulse generator's signal into usable form within the drive. Again, with costs coming down and because this ac drive is so similar to its open-loop version (power bridges, protective circuits, and so on are the same), this open-loop/closed-loop configurable product is becoming a reality.

The next requirement of an ac drive will be regeneration. At the very least, this capability will have to equal that of the regenerative dc drive. Without this capability, the ac drive is still unable to perform the holdback and braking functions of a multiple-section line, which are currently provided by traditional dc drives. Again, this requirement involves extra costs. An ac drive already has an extra power bridge over its dc counterpart in its inverter section. To make the ac drive regenerative, another complement of power devices has to be added to the converter section to allow for regen energy to have a path back onto the supply line. In addition, the control for this function has to be incorporated into the drive; however, with software and microprocessor capability, this is not as much a cost burden as the extra set of power devices. A simplified version of this power scheme for an ac drive is shown in figure 13-1. Once this obstacle is overcome, a huge issue in the "which is better, AC or DC?" argument will be gone.

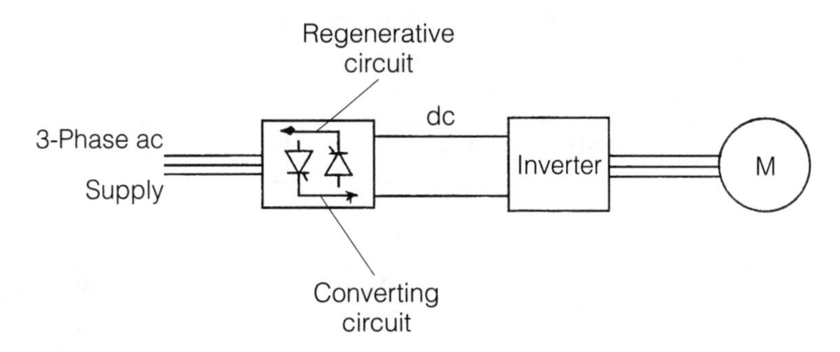

■ **13-1** *Regenerative ac drive bridge scheme.*

So, which *is* better, ac or dc drives and motors? The operative words in the question are not ac or dc, but rather, drives and motors. The package has to be compared because one motor might be better suited for an application than the other. The motor selected automatically determines the type of drive used (unless that ac/dc configurable drive is available). Throughout this book we have touched upon this ac-or-dc question. Many issues enter into the decision: the type of application, the motor environment, the performance expected, the voltage available and so on. But the lingering factor is simply: user choice! If a person has grown up with and had good experiences with one type versus the other, then many times that person will stick with that technology.

It is a fact, however, that ac electronic drives are displacing dc electronic drives in many applications. This displacement is not as dramatic as many people make it out to be. Nonetheless, there is a gradual trend to use ac over dc drives and motors in many applications. Because this is happening, much of electronic drive future development is being centered around the ac version. With the prevalence of the ac motor, compact sizes available, and energy savings an omnipresent issue, ac electronic drives are getting good publicity. Future drive technology development will focus much of the efforts around ac products.

For instance, in the automobile industry, ac drives are being considered as the choice power controller in electric vehicles. With environmental concerns, and with the energy crisis still in the back of our minds, electric automobiles are slowly emerging as a viable alternative to the gasoline-powered car. First, a look at the mechanics of the electric car and the inverter's role. As shown in figure 13-2, an electric car is made up of certain basic components: an ac inverter, batteries to store dc electrical energy, and an electric ac motor. While this technology has a ways to go, it is taking an existing technology in ac drives and exploiting it.

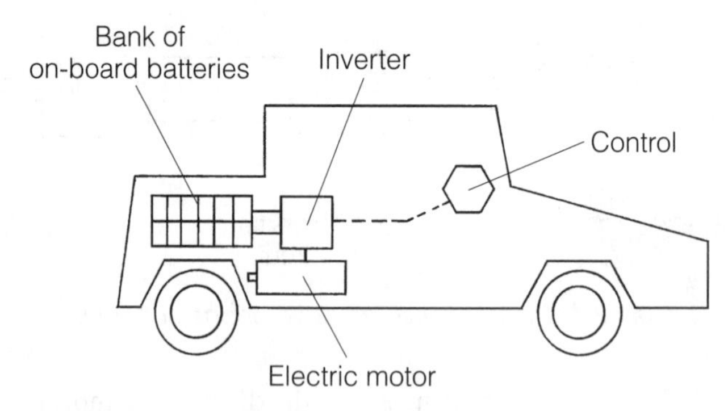

■ **13-2** *The future of electric cars.*

Some of the issues relative to electric automobiles include safety and packaging. Safety issues include addressing vehicle accidents where passengers and rescuers can be electrocuted by batteries and broken cables. Somehow with all the software and fault-handling capability of ac drives, these issues should be resolved. Second, packaging of several batteries within an automobile can add tremendous amounts of weight. Perhaps with advances in superconductivity and other relative industries, these issues can be overcome, and the electric car will become a viable alternative.

Other industries' developments and other factors continue to affect the electronic drive's market and technology. With superconductive materials someday in our future, how will this change the drive industry? Microprocessor advances and power semiconductor breakthroughs are definitely going to have more immediate impacts on the drive products' development. Other factors that will have an effect on electronic drives will be the Clean Air Act(s), future harmonic distortion regulations, and energy rebates. The effects will be in building the product cost-effectively to address market and application needs as these peripheral rules, codes, and regulations dictate. Competition within the drives' industry will spawn new developments.

The Clean Air Act was enacted several years ago and has caused ac drives to be a consideration in certain facilities where adherence to the Act is required. Factory and plant emissions have to be controlled in some way. These methods usually involve air handlers, scrubbers, incinerators, and other devices to clean exhaust air before it leaves the facility. This means that fans and pumps are used quite often, and therefore, variable-frequency drives are needed. This Clean Air Act has had a substantial impact on ac drive growth.

Likewise, IEEE has established guidelines for harmonic distortion limits for industrial and commercial facilities. This has caused the drive manufacturer to change their product design to meet these guidelines. This might mean the addition of hardware or complete design revamp as far as the manufacturer is concerned. However, as far as the end user is concerned, meeting the IEEE actual limits when applying the drives is all that matters. (Much of this was discussed at length in the previous chapter.) One prediction about drives and harmonics is that there will be drive units developed soon that will have a pure sine wave going in and a pure sine wave going out, and no harmonic distortion will be present.

Energy rebates are the best way to get a user to purchase and install drive equipment. If the price is lowered so much that the cost of the product no longer becomes an issue, and the energy justification (payback) is also there, then these installations will continue to happen. Energy rebates furnished by the electric utilities will come and go. But it is almost safe to assume that not too many more new power plants are going to be built over the next decade. Environmental red-tape and sky-rocketing costs, along with the amount of time (several years) required to bring a new power plant online, prohibits the thought. The better scenario is to conserve in any way possible, and ac electronic drives have proven, dramatically, that they save energy.

Earlier it was stated that the electric automobile is one major electronic drive application that is gaining ground. Another is magnetic levitation. Mag-lev is the latest in high-speed passenger train technology. Although it has been around for some time, mag-lev is starting to get impetus probably due in part to energy conservation and the trend toward mass transit. The magnetic levitation train works on the principle of linear induction. The train and its track are likened to the standard induction motor: the train is like the rotor, and the track is like the stator (figure 13-3). Utilizing this technology means that ac electronic drives, usually in higher horsepowers, are necessary in some form to provide ac to dc and dc to ac conversion and inversion.

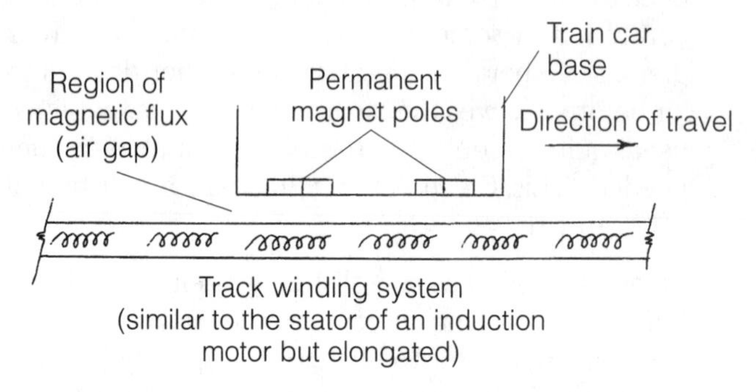

13-3 *The magnetic levitation (mag-lev) train.*

Size of drives, costs, and packaging are three areas where future product development will occur. Fifteen years ago, a 5-HP electric drive was the same physical size as a 200-HP drive today. Microprocessors, reduction in other component sizes, and the ability to get more amperage through smaller power semiconductors have allowed for the size reduction of the drive package. Smaller, more efficient power semiconductors mean that less area in the drive must be dedicated to heat dissipation and, again, allow for a smaller design.

New methods for handling and dealing with heat are also being developed. A cool drive is a happy drive and getting the heat away lengthens its life and improves its power capabilities.

All of these product developments also have cost reduction in mind. A more expensive drive with more capability will not usually sell. A less expensive drive, even with less capability, will almost always sell. A less expensive drive with more capability will always sell!

Other industry trends include physically placing or packaging the drive with the motor. Figure 13-4 shows a motor with the drive mounted on top. At present, this is not the accepted practice, and obviously, there have to be concerns whether this approach will even work at all. Motor heating will contribute to drive heating. Motor and machine vibration have to be addressed. However, this technique is being explored by those drive manufacturers who also manufacture electric motors. This development will require a great deal of proof in the field to convince engineers and users that

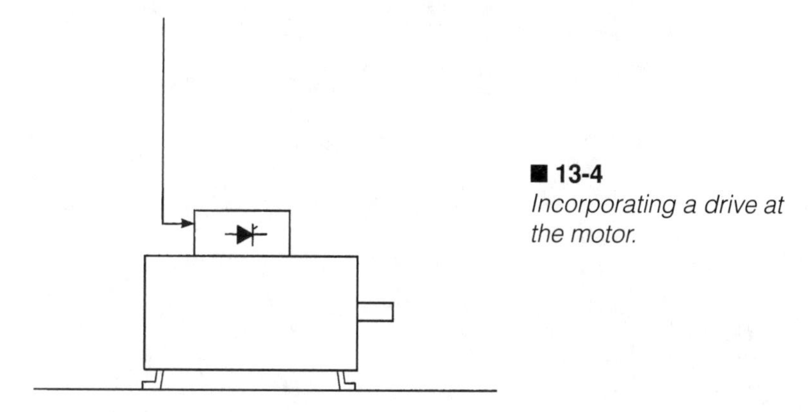

■ **13-4**
Incorporating a drive at the motor.

the approach will work. Additionally, there will be severe limitations on the actual size and horsepower drive that can be applied in this manner.

As we have seen, many factors and issues are involved in continuing the development of the electronic drive. Drives, both ac and dc, along with their servo and stepper brethren, will continue to proliferate. Wherever there is a need, there will be a product.

A

Standards organizations and addresses

American Institute of Motion Engineers (AIME)
Kohrman Hall
Western Michigan University
Kalamazoo, MI 49008
(616) 387-6533

American National Standards Institute (ANSI)
11 West 42nd Street
New York, NY 10036
(212) 642-4900

Standards Council of Canada (CSA)
1200-45 O'Connor
Ottawa, Ontario K1P6N7
(613) 328-3222

Electrical Apparatus Service Association, Inc. (EASA)
1331 Baur Boulevard
St. Louis, MO 63132

Electronic Industries Association (EIA)
2001 Pennsylvania Avenue
Washington, DC 20006-1813
(202) 457-4919

ETL Testing Laboratories, Inc. (ETL)
Industrial Park, P.O. Box 2040
Cortland, NY 13045
(607) 753-6711

Institute of Electrical & Electronic Engineers (IEEE)
445 Hoes Lane
Piscataway, NJ 08854
(908) 562-3803

National Electrical Manufacturers Association (NEMA)
2101 L Street, NW
Washington, DC 20037
(202) 457-8400

National Fire Protection Association (NFPA)
1 Batterymarch Park
P.O. Box 9101
Quincy, MA 02269-9101

National Institute of Standards & Technology (NIST)
Building 221/A323
Gaithersburg, MD 20899
(301) 975-2208

National Standards Association (NSA)
1200 Quince Orchard Boulevard
Gaithersburg, MD 20878
(800) 638-8094

Robotic Industries Association (RIA)
900 Victors Way, P.O. Box 3724
Ann Arbor, MI 48106
(313) 994-6088

Society of Manufacturing Engineers (SME)
1 SME Drive, P.O. Box 930
Dearborn, MI 48121
(313) 271-1500

Underwriters Laboratories (UL)
333 Pfingsten Road
Northbrook, IL 60062
(708) 272-8800

B

Common symbols used in drive diagrams

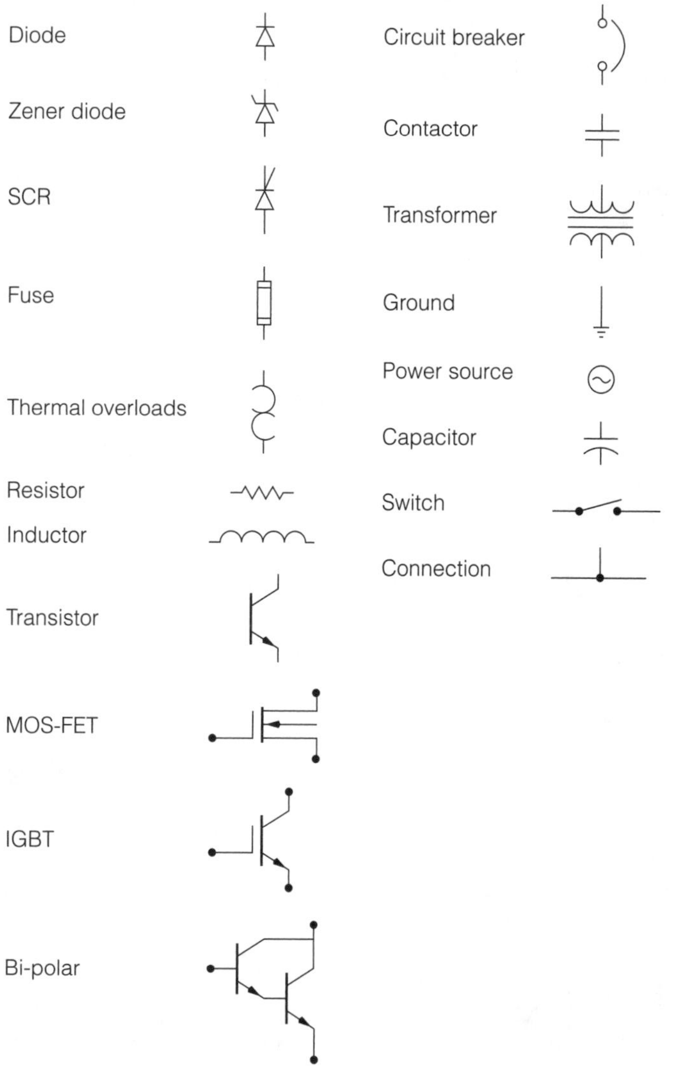

Diode

Zener diode

SCR

Fuse

Thermal overloads

Resistor

Inductor

Transistor

MOS-FET

IGBT

Bi-polar

Circuit breaker

Contactor

Transformer

Ground

Power source

Capacitor

Switch

Connection

Bibliography and suggested further reading

Bewley, L.E. *Travelling Waves on Transmission Systems*. 2nd ed. New York: John Wiley and Sons, 1951.

Boylestad, Robert L. *Introductory Circuit Analysis*. 5th ed. Columbus, OH: Charles E. Merrill, 1982.

Cummings, *DC Motors, Speed Controls, Servo Systems*, Hopkins, MN: Electrocraft Corporation, 1980.

Dulin, John J.; Veley, Victor F.C.; and Gilbert, John. *Electronic Communications*. Blue Ridge Summit, PA: McGraw-Hill, 1990.

Early, Murray, and Caloggero, *National Electrical Code Handbook*. 5th ed. Quincy, MA: NFPA, 1990.

Fitzgerald, A.E., D. Kingsley, Jr., and A. Kusko. *Electric Machinery: The Processes, Devices, and Systems of Electro-Mechanical Energy Conversion.* 3rd ed. IEEE, 1971.

Gottlieb, Irving M., *Power Supplies, Switching Regulators, Inverters, and Converters.* 2nd ed. Blue Ridge Summit, PA: McGraw-Hill, 1994.

IEEE Standards 519-1981 and 519-1992. "IEEE Guide for Harmonic Control and Reactive Compensation of Static Power Converters."

Lowden, Eric. *Practical Transformer Design Handbook*. 2nd ed. Blue Ridge Summit, PA: McGraw-Hill, 1989.

Spitzer, David W. *Variable-Speed Drives: Principles and Applications for Energy Cost Savings*. 2nd ed. Instrument Society of America, 1990.

Tomal, Daniel R. and Neal S. Widmer. *Electronic Troubleshooting*. Blue Ridge Summit, PA: McGraw-Hill, 1993.

Underwriters Laboratories, *UL/ANSI Standard #508: Industrial Control Equipment*, May 1994, Northbrook, IL.

Williams, B.W. *Power Electronics: Devices, Drives, Applications, and Passive Components*. 2nd ed.

Index

A

ac drives, 1, 7, 135-188, **137**
- acceleration, 167-168
- affinity laws, 153-155
- applications for ac drives, 136-137, 158-167, 187-188
- automatic restart, 168
- auxiliary blower at motor, 160
- base speed of ac motors, 149
- brake horsepower calculations, 155-156
- braking, 161
- bypass for ac drives, 187-188, **188**
- cable sizing, 161
- chopper diode rectifier, 143-144, **144**
- circuitry, 160
- constant torque applications, 157-158, **158**, **159**, 160
- contactors, input and output, 166-167
- converter bridge or section, 137, **137**
- cooling, 163
- cost calculations for ac drives, 152-157
- critical frequencies, 169, **169**
- current source inverter (CSI), 139, **140**, 146, **147**
- cycloconverters, 186-187, **186**
- dc bus or filter, 138, **138**
- dc drive replacement with ac, 161-162, 341-342
- deceleration, 168
- digital vs. analog control, 160
- diode rectifiers, 143-144, **143**, **144**
- distance between drive and motor, 165-166
- earth ground leakage, 177
- efficiency, 163
- encoder feedback, 160
- energy savings with electronic ac drives, 152-157
- fault logging, 169, **170**
- flux control, 150-151
- flux vector ac drives, 178-184, **179**
- full-load speed, 148
- fuses blown, 177
- ground faults, 162
- grounding, 161
- harmonic distortion, 162
- history and development, 135-136
- horsepower vs. voltage, 162-163
- humidity and moisture, 165
- inrush currents, 161
- insulated gate bipolar transistor (IGBT) inverters, 145, **146**
- inverter section, 138-140, **139**, **140**, 145-147
- load commutated inverters (LCI), 184-185, **185**
- location of drive and motor, 161
- loss of automatic signal, 168
- low line input problems, 176
- magnetizing current in ac motor, 148
- maintenance, 171-172, 187-188
- MOSFET inverters, 145
- motor will not run, 176
- multiple motor operations, 162
- overcurrent problems, 176
- overfrequency operations, 151
- overload problems, 176
- overspeeding, 151
- overtemperature problems, 176
- overvoltage problems, 177
- poles in ac motor, 148
- power dip problems, 176
- power factors, 162
- power loss ride-through, 169-170
- power supplies, 161
- pulse amplitude modulated (PAM) drives, 140, 141-143, **142**
- pulse width modulated (PWM) drives, 140, 141, **141**, **142**
- reactors, input and output, 167
- rectification, 143-145
- regeneration, 161
- selecting the ac drives, 158-167
- short circuits, 162, 177
- silicon controlled rectifier (SCR), 144-145, **144**
- single-phase input to three-phase drive, 162
- slip compensation, 170-171
- slip, ac motors, 148-150

Illustrations are in **boldface**.

ac drives, *continued*
speed problems, fluctuations, 177
speed range, 160
spinning load pick up, 171
stall prevention, 171, 176-177
standing wave conditions, 165-166
synchronous speed calculations, 147-148, **148**
transistor trips, 176-177
traverse of P jump, 169, **170**
troubleshooting and repairs, 172-175
undervoltage problems, 176
variable torque applications, 157-158, **158**, 160
variable voltage inverter (VVI), 139, **139**
ventilation, 163-164, **164**
voltage source inverter (VSI), 139, **139**
volts per hertz ratio, 150, **151**
ac induction motor, 70, **70**, 84-85, **85**
accelerating torque, 46
affinity laws, ac drives, 153-155
aftermarket add-ons, 8
alternating current (ac), 26, 29-30
alternating current drives (*see* ac drives)
amplitude, 30-31, **30**, **31**
analog signal, 237, **237**
applications for electronic drives, 2-3, **4**, 279-311, **312**, 339
aftermarket add-ons, 8
automobiles, 342-343, **342**
batch centrifuges drive application, 296-298
belt conveyor drive application, 292-293, **293**
box palletizer drive application, 308-310, **309**
centrifugal fan application, 282-283, **282**, **283**
chain conveyor drive application, 293-294, **294**
compressor drive application, 305-306, **306**
cooling tower drive application, 290-291, **290**
cranes and hoists drive applications, 294
electric cars, 342-343, **342**
extruder drive application, 281-282, **281**
fan applications, 282-283, 283-284, **282**, **283**
glass furnace drive applications, 300-301, **301**
HVAC drive application, 282-283, **283**
lead screws drive applications, 299-300, **299**
magnetic levitation train, 343-344, **344**
marketplace, 7-8
mixers drive application, 295-296, **296**
motion control, 22-23
multiple drives tied to common bus, 303-305
myths, misconceptions, misapp, 5-7
presses, drill, punch, and brick, 298-299
pump drive applications, 284-290
centrifugal pump, 284
parallel pump, variable frequency drive, 287
positive displacement (PD) pumps, 288-289
roll grinding drive application, 291-292, **291**
soft-start S curve, 310, **311**
tension control, 306-308, **307**
web control system, multiple-section, 306-308
winders, drive applications, 301-303, **302**
automation engineers, 266
automotive drive applications, electric cars, 342-343, **342**
autotransformers, 36

B

back EMF in dc motors, 94
backlash, 56, **57**
ball and lead screws (*see also* lead screws), 61-63, **62**
batch centrifuges drive application, 296-298, **297**
bearings, feedback systems, 254
belt conveyor drive application, 292-293, **293**
bipolar, 39
"black box" defined, 1
box palletizer drive application, 308-310, **309**
brake horsepower (BHP), 45, 155-156
braking, 6, 59-60, 271-274, **273**
common bussing, 272-273
dc injection braking, 273-274
dynamic braking, 272, **273**
inertia, 51-54, **53**
regenerative braking, 116-117, 271-274, **273**
breakaway torque, 46
bypass for ac drives, 187-188, **188**

C

cabling (*see* wiring and cabling)
capacitor induction motor, 87
capacitors and capacitance, 33-34, 221
commutation capacitors, 34
central processor unit (CPU), 236, **236**
centrifugal fan application, 282-283, **282**, **283**
centrifuges drive application, 296-298, **297**
chain conveyor drive application, 293-294, **294**
chains and sprockets, 58-59
chopper drives, dc, 121, **121**
circuit breakers, 42, 332
closed-loop control, 246-248, **247**
clutch systems, 59-60
inertia, 51-54, **53**

common bussing, 272-273
communications, drive communications, 241-243
CANs, 242
LANs, 242
commutation, 95
commutation capacitors, 34
commutation notches, 321, **322**
compressor drive application, 305-306, **306**
compressors for pneumatic systems, 64-65
configurable drives, 340
control boards, 2
control wiring, 216-217
control, drive control, 234-237, **235**, **236**, **237**, 238-241
automatic mode control, 238-239
communications, 241-243
feedback systems, 240-241
flux vector drives, 240-241
local control schemes, 238
preset speed control, 239-240, **240**
proportional integral derivative (PID) controller, 239, 256
ramping or acceleration, 241
setpoint, 239
controller area networks (CANs), 242
converter bridge or section, ac drives, 1, 137, **137**
conveyor drive application
belt conveyor, 292-293, **293**
chain conveyor, 293-294, **294**
servo drive systems, 200, **205**
cooling, 6, 70, 71-77, 344
cooling tower drive application, 290-291, **290**
coordinated drive machine control, 268-269
cost effectiveness of electronic drives, 2-3, **4**, 8, **9**, 339
counter electromotive force (CEMF), 248
couplings, 60-61
cranes and hoists drive applications, 294, **295**
current control, 14
current or amperage, 27
current source inverter (CSI), ac drives, 139, **140**, 146, **147**
current transformers (CTs), 248, **248**
cycle (*see* period or cycle of sine wave)
cycloconverters, 186-187, **186**

D

dancer feedback system, **258**
dc brush-type motor, 70, **70**
dc bus or filter, ac drives, 138, **138**
dc drives, 1, 7, 103-133
ac drive replacements, 161-162, 341-342
ac vs. dc drives, 132-133
applications for dc drives, 124, 132-133
armature current curve, 104, **105**
armature voltage curve, 104, **104**
chopper drives, 121, **121**
contactor will not enable, 132
current feedback, 112
current limits, 123-124
diagnostics for dc drives, 122
digital dc drives, 133
digital encoder for feedback, 113-114, **114**
drive power bridges (dc), 107-110
drive speed regulation, 122-123
equivalent circuit, 104, **104**
features of dc drives, 121-124
feedback arrangements (dc), 112-115, **113**
field economy in dc drives, 122
field loss in dc drives, 122, 131
field weakening in dc drives, 121-122
form factors, 111
four-quadrant control scheme, 109-110, **110**
full-wave rectification, 105, **106**
fuses and protection, 126-127
fuses blown, 131-132
half-wave rectification, 104-105, **105**
low line input problems, 130
M or loop contactors, 117, **118**
maintenance, 125
nonregenerative action of dc drives, 107
overcurrent or overload problems, 131
overshoot, 123, **123**
overtemperature problems, 131
power dip problems, 130
regenerative braking, 116-117
regenerative vs. nonregenerative drives, 115-117, **116**
regulation in dc drives, 123
replacement parts, 126-127
ripple, 111
selecting dc drives, 124
short circuits, 131
silicon controlled rectifiers (SCR), testing, 128-130
spares for repair of dc drives, 126-127
speed feedback, 112, **113**
speed problems, fluctuations, 132
speed range in dc drives, 123
speed scaling, 118-119
stacking or addition of power bridge, 109
stall trips, 132
tachometers, 113, **114**, 113, 118-119, **119**, 131

dc drives, *continued*
three-phase full-wave rectification, 105-106, **106**
three-phase half-wave rectification, 105, **106**
thyristor or SCR-based dc drives, 107
traction drives, 119-121
transistorized dc drives classification, 107
troubleshooting and repairs, 126-130
two-quadrant control scheme, 107, 109, **109**
undervoltage problems, 130
web converting line, regenerative systems, 116, **117**
dc injection braking, 273-274
dc motors, 70, **70**, 89-96, **90**
Delta transformer system, 227, **227**
digital encoder for feedback, 113-114, **114**
digital input/output control, 14-15
digital signal, 237, **237**
diode rectifiers, 143-144, **143**, **144**
diodes, 40, **40**, 221
zener diode, 40, **41**
direct current (dc), 26, 30
direct current drives (*see* dc drives)
distortion (*see* harmonic distortion)
downtime log, **276**
drill presses, punch, and brick, 298-299, **298**
drives defined, 1
drivetrain components, **44**
duty cycles, 60
dynamic braking, 272, **273**
dynamic torque, 46

E

eddy current, 329
eddy current clutches, 5, 17-22, **18**
efficiency calculations, 28-29
efficiency of electronic drives, 2, 65-66
electric cars, drive applications, 342-343, **342**
electric motors (*see* motors, electric)
electrical and electronic, 26-42
alternating current (ac), 26, 29-30
amplitude, 30-31, **30**, **31**
capacitors and capacitance, 33-34
circuit breakers, 42
commutation capacitors, 34
current or amperage, 27
diodes, 40, **40**
direct current (dc), 26, 30
efficiency calculations, 28-29
electron flow to produce electricity, 26
formulas used in electrical work, 27-30
Fourier analysis, 28
frequency standards for electrical service, 26
frequency, 30-31, **30**, **31**
fundamental or sine wave, 30, **30**
fuses and circuit protection, 40-42
gate-turn-off thyristors (GTO), 39
half-cycle of sine wave, 31
harmonic distortion, 28
horsepower calculations, 28, 29
inductors and inductance, 35-37
La Place transformation, 28
magnetism and electromagnetism, 26
mechanical power from electrical power, 42-68
National Electric Code Handbook, 41, 42
Ohm's law, 27-28, **27**
outages of electricity, 31
period or cycle of sine wave, 31
power factor calculations, 29
RC networks, 33, **34**
rectification of electrical power, 28, 38
relays, 34-35
resistors and resistance, 27, 32-33, **33**
semiconductors, power, 37-39
short circuits, 41-42, **41**
silicon controlled rectifiers (SCR), 38-39
sine wave or fundamental, 30, **30**
square wave, **31**
symbols used in drive diagrams, 349
thryistors, 38
thyratrons, 38
transformers, 35-37, **35**
transistors, 39-40
triangular wave, **31**
usage of power, 28
vacuum tubes or electron tubes, 37-38
voltage, 27
voltage standards for electrical service, 26
zener diode, 40, **41**
electrical coupling, 329
electromagnetic interference (EMI), 329-330
electromagnetic noise, 224
electron flow to produce electricity, 26
electrostatic noise, 224
enclosures
electrical motor enclosures, 72-77
electronic drives, 230-233, **232**
encoders, optical encoders, 250-254, **250**, **251**, **252**
explosion-proof motors, 77
extruder drive application, 281-282, **281**

F

fan applications, 282-283, 283-284, **282**, **283**
fan system of motion control, 20-21, **21**, **22**
Fast Fourier Transforms (FFT), 324

feedback methods, 240-241, 248-250
bearings, 254
dancer feedback system, **258**
magnetic pickups, 255-256, **256**
optical encoders, 250-254, **250**, **251**, **252**
peripheral feedback and control devices, 256-257
pressure sensors, 258-259
proportional integral derivative (PID) controller, 256
resolvers, 254-255, **255**
rotary variable differential transformer (RVDT), **258**
temperature sensors, 258-259
thermoelectric devices, 259
transducers, 257-258, **258**
field effect (FET), 39
field wound dc motors, 92
filters, 336-337, **336**
line filters, 215
tuned filters, 318-319, **319**
fluctuations in power, 331-332
fluid coupling control, 18-19, **19**
fluid-based speed variator, 19-20, **20**
flux, 92
flux vector ac drives, 178-184, **179**, 240-241
earth ground leakage, 184
external fault conditions, 183
fuses blown, 184
low line input problems, 183
motor will not run, 183
open- and closed-loop, 340
overcurrent problems, 183-184
overload problems, 183-184
overtemperature problems, 183
overvoltage problems, 184
power dip problems, 183
relays tripping, 184
short circuits, 184
slow speed, 182-183
speed problems, fluctuations, 184
stalling motor, 184
transistor tripping, 184
troubleshooting and repairs, 182-184
undervoltage problems, 183
flywheel effect, 52, **53**
force, 43
form factors, 37, 94-95
dc drives, 111
formulas used in electrical work, 27-30
Fourier analysis, 28, 314
frame standards for electric motors, 83, **84**
“freak” (frequency) drive, 1
frequency standards for electrical service, 26
frequency, 30-31, **30**, **31**
friction, 54
full-wave rectifier, 105, **106**
fundamental or sine wave, 30, **30**
fuses and circuit protection, 40-42, 332
future developments in electronic drives, 339-345
ac drive replacements, 341-342
automotive drive applications, electric cars, 342-343, **342**
configurable drives, 340
cooling systems, 344
dc drive replacement with ac, 341-342
drive incorporated into motor, 344, **345**
flux vector drives, open- and closed-loop, 340
magnetic levitation (maglev) trains, 343-344, **344**
regeneration, 340-341, **341**
voltage-configurable drives, 340

G

gate firing boards, 2
gate-turn-off thyristors (GTO), 39
gears, **44**, 54-57, **56**
backlash, 56, **57**
gear reduction, 5, **6**, 52, **53**, 54-57, **56**
gearboxes, 55
helical gears, 56-57
inertia, 52, **53**
mechanical advantage, 55
planetary gears, 57
servo drive systems, 199-201, **200**, **202-203**
stage in gears, 55-56
switchgear, 233-234
worm gear reducer, 56
gears, 44, **44**
glass furnace drive applications, 300-301, **300**
ground loops, 219, **219**, 330-331, **331**
noise suppression, 224
grounding, 330-331, **331**
wiring and cabling, 217-220, **218**, **219**, **220**

H

half-cycle of sine wave, 31
half-wave rectifier, 104-105, **105**
Hall-effect sensors, 248-249, **249**
harmonic distortion, 28, 223, 313-324
cable derating graphs, 320, **320**
commutation notches, 321, **322**
current limits, 321, **323**
Fast Fourier Transforms (FFT), 324
Fourier analysis, 314
line reactors, 320
linear loads, 317-321, **318**
measuring harmonic distortion, 323-324
noise, 314
nonlinear loads, 317-321
point of common coupling (PCC), 319, **319**
power factors, 326-328, **327**, **328**
radio frequency interference (RFI), 325-326

harmonic distortion, *continued*
short circuit method of analysis, 316
total demand distortion (TDD), 315
total harmonic distortion (THD), 315-316, **316**
tuned filters, 318-319, **319**
harmonics (*see* harmonic distortion)
hazardous areas, motors, electric, 77
heatsinks, 1
helical gears, 56-57
horsepower, 5, 28, 29, 43, 44-45, 47, 48
brake horsepower (BHP), 45
linear, 47
speed, torque, and horsepower relationships, 48-51, **49**, **50** **51**
HVAC drive application, 282-283, **283**
hydraulic systems, 63-64

I

induction motor, 70, **70**, 84-85, **85**
inductors and inductance (*see also* transformers), 35-37
inertia, 51-54, **53**
flywheel effect, 52, **53**
gear reduction, 52, **53**
mass and inertia, 53
reflected inertia, 52
input/output cards, 2
input/output control, digital, 14-15
input/output signals, 234-237, **235**
analog vs. digital, 237, **237**
insulated gate bipolar (IGBT), 39
insulation ratings, induction motors, 85, **86**
integration, drive integration, 261-278
automation engineers, 266
braking, 271-274, **273**
checklist or spreadsheet for integration, 265
coordinated drive machine control, 268-269
downtime log, **276**
LANs and WANs, 264
machine systems and applications, 266-267
multiple drives for one machine, 262, **263**
multiple drives, multiple machines, 262, **263**
programmable logic controller (PLC), 264
regeneration, 271-274, **273**
regenerative braking, 271-274, **273**
remote control, 269-271, **270**
retrofitting, 265
servicing, service contracts, 277-278
single drive, multiple machines, 262, **264**
spare parts, inventories for repair, 277-278
tension control, 267-268, **268**
training programs, 275-277
troubleshooting, 274-275
turnkey systems, 261-262
warranties, 277-278
interference (*see* radio frequency interference (RFI))
inverter section, ac drives, 1, 138-140, **139**, **140**, 145-147
isolation transformers, 36, 226
isolation, optical isolation, 335-336, **336**

K

K factor in transformers, 36-37, 229-230

L

La Place transformation, 28
lead screw systems, servo drive systems, 200, **205**
line filters, 215
line reactors, 215, **215**, 228, 320
linear motor systems, 206-207, **206**
load commutated inverters (LCI), 184-185, **185**
loads, 6, 7
local area networks (LANs), 242, 264
local drive control, 238

M

M or loop contactors, dc drives, 117, **118**
M-G set (*see* motor-generator set)
machine systems and applications, 266-267
magnetic levitation (maglev) trains, 343-344, **344**
magnetic noise, 224
magnetic pickups, feedback systems, 255-256, **256**
magnetism and electromagnetism, 26
marketplace for electronic drives, 7-8
mass and inertia, 53
mechanical systems, 42-68
ball and lead screws, 61-63, **62**
brake horsepower (BHP), 45
braking, 59-60
chains and sprockets, 58-59
clutches, 59-60
couplings, 60-61
drivetrain components, **44**
efficiency, 65-66
flywheel effect, 52, **53**
force, 43
friction, 54
gears, gearboxes, gear reduction, 54-57, **56**
horsepower, 43, 44-45
hydraulic systems, 63-64
inertia, 51-54, **53**
maintenance, 67-68
noise, 65
pneumatic systems, 64-65
power, 43

power transmission (PT), 42
revolutions per minute (rpm), 47
roller chains, 58-59
service factors, 66-67
speed, 47
speed, torque, and horsepower relationships, 48-51, **49**, **50**, **51**
torque, 43-44, 45-48, **45**
V-belts, 57-58, **58**
variable-speed pulleys, 57-58, **58**
windage and friction, 54
work, 43-44, **43**
memory, 236-237, **236**
metal oxide varistors (MOVs), 221
misapplication of electronic drives (*see* applications for electronic drives)
mixers drive application, 295-296, **296**
MOSFETs, 39-40
motion control, 11-23
current control, 14
digital input/output control, 14-15
eddy current clutches, 17-22, **18**
electrical motion control, 15
electronic drive applications, 22-23
fan system of motion control, 20-21, **21**, **22**
fluid coupling control, 18-19, **19**
fluid-based speed variator, 19-20, **20**
hydraulic motion control, 15-16
methods of motion control, 15-16
motor-generator (M-G) sets, 16-17, **17**
pneumatic motion control, 15-16
position control, 14
pump system of motion control, 20-21, **22**
reengineering mechanical systems for electronic drives, 12
speed control, 12-13, **13**
standardization, 15
steam-actuated motion control, 15
torque control, 13-14
transmissions, 20, **20**
V-belt drive, 18, **19**, 57-58, **58**
variable-pitch pulleys, 18
variable-speed control, 16
variable-speed pulleys , 57-58, **58**
motor driver, 1
motor-generator (M-G) sets, 1, 16-17, **17**
motors, electric, 69-101, 344
ac induction motor, 70, **70**
ac motors, 83-89
applications for electric motors, 69-71, 100-101
armature of dc motors, 91
auxiliary blower for cooling, 75, **76**
back EMF in dc motors, 94
bearings in motor, 80
capacitor induction motor, 87
commutation, 95
controllers for electric motors, 100-101
cooling and heatsinks, 70, 71-77
current to winding ratio, dc motors, 91
dc brush-type motor, 70, **70**
dc motors, 89-96, **90**
demagnetization of dc motors, 92
drip proof fully guarded separately ventilated (DPFG-SV) unit, 76
drive incorporated into motor, 344, **345**
efficiency ratings, 101
enclosure types for cooling, 72-77
equivalent circuit for dc motor, 90, **90**
explosion-proof motors, 77
field weakening in dc motors, 92, **93**
field wound dc motors, 92
flux, 92
form factors, 94-95
frame standards, 83, **84**, 91
hazardous area applications, 77
history and development, 69
induction ac motors, 84-85, **85**
insulation ratings, induction motors, 85, **86**
life expectancy vs. design, 79-80
nameplate for motor, 89, **89**, 95, **95**
NEMA ratings for motors, 80-83, **81**, **82**, 111
open drip proof (ODP) enclosure, 72, **72**
parallel or shunt wound dc motors, 91, 94
permanent magnet dc motors, 91-92, **93**
polyphase induction motor, 84, 87
protecting the motor, 78-80
repulsion type motors, 88, 90
rotor bar, 70, **71**
rotor, 69, **70**
series wound dc motor, 91, 92, 94, **94**
service factors and sizing, 75
servomotors, 96-99, **97**, **98**
shaded pole induction motor, 87
shunt wound dc motor, 94, **94**
size of motor for application, 75
slip, induction motors, 85
sparks from dc brush-type motors, 77
speed controller, 78, **78**
split phase induction motor, 87, **87**
squirrel cage induction motor, 84, 86, **86**

motors, electric, *continued*
starters, 233, **234**
stator, 69, **70**
stepper motors, 99-100, **99**
switchgear, 233-234
synchronous induction motor, 87-88
thermal overload protection, 78, **79**
thermostat, 79, **79**
three-phase induction motors, 84
torsional analysis for motor sizing, 88-89
totally enclosed air over (TEAO) enclosure, 73, 73, **74**
totally enclosed fan cooled (TEFC) enclosure, 73, **73**
totally enclosed nonventilated (TENV) enclosure, 72, **73**
totally enclosed unit cooled (TEUC) enclosure, 73, **74**, 75
totally enclosed water to air cooled(TEWAC) enclosure, 73, **74**
usage of electricity by motors, 101
windings in motor, 79-80
wound rotor induction motor, 87-88

N

nameplate for motor, 89, **89**, 95, **95**
National Electric Code Handbook, 41, 42
NEMA ratings for motors, 80-83, **81**, **82**, 111
noise (*see also* harmonic distortion), 65, 222-225, 328-329
eddy current, 329
electrical coupling, 329
electromagnetic interference (EMI), 329-330
emission, 328
ground loops, 224
harmonic distortion, 314
immunity, 328
power supplies, 332-334
radio frequency interference (RFI), 329-330
sources of electrical noise, 224
thermal noise emission, 329
nonregenerative drives, dc drives, 107, 115-117, **116**

O

Ohm's law, 27-28, **27**
open-loop control, 246-248, **246**
optical encoders, 250-254, **250**, **251**, **252**
optical isolation, 335-336, **336**
outages of electricity, 31
overhead cranes and hoists drive applications, 294, **295**
overshoot, servo drive systems, 198, **198**

P

parallel or shunt wound dc motors, 91, 94
peak inverse voltage (PIV), 38
peak reverse voltage (PRV), 38
period or cycle of sine wave, 31
permanent magnet dc motors, 91-92, **93**
PIV (*see* transmissions)
planetary gears, 57
pneumatic systems, 64-65
point of common coupling (PCC), 319, **319**
polyphase induction motors, 84, 87
position control, 14
positive displacement (PD) pumps, 288-290, **289**
potential transformers (PTs), 248
power, 28, 29, 43
power factors, 29, 326-328, **327**, **328**
power supplies (*see also* harmonic distortion; noise suppression; wiring and cabling)
circuit breakers, 332
disturbances to electrical power, 332-334
fluctuations in power, 331-332
fuses, 332
noise, 332-334
outages, 331-332
shielding, 334, **335**
spikes, 332
surges, 332
power transmission (PT) (*see* mechanical systems)
preset speed control, 239-240, **240**
presses, drill, punch, and brick, 298-299, **289**
pressure sensors, 258-259
programmable logic controller (PLC), 5-6, 264
proportional integral derivative (PID) controller, 239, 256
pulse amplitude modulated (PAM) drives, 140, 141-143, **142**
pulse width modulated (PWM) drives, 140, 141, **141**, **142**
pump drive applications, 284-290
centrifugal pump with bypass, 284-285, **285**
parallel pump, variable frequency drives, 287-288, **288**
positive displacement (PD) pumps, 288-290, **289**
specific gravities for common liquids, **285**

R

rack-and-pinion systems, servo drive systems, 199, **204**
radio frequency interference (RFI) 325-326, 329-330

ramping or acceleration, control, drive control, 241
RC networks, 33, **34**, 221, **336**
rectification, 1, 28, 38, 142-143
reengineering concepts, 12
regeneration, 271-274, **273**, 340-341, **341**
regenerative braking, 116-117, 271-274, **273**
regenerative vs. nonregenerative dc drives, 115-117, **116**
relays, 34-35
remote control, 269-271, **270**
repulsion-type motors, 88, 90
resistors and resistance, 27, 32-33, **33**, 221
  parallel configuration, 32-33, **33**
  RC networks, 33, **34**
  series configuration, 32-33, **33**
resolvers, 254-255, **255**
retrofitting for electronic drives, 265
revolutions per minute (rpm), 47
ringing, in servo drive systems, 198
ripple, dc drives, 111
robotic control
  lead screws drive applications, 299-300, **299**
  servo drive systems, 201
roll grinding drive application, 291-292, **291**
roller chains, 58-59
rotary variable differential transformer (RVDT), **258**
rotor bar, 70, **71**
rotor, 69, **70**
running or process torque, 46

S

S curve for soft starts, 310, **311**
semiconductors, power, 37-39
series-wound dc motor, 91, 92, 94, **94**
service factors, 60, 66-67
servicing, service contracts, 277-278
servo drive systems, 7
  amplifiers, servo amplifiers, 191
  applications, 207-208
  bandwidth or response factor, 196
  components of servo drive systems, 195
  conveyor systems, 200, **205**
  derivative gain, 197, **198**
  design and selection criteria, 199, **202-203**
  feedback devices, 196-197
  gain, 197, **198**
  gearbox and gearing, 199-201, **200**, **202-203**
  integral gain, 197, **198**
  lead screw systems, 200, **205**
  motors, servo motors, 191
  multiple servo motor control, 194-195, **195**
  overshoot, 198, **198**
  position control, 193-194, **194**, 196
  proportional gain, 197, **198**
  rack-and-pinion systems, 199, **204**
  ringing, 198
  robotic control, 201
  speed control, 190-191, **192**, 196
  stability of system, 197
  timing belts, 199-200, **204**
  torque and power requirements, 190
  torque control, 191, **192**, 193, 196
  tuning the system, 197-198
  velocity control, 193, **193**, 196
  velocity gain, 197, **198**
servomotors, 96-99, **97**, **98**
setpoint, 239
shaded pole induction motor, 87
shielding, 334, **335**
  wiring and cabling, 220-221, **221**
short circuits, 41-42, **41**
signal conditioners, 336-337
signal wiring, 216-217
silicon controlled rectifier (SCR), 1, 38-39, **129**, **130**
  ac drives, 144-145, **144**
  dc drives, 107
  peak inverse voltage (PIV), 38
  peak reverse voltage (PRV), 38
  testing, 128-130
sine wave or fundamental, 30, **30**, 314
slip, ac motors, 85, 148-150
soft-start S curve, 310, **311**
specific gravities for common liquids, **285**
speed, 47
  speed, torque, and horsepower relationships, 48-51, **49**, **50**, **51**
speed control, 12-13, **13**, 248-250
  motors, electric, 78, **78**
speed scaling, 118-119
spikes or voltage transients, 214-215, **214**, 332
spindle drives, 207
split phase induction motor, 87, **87**
square wave, **31**
squirrel cage induction motor, 84, 86, **86**
standardization
  motion control, 15
  organizations for standards, addresses, 347
standing wave conditions, 165-166
star grounding system, 219, **220**
start-up, 7
starters, motor, 233, **234**
static vs. dynamic torque, 46
stator, 69, **70**
step-up/step-down transformers, 36, 226-227
stepper drives, 201, **206**
  applications, 207-208
  linear motor systems, 206-207, **206**
stepper motors, 99-100, **99**

surges, power surges, 332
switchgear, 233-234
symbols used in drive diagrams, 349
synchronous induction motor, 87-88

T

tachometers, 113, **114**, 118-119, **119**
temperature sensors, 258-259
tension control, 267-268, **268**, 306-308, **307**
thermal noise emission, 329
thermal overload protection, motors, electric, 78, **79**
thermoelectric devices, 259
thermostat, motors, electric, 79, **79**
three-phase induction motors, 84
thyratrons, 38
thyristor drives, 1, 38
  dc drives, 107
  gate-turn-off thyristors (GTO), 39
timing belts, servo drive systems, 199-200, **204**
torque, 13-14, 43-44, 45-48, **45**
  accelerating torque, 46, 47
  breakaway torque, 46
  calculating torque, 46-47
  linear vs. rotary torque, 47
  running or process torque, 46
  speed, torque, and horsepower relationships, 48-51, **49**, **50**, **51**
  static vs. dynamic torque, 46
  wrenches, torque wrenches, 48
torque control, 13-14
torque wrenches, 48
torsional analysis for motor sizing, 88-89
total demand distortion (TDD), 315
total harmonic distortion (THD), 315-316, **316**
traction drives, dc, 119-121
training programs, 275-277
trains, magnetic levitation (maglev) trains, 343-344, **344**
transducers, 257-258, **258**
transformers, 35037, **35**, 215, **215**, 225-230, **226**, **227**
  autotransformers, 36
  coil and turns ratio, 35-36
  cooling, 229
  current transformers (CTs), 248, **248**
  dc bus chokes, 228
  Delta transformer system, 227, **227**
  form factor, 37
  isolation transformers, 36, 226
  K factor in transformers, 36-37, 229-230
  line reactors, 228
  linear and nonlinear loads, 228-229
  potential transformers (PTs), 248
  ratings, 36
  rotary variable differential transformer (RVDT), **258**
  sizing transformers, 229
  step-up/step-down transformers, 36, 226-227
  weight of unit vs. heating, K-factor, 36-37
  Wye transformer system, 227, **227**
transients, voltage spikes, 214-215, **214**
transistors, 39-40
  bipolar, 39
  dc drives, 107
  field effect (FET), 39
  insulated gate bipolar transistor (IGBT), 39, 145, **146**
  MOSFETs, 39-40, 145
transmissions, 20, **20**
triangular wave, **31**
troubleshooting general drive problems, 274-275
tuned filters, 318-319, **319**
turnkey systems, 261-262

U

use of electronic drives (*see* applications for electronic drives)

V

V-belt drive, 18, **19**, 57-58, **58**
vacuum tubes or electron tubes, 37-38
variable frequency drive (VFD), 1, 7
variable speed drive (VSD), 1
variable voltage inverter (VVI), ac drives, 139, **139**
variable-pitch pulleys, 18, **19**
variable-speed control, 16
variable-speed pulleys, 57-58, **58**
variator, fluid-based speed, 19-20, **20**
VF drive, 1
voltage, 27
voltage source inverter (VSI), ac drives, 139, **139**
voltage standards for electrical service, 26
voltage transients or spikes, 214-215, **214**
voltage-configurable drives, 340
volts-per-hertz drive, 1

W

warranties, 277-278
web control system, multiple-section, 306-308, **307**
web converting line, regenerative systems, 116, **117**
wide area networks (WANs), 264
windage and friction, 54
winders, drive applications, 301-303, **302**, **303**, **304**
wiring and cabling, 209, 210-225
  air gap flux rate, 212-213
  aluminum vs. copper wire, 213-214
  American Wire Gauge (AWG) sizes, 213

bundling cables, 217
cable derating graphs, 320, **320**
control wiring, 216-217
current-carrying capacities, 213
diameter, resistance, weight (gauge) of wiring, 210, **211**, **212**
digital vs. analog signal wiring, 217
filters, 215, 336-337, **336**
ground loops, 219, **219**
grounding, 217-220, **218**, **219**, **220**
line reactors, 215, **215**
National Electrical Code (NEC) standards, 210, **211**, **212**
noise suppression, 222-225
optical isolation, 335-336, **336**
resistance values of wire, 213
shielding, 215, 220-221, **221**, 334, **335**
signal conditioners, 336-337
signal wiring, 216-217
spikes or voltage transients, 214-215, **214**
star grounding system, 219, **220**
transformers, 215, **215**, 225-230, **226**, **227**
voltage boost, 210-211
work, 43-44, **43**
worm gear reducer, 56
wound rotor induction motor, 87-88
wrenches, torque wrenches, 48
Wye transformer system, 227, **227**

Z

zener diode, 40, **41**